ORBITAL INTERACTIONS
IN CHEMISTRY

ORBITAL INTERACTIONS IN CHEMISTRY

THOMAS A. ALBRIGHT
Department of Chemistry
University of Houston

JEREMY K. BURDETT
Department of Chemistry
The University of Chicago

MYUNG-HWAN WHANGBO
Department of Chemistry
North Carolina State University

A Wiley-Interscience Publication

JOHN WILEY & SONS

New York • Chichester • Brisbane • Toronto • Singapore

Copyright © 1985 by John Wiley & Sons, Inc.

All rights reserved. Published simultaneously in Canada.

Reproduction or translation of any part of this work
beyond that permitted by Section 107 or 108 of the
1976 United States Copyright Act without the permission
of the copyright owner is unlawful. Requests for
permission or further information should be addressed to
the Permissions Department, John Wiley & Sons, Inc.

Library of Congress Cataloging in Publication Data:

Albright, Thomas A.
  Orbital interactions in chemistry.

  "A Wiley-Interscience publication."
  Incudes index.
    1. Molecular orbitals.   I. Burdett, Jeremy K.,
1947–   II. Whangbo, Myung-Hwan.   III. Title.

QD461.A384   1984   541.2'2   84-15310
ISBN 0-471-87393-4

Printed in the United States of America

10 9 8 7 6 5 4 3 2

Foreword

This book provides a way of comprehending the geometry and reactivity of all molecules, and it also serves as a guide to understanding the molecular orbital calculations, now so easily done, on these molecules.

Curious, isn't it, that one should have to talk about understanding calculations? For if those calculations are correct, if they predict the observable that interests us, then we, through the computer we command and the computer program we have written, should truly be in the position of understanding the molecule. Yet science is not so simple. First the "correct" calculation is achieved only with immense effort, through a layering of numerical expansions and approximations that makes us move away from chemical and physical understanding to numerical applied mathematics. The numbers obscure the chemistry—worse, even, is the fact that we must deal with psychologically addictive computers. Second, each science, and chemistry in particular, generates it's own modalities of understanding. The complexity of the phenomena causes questions to be asked at a certain level. For instance, we have found the classification of functional groups as π donors or acceptors useful in both organic and inorganic chemistry. So when we look at the rotational barrier or reactivity of an $Fe(CO)_4$ complex of a substituted olefin we naturally ask how that activation energy is affected by putting on electron-withdrawing or -accepting substituents on the olefin. And we do not care as much whether the barrier to rotation is x kcal/mol for $R = CH_3$ but $x + 1.5$ when $R = OCH_3$ as we do about the trend with change in the electronic nature of the substituent. The question is posed at a certain level of complexity, and understanding is posited when the answer is given at the same, chemical level of complexity.

What Tom Albright, Jeremy Burdett, and Myung-Hwan Whangbo have provided us with is that appropriate intermediate level of thinking—a way of analyzing molecules and of calculations on molecules in a theoretically consistent framework that allows a chemist to understand calculations in the natural language of chemistry. There are a number of unique features in this book. First, there is the consistent use of the fragment picture—the building up of more complicated molecules from simpler ones, of molecules from their pieces. In this building-up process the natural language is that of perturbation theory. Second, an important aspect of their approach is the constant and crucial inclusion of overlap in the interaction picture. A third feature that distinguishes this work is its breadth—the taking into consideration and treatment, in one and the same language, of structure and reactivity problems in the organic and inorganic realm, and even in the solid state. I agree with the authors that these disciplines should not be treated as separate entities—molecules are molecules, and it is only the prejudices of our education that make us sort them out into boxes containing those we care to think about and therefore call interesting and those that we think are dull because we are afraid of them. The authors show us that frontier orbital ideas and simple concepts of symmetry and bonding are applicable to understanding all molecules.

The last feature that characterizes this book is its pedagogical concern and effectiveness. It would seem obvious that if we understand something we must be able to communicate it, to teach it, for we have taught ourselves a fact, an explanation, a truth. If we could teach ourselves (we call that research) should it not be easier to teach other, more clever people? Unfortunately, as we all know, it is not so simple, and expository skill and clarity do not always go hand in hand with scientific achievement. But this book does it all—it not only shows us how to understand molecules simply and economically, but also explains and teaches us how to explain.

ROALD HOFFMANN

Cornell University
June 1984

Preface

Molecular orbital theory has now come of age. It is used to enable us to understand a wide range of physical properties of molecules, their reactivities, and the pathways taken by chemical reactions. Over the years, since its conception in the 1920s and 1930s, the theory has been developed along two very different lines by largely separate groups of people. The first, the *ab initio* group, attempts to solve the molecular electronic Schrödinger equation to what might be called chemical accuracy. The second, the semiempirical group, does not try to evaluate the many molecular integrals needed in all but the simplest molecules and makes extensive use of experimental information to parametrize many of the integrals. Because the calculations performed by the first group are invariably expensive, and those of the second usually not, very different philosophies concerning their approach to chemical problems exist between the two groups. It is an unfortunate truism that proponents of these two different schools often react negatively to each other, although the combined efforts of both have resulted in dramatic progress in our understanding of molecular electronic structure.

If we wish to compute accurately the equilibrium geometry of a molecule or solid, then there is at present no substitute for a quality calculation. First-principles calculations on molecules containing up to half a dozen first-row atoms, with generally exceedingly good matching of theoretical and experimental bond lengths and angles, vibrational frequencies, and so on, are now possible. As the molecules become larger, approximations have to be made to make the computational problem tractable; consequently, the errors increase. Thus, many *ab initio* calculations on larger systems are really only approximate ones. There are still unfortunate surprises even at the small molecule level. In the area of the solid state (which might be called the giant molecule level), structural prediction by calculation has only been achieved for very simple systems with small unit cells. When looking at bond dissociation energies of molecules, or cohesive energies of solids, again good calculations are necessary. These are also needed when bonds are stretched, as in chemical reactions. Configuration interaction is very important in quantitative terms if accurate activation parameters are to be calculated. The huge computational requirements involved here have restricted good calculations to the area of small molecules.

From what we have said then, the *ab initio* route is the one to take if one is concerned with the numerology of chemistry, although for practical reasons a semi-empirical approach may be necessary. The size of the molecule that can be handled by any method depends (a) upon how good we want the calculation to be for it to be useful and (b) upon the nature of the variable to be computed. Equilibrium geometries, for example, are obtained much more easily and accurately than activation energies and the geometries of transition states as mentioned above.

This book takes the problem one step further. We shall study in some detail the mechanics behind the molecular orbital level structures of molecules. We shall ask why these orbitals have a particular form and are energetically ordered in the way that is generated by a computer calculation, whether it is generated by an *ab initio* or semiempirical technique. Furthermore, we want to understand in a qualitative or semiquantitative sense what happens to the shape and energy of orbitals when the molecule distorts or undergoes a chemical reaction. These models are useful to the chemical community. An experimentalist must have an understanding of why molecules of concern react the way they do, as well as what determines their molecular structure and how this influences reactivity. So, too, it is the duty and obligation of a theorist to understand why the numbers from a calculation come out the way they do. Models in this vein must be simple. The ones we use here are based on concepts such as symmetry, overlap, and electronegativity. The numerical and computational aspects of the subject are, in fact, deliberately deemphasized. In other words, the goal of this approach is the generation of global ideas that will lead to a qualitative *understanding* of electronic structure.

An important aim of the book is then to show how common orbital situations arise throughout the whole chemical spectrum. For example, there are isomorphisms between the electronic structure of CH_2, $Fe(CO)_4$, and $Ni(PR_3)_2$ and between the Jahn–Teller instability in cyclobutadiene and the Peierls distortion in solids. These relationships will be highlighted, and to a certain extent, we have chosen problems that allow us to make such theoretical connections across the traditional boundaries between the subdisciplines of chemistry.

Qualitative methods of understanding molecular electronic structures are based either on valence bond theory promoted largely by Linus Pauling or delocalized molecular orbital theory following the philosophy suggested by Robert Mulliken. The orbital interaction model that we shall use in our book, which is based upon delocalized molecular orbital theory, was largely pioneered by Roald Hoffmann and Kenichi Fukui. This is one of several models that can be employed to analyze the results of computations. This model is simple and yet very powerful. Although chemists are more familiar with valence bond and resonance concepts, the delocalized orbital interaction model has many advantages. In our book we will often point out links between the two viewpoints.

There are roughly three sections in this book. The first develops the models we will use in a formal way and serves as a review of molecular orbital theory. The second covers the organic–main group areas with a diversion into solids. Typical concerns in the inorganic–organometallic fields are covered in the third section along with cluster chemistry. Each section is essentially self-contained, but we hope that the organic chemist will read on further into the inorganic–organometallic chapters and vice versa. For space considerations, many interesting problems were not included. We have attempted to treat those areas of chemistry that can be appreciated by a general audience. Nevertheless, the strategies and arguments employed should cover many of the structure and reactivity problems that one is likely to encounter. We hope that readers will come away from this work with the idea that there is an underlying structure to all of chemistry and that the conventional divisions into organic, inorganic, organometallic, and solid state are largely artificial. Introductory material in quantum mechanics along with undergraduate organic and inorganic chemistry constitutes the necessary background information for this book.

It is impossible to list all the people whose ideas we have borrowed or adapted in this book. We do, however, owe a great debt to a diverse collection of chemists who have gone before us and have left their mark upon particular chemical problems. The genesis of this book came about when the three of us worked at Cornell University with Roald Hoffmann. This was an enjoyable and very exciting time for us. We learned a great deal from Roald Hoffmann about how to think and understand chemical problems on the basis of orbital interactions. We must also thank the Sloan and Dreyfus Foundations, who provided vital support for us during the early years of our academic careers. The manuscript was carefully and critically read by Timothy Hughbanks, Sung-Kwan Kang, Jesus Lopez, Jerome Silvestre, and Dao-Xin Wang. A very skillful rendering of the diagrams was performed by Moon Vanko and S.-H. Kang. The typing of the manuscript was expertly done by Natalie Mosley, Nancy Trombetta, and Joyce Weatherspoon. We would like to thank our wives Jin-Ok, Judy, and Roberta, and our children, Albert, Alexander, Harry, Jennifer, Meredith, and Rufus for their patience and moral support.

THOMAS A. ALBRIGHT
JEREMY K. BURDETT
MYUNG-HWAN WHANGBO

Houston, Texas
Chicago, Illinois
Raleigh, North Carolina
November 1984

Contents

CHAPTER ONE

Atomic and Molecular Orbitals

1.1. INTRODUCTION

The goal of this book is to show the reader how to work with and understand the molecular orbital structure of molecules and solids. It is not our intention to present a formal discussion of the tenets of quantum mechanics or to discuss the methods and approximations used to solve the molecular Schrödinger equation. There are several excellent books[1-5] which do this and many "canned" computer programs which are readily available to carry out the numerical calculations at different levels of sophistication. The real challenge, and the motivation behind this volume is to be able to understand where the numbers generated by such computations actually come from. The first part of the book contains some mathematical material on which we will be able to build a largely qualitative discussion of molecular orbital structure. We will see how the level structure of complex molecules may be constructed and how these orbitals change as a function of a geometrical perturbation or as a result of the presence of a second molecule as in a chemical reaction. We shall see that many concepts and results together form a common thread which enables different fields of chemistry to be linked in a satisfying way.

1.2. ATOMIC ORBITALS

The molecular orbitals (MOs) of a molecule are usually expressed as a linear combination of the atomic orbitals (LCAO) centered on its constituent atoms, as we shall see later. These atomic orbitals have the form shown in equation 1.1. This is

$$\chi(r, \theta, \phi) = R(r) Y(\theta, \phi) \tag{1.1}$$

TABLE 1.1

Orbital Type		Expression for Y
s	1	1
p_x	x/r	$\sin\theta\cos\phi$
p_y	y/r	$\sin\theta\sin\phi$
p_z	z/r	$\cos\theta$
$d_{x^2-y^2}$	$(x^2-y^2)/r^2$	$\sin\theta\cos\theta\cos 2\phi$
d_{z^2}	$(3z^2-r^2)/r^2$	$3\cos^2\theta - 1$
d_{xy}	xy/r^2	$\sin\theta\cos\theta\sin 2\phi$
d_{xz}	xz/r^2	$\sin^2\theta\cos\phi$
d_{yz}	yz/r^2	$\sin^2\theta\sin\phi$

a simple product of a function, $R(r)$, which only depends upon the distance, r, of the electron from the nucleus, and a function $Y(\theta,\phi)$ which contains all the angular information needed to describe the wavefunction. The Schrödinger wave equation may only be solved exactly for one-electron atoms (e.g., H, Li^{2+}) in which case analytical expressions for R and Y are found. For many-electron atoms, the angular form of the atomic orbitals is the same as for the one-electron atom (shown in Table 1.1) but now the radial function $R(r)$ must be approximated in some way as will be shown below.

Figure 1.1a shows a plot of the wavefunction χ for an electron in a $1s$ orbital, as a function of distance from the nucleus. This has been chosen to be the x axis of an arbitrary coordination system. With increasing x, χ sharply decreases and becomes negligible outside a certain region indicated by the dashed lines. The boundary sur-

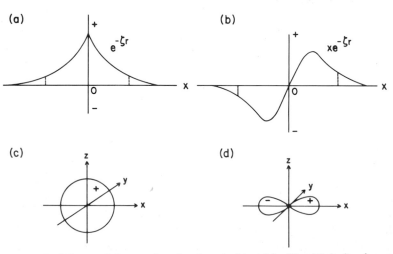

FIGURE 1.1. Radial part of the wavefunction for a $1s$ (a) and $2p$ (b) orbitals showing an arbitrary cutoff beyond which $R(r)$ is less than some small value. The surface in three dimensions defined by this radial cutoff is shown in (c) for the $1s$ orbital and in (d) for the $2p$ orbital.

face of the s orbital, outside of which the wavefunction has some critical (small) value, is shown in Figure 1.1c. The corresponding diagrams for a $2p_x$ orbital are shown in Figures 1.1b and 1.1d. Note that the wavefunction for this p orbital changes sign when $x \to -x$. It is often more convenient to represent the sign of the wavefunction by the presence or absence of shading of the orbital lobes as in 1.1 and 1.2.

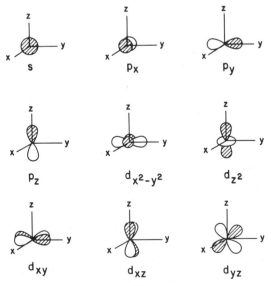

The characteristic features of s, p, and d orbitals using this convention are shown in Figure 1.2 where the positive lobes have been shaded. By squaring the wavefunction and integrating over a volume element the probability of finding an electron within that element is found. So there is a correspondence between the pictorial representations in Figure 1.2 and the electron density distribution in that orbital. In particular, the probability function or electron density is exactly zero for the p_x orbital at the nucleus $(x = 0)$. In fact, the wavefunction is zero at all points on the yz plane. This is the definition of a nodal plane. In general, an s orbital has no such angular nodes, a p orbital one node, and a d orbital two. For the d_{z^2} orbital there are two nodal cones as shown in 1.3 at an angle $\theta = 54.73°$. In general, an atomic wavefunction

FIGURE 1.2.   Atomic s, p, and d orbitals drawn using the shading convention described in the text.

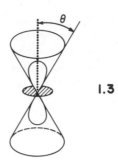

1.3

with quantum numbers n, l, m will have $n - 1$ nodes altogether, of which l are angular nodes, and, therefore, $n - l - 1$ are radial nodes. (Sometimes it is stated that there are n nodes altogether. In this case the node which always occurs as $r \rightarrow \infty$ is included in the count.)

As mentioned above the radial function $R(r)$ for many-electron atoms needs to be approximated in some way. The atomic orbitals most frequently employed in molecular calculations are Slater type orbitals (STOs) and Gaussian type orbitals (GTOs). Their mathematical form makes them relatively easy to handle in machine calculations. An STO with principal quantum number n is written as

$$\chi(r, \theta, \phi) \propto r^{n-1} \exp(-\zeta r) \cdot Y(\theta, \phi) \tag{1.2}$$

where ζ is the orbital exponent. ζ is obtained by applying the variational theorem to the atomic energy evaluated using the wavefunction of equation 1.2. This theorem tells us that an approximate wavefunction will always overestimate the energy of a given system. So minimization of the energy with respect to the variational parameter ζ will lead to determination of the best wavefunction of this type. Slater has compiled a set of rules which reproduce in general the variation of ζ across the Periodic Table.[6] Sometimes we may wish to be more exact and choose a double zeta basis set for our molecular calculation made up of wavefunctions of the type

$$\chi \propto r^{n-1} [c_1 \exp(-\zeta_1 r) + c_2 \exp(-\zeta_2 r)] \cdot Y \tag{1.3}$$

where now the atomic energy has been minimized with respect to ζ_1 and ζ_2. Often it is found that observables such as molecular geometry are best reproduced by *ab initio* calculations if "polarization" functions are added to the basis set. For example, we might choose a wavefunction for the $2p$ orbital of the carbon atom as

$$\chi = c_1 \chi(2p) + c_2 \chi(3d) \tag{1.4}$$

where the $3d$ wavefunction has an angular function, Y, corresponding to a d orbital and the radial part of equation 1.2 for $n = 3$. Commonly p functions are added to the basis set for hydrogen atoms. These polarization functions will lower the total energy calculated for the molecule according to the variation principle and their inclusion may lead to a better matching of observed and calculated geometries. However, these polarization functions do not generally mix strongly into the occupied molecular orbitals and are not chemically significant. Even in so-called "*ab initio*" calculations, therefore, there is considerable choice as to the basis set (equations 1.2–1.4) and indeed of the exponents, ζ, themselves. In practice the details of the basis set chosen for a given problem rely heavily on previous experience.

A general expression for a Gaussian type orbital (GTO) is

$$\chi(x, y, z) \propto x^i y^j z^k \exp(-\alpha r^2) \qquad (1.5)$$

where i, j, k are positive integers or zero. Here α is the orbital exponent. Orbitals of s, p, and d type result when $i + j + k = 0, 1, 2$, respectively. For example, a p_x orbital results for $i = 1$ and $j = k = 0$. The one major difference between STOs and GTOs is shown in **1.4** and **1.5**. Unlike GTOs, STOs are not smooth functions at the

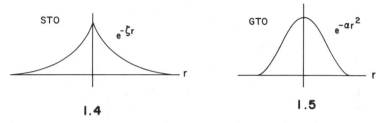

1.4   **1.5**

origin like the hydrogenic functions. The great convenience of GTOs, however, lies in the fact that evaluation of the molecular integrals needed in *ab initio* calculations is performed much more efficiently if GTOs are used. In practice, the functional behavior of an STO is simulated by a number of GTOs with different orbital exponents (equation 1.6)

$$\exp(-\zeta r) \simeq c_1 \exp(-\alpha_1 r^2) + c_2 \exp(-\alpha_2 r^2) + \cdots \qquad (1.6)$$

where GTOs with large and small exponents are designed to fit the center and tail portions, respectively, of an STO. If n GTOs are used to fit each STO, then the atomic wavefunctions are of STO-nG quality, using the terminology in current usage.

1.3. MOLECULAR ORBITALS

For a molecule with a total of m atomic basis functions $\{\chi_1, \chi_2, \cdots, \chi_m\}$, there will be a total of m resultant molecular orbitals constructed from them. For most purposes these atomic orbitals can be assumed to be real functions and normalized (equation 1.7) such that the probability of finding an electron in χ_μ when integrated over all space is unity. Here χ_μ^* is the complex conjugate of χ_μ. In equation 1.8 we show an alternative, useful, way of writing such integrals. The molecular orbitals of

$$\int \chi_\mu^* \chi_\mu \, d\tau = \int \chi_\mu^2 \, d\tau = \int \chi_\mu^2 \, dx \, dy \, dz = 1 \qquad (1.7)$$

$$\int \chi_\mu^* \chi_\mu \, d\tau = \langle \chi_\mu | \chi_\mu \rangle \qquad (1.8)$$

a molecule are usually approximated by writing them as a linear combination of atomic orbitals (LCAO) such that

$$\psi_i = c_{1i}\chi_1 + c_{2i}\chi_2 + \cdots + c_{mi}\chi_m$$

$$= \sum_\mu c_{\mu i}\chi_\mu \qquad (1.9)$$

where $i = 1, 2, \cdots, m$. These MOs are normalized and orthogonal (i.e., orthonormal), namely

$$\langle \psi_i | \psi_j \rangle = \int \psi_i^* \psi_j \, d\tau = \delta_{ij} \qquad (1.10)$$

where $\delta_{ij} = 1$ if $i = j$ and $\delta_{ij} = 0$ if $i \neq j$. Note that the sum in equation 1.9 runs over all the atomic orbitals of the basis set. The $c_{\mu i}$ are called the molecular orbital coefficients. They may be either positive or negative and the magnitude of the coefficient is related to the weight of that atomic orbital in the molecular orbital. Equation 1.9 is perhaps at first sight the most frightening aspect of delocalized molecular orbital theory. For a molecule of any reasonable size, this obviously represents quite a large sum. In actual fact, not all of the $c_{\mu i}$ will be significant in a given molecular orbital ψ_i. Some will be exactly zero, forced to be so by the symmetry of the molecule. Generally the more symmetric the molecule the larger the number of $c_{\mu i}$ which are zero. Furthermore, symmetry requirements often dictate relationships (sign and magnitude) between orbitals on different atoms. We shall devote a considerable amount of effort to provide simple ways to understand how and why the orbital coefficients in the molecular orbitals of molecules and solids turn out the way they do.

The molecular orbital coefficients $c_{\mu i}$ ($i, \mu = 1, 2, \cdots, m$) which specify the nature, and hence, energy of the orbital ψ_i, are determined by solving the eigenvalue equation of the effective one-electron Hamiltonian, H^{eff}, associated with the molecule (equation 1.11):

$$H^{\text{eff}} \psi_i = e_i \psi_i \qquad (1.11)$$

What H^{eff} is exactly we shall leave for the moment and discuss more fully in Chapter 8. The resultant energy e_i measures the effective potential exerted on an electron located in ψ_i. This molecular orbital energy is the expectation value of H^{eff}, that is,

$$e_i = \frac{\int \psi_i H^{\text{eff}} \psi_i \, d\tau}{\int \psi_i^2 \, d\tau} = \frac{\langle \psi_i | H^{\text{eff}} | \psi_i \rangle}{\langle \psi_i | \psi_i \rangle} \qquad (1.12)$$

$$= \langle \psi_i | H^{\text{eff}} | \psi_i \rangle \qquad (1.13)$$

Given two atomic orbitals χ_μ and χ_ν centered on two different atoms, the overlap integral $S_{\mu\nu}$ is defined as

$$S_{\mu\nu} = \langle \chi_\mu | \chi_\nu \rangle \qquad (1.14)$$

Its origin is clear from the spatial overlap of the two wavefunctions in **1.6** where we have chosen two $1s$ orbitals from Figure 1.1 as examples. **1.7** shows an alternative representation in terms of two orbital lobes. For the purposes of clarity this is better

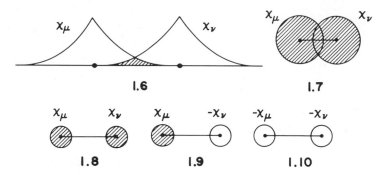

written as in **1.8**. According to the sign convention of **1.1**, the overlap integrals in
1.9 and **1.10** are given by equations 1.15 and 1.16, respectively. This simply shows

$$\langle \chi_\mu | -\chi_\nu \rangle = (-1) \langle \chi_\mu | \chi_\nu \rangle = -S_{\mu\nu} \tag{1.15}$$

$$\langle -\chi_\mu | -\chi_\nu \rangle = (-1)^2 \langle \chi_\mu | \chi_\nu \rangle = S_{\mu\nu} \tag{1.16}$$

that the overlap integral between two orbitals is positive when lobes of the same
sign overlap and negative when the two lobes have opposite signs. Figure 1.3 shows
pictorially some of the various types of overlap integrals that are encountered in
practice. The σ type overlaps shown in Figures 1.3a–1.3d contain no nodes along
the internuclear axis, the π type overlaps (Figure 1.3e–1.3g) are between orbitals
with one nodal plane containing this axis, and those of δ type (Figure 1.3h–1.3i)
contain two such nodal planes. The variation of the overlap integral with the distance
between the two atomic centers depends in detail upon the form of $R(r)$ chosen in
equation 1.1, but clearly will approach zero at large internuclear distances. When
the two interacting orbitals are identical, the overlap integral will be unity when the
separation is zero as shown by equation 1.7 for this hypothetical example. A complete S vs. r curve for the case of two $1s$ orbitals is shown in **1.11**. It may be readily
seen from Figure 1.1 that the overlap between an s orbital and a p orbital at $r = 0$ is
identically zero, as shown in **1.12**. The angular dependence of the overlap integral

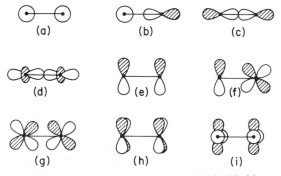

FIGURE 1.3.   Types of overlap integrals between atomic orbitals, (a)–(c) correspond to σ overlap, (e)–(g) correspond to π overlap and (h), (i) correspond to δ overlap.

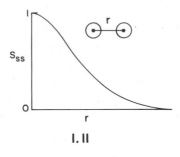

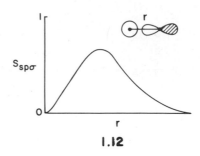

1.11   1.12

follows immediately from the analytic form of $Y(\theta, \phi)$ in equation 1.1. We can often write the overlap integral as in equation 1.17:

$$S_{\mu\nu} = S_{\mu\nu}(\lambda, r)\, f(\text{angular geometry}) \qquad (1.17)$$

$S_{\mu\nu}(\lambda, r)$ depends upon the distance between the two orbitals and the nature, $\lambda = \sigma$, π, or δ, of the overlap between them. It is also, of course, strongly dependent upon the identity of the atoms on which the orbitals μ and ν are located. The angular geometry dependent term is independent of the nature of the atoms themselves and only depends upon the description (s, p, or d) of the two orbitals.[6] The angular variation of some of the more common types of overlap integral are shown in Figure 1.4.

The energy of interaction associated with two overlapping atomic orbitals χ_μ and χ_ν is given by

$$H_{\mu\nu} = \langle \chi_\mu | H^{\text{eff}} | \chi_\nu \rangle \qquad (1.18)$$

The diagonal element $H_{\mu\nu}$ (when $\mu = \nu$ in equation 1.18) refers to the effective potential of an electron in the atomic orbital χ_μ. The off-diagonal element $H_{\mu\nu}$ can be approximated by the equation

$$H_{\mu\nu} = K(H_{\mu\mu} + H_{\nu\nu})\, S_{\mu\nu} \qquad (1.19)$$

which is known as the Wolfsberg–Helmholz formula. (K is a proportionality constant.) Since the $H_{\mu\nu}$ are negative quantities, $H_{\mu\nu} \propto -S_{\mu\nu}$ which implies that the interaction energy between two orbitals is negative (i.e., stabilizing) when their overlap integral is positive. There are a number of ways to compute $H_{\mu\nu}$, depending upon the level of approximation. The important result, however, is that, whatever the exact functional form, there is a direct relationship between $H_{\mu\nu}$ and $S_{\mu\nu}$.

The overlap integral, $S_{\mu\nu}$, and the interaction integral $H_{\mu\nu}$ (often called the resonance integral or hopping integral) are symmetric such that $S_{\mu\nu} = S_{\nu\mu}$ and $H_{\mu\nu} = H_{\nu\mu}$. (This second equality arises because of the Hermitian properties of the Hamiltonian.) For an arbitrary function ψ_i (equation 1.9), the integrals needed in equation 1.12 may be written as

$$\langle \psi_i | \psi_i \rangle = \left\langle \sum_\mu \chi_\mu c_{\mu i} \,\middle|\, \sum_\nu \chi_\nu c_{\nu i} \right\rangle$$

$$= \sum_\mu \sum_\nu c_{\mu i} S_{\mu\nu} c_{\nu i} \equiv A \qquad (1.20)$$

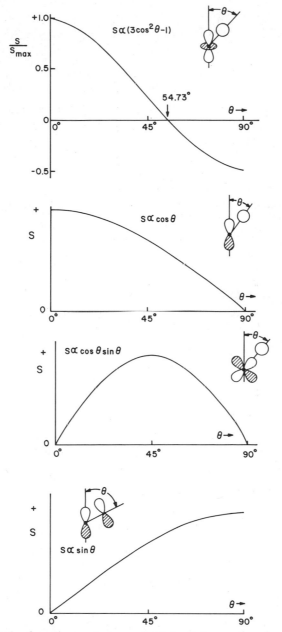

FIGURE 1.4. Angular dependence of the overlap integral for some commonly encountered pairs of atomic orbitals.

and

$$\langle \psi_i | H^{\text{eff}} | \psi_i \rangle = \left\langle \sum_\mu \chi_\mu c_{\mu i} \left| H^{\text{eff}} \right| \sum_\nu \chi_\nu c_{\nu i} \right\rangle$$

$$= \sum_\mu \sum_\nu c_{\mu i} H_{\mu\nu} c_{\nu i} \equiv B \qquad (1.21)$$

If ψ_i is an eigenfunction of H^{eff}, then it will be normalized and equation 1.12 will result, but for an arbitrary ψ_i, A will not be equal to unity. From equation 1.12, the energy e_i is given by

$$e_i = \frac{\langle \psi_i | H^{\text{eff}} | \psi_i \rangle}{\langle \psi_i | \psi_i \rangle} = \frac{B}{A} \qquad (1.22)$$

According to the variational theorem, the coefficients $c_{\mu i}$ are chosen such that the energy is minimized, that is,

$$\frac{\partial e_i}{\partial c_{1i}} = \frac{\partial e_i}{\partial c_{2i}} = \cdots = \frac{\partial e_i}{\partial c_{mi}} = 0 \qquad (1.23)$$

For any coefficient $c_{\kappa i}$ ($\kappa = 1, 2, \cdots, m$)

$$\frac{\partial e_i}{\partial c_{\kappa i}} = \frac{\partial}{\partial c_{\kappa i}} \left(\frac{B}{A} \right) = \frac{1}{A} \frac{\partial B}{\partial c_{\kappa i}} - \frac{B}{A^2} \frac{\partial A}{\partial c_{\kappa i}}$$

$$= \frac{1}{A} \left(\frac{\partial B}{\partial c_{\kappa i}} - e_i \frac{\partial A}{\partial c_{\kappa i}} \right)$$

$$= 0 \text{ from equation 1.23} \qquad (1.24)$$

Therefore,

$$\frac{\partial B}{\partial c_{\kappa i}} - e_i \frac{\partial A}{\partial c_{\kappa i}} = 0 \qquad (1.25)$$

Since the indices μ and ν in equations 1.20 and 1.21 are only used for summation

$$\frac{\partial A}{\partial c_{\kappa i}} = \sum_\nu S_{\kappa\nu} c_{\nu i} + \sum_\mu S_{\mu\kappa} c_{\mu i}$$

$$= 2 \sum_\mu S_{\kappa\mu} c_{\mu i} \qquad (1.26)$$

Similarly,

$$\frac{\partial B}{\partial c_{\kappa i}} = 2 \sum_\mu H_{\kappa\mu} c_{\mu i} \qquad (1.27)$$

Combining equations 1.25–1.27,

$$\sum_\mu H_{\kappa\mu} c_{\mu i} - e_i \sum_\mu S_{\kappa\mu} c_{\mu i} = \sum_\mu (H_{\kappa\mu} - e_i S_{\kappa\mu}) c_{\mu i} = 0 \qquad (1.28)$$

Here as a reminder, i indexes the molecular orbital level while μ, ν, and κ index the μth, νth, and κth atomic orbitals, respectively, in the LCAO expansion of equation 1.9. Equation 1.28 is satisfied for $\kappa = 1, 2, \cdots, m$ and the explicit form of these m equations is

$$(H_{11} - e_i S_{11})\, c_{1i} + (H_{12} - e_i S_{12})\, c_{2i} + \cdots + (H_{1m} - e_i S_{1m})\, c_{mi} = 0$$

$$(H_{21} - e_i S_{21})\, c_{1i} + (H_{22} - e_i S_{22})\, c_{2i} + \cdots + (H_{2m} - e_i S_{2m})\, c_{mi} = 0$$

$$\vdots \qquad\qquad \vdots \qquad\qquad \vdots$$

$$(H_{m1} - e_i S_{m1})\, c_{1i} + (H_{m2} - e_i S_{m2})\, c_{2i} + \cdots + (H_{mm} - e_i S_{mm})\, c_{mi} = 0$$

$$(1.29)$$

A well-known mathematical result from the theory of such simultaneous equations requires the following determinant to vanish.

$$
\begin{vmatrix}
H_{11} - e_i S_{11} & H_{12} - e_i S_{12} & \cdots H_{1m} - e_i S_{1m} \\
H_{21} - e_i S_{21} & H_{22} - e_i S_{22} & \cdots H_{2m} - e_i S_{2m} \\
\vdots & \vdots & \vdots \\
H_{m1} - e_i S_{m1} & H_{m2} - e_i S_{m2} & \cdots H_{mm} - e_i S_{mm}
\end{vmatrix} = 0
$$

$$(1.30)$$

Solution of the polynomial equation that results from expansion of the determinant 1.30 provides m orbital energies e_i $(i = 1, 2, \cdots, m)$ which, according to the variational theorem, are a set of upper bounds to the true orbital energies. Written in matrix notation, equation 1.30 becomes

$$\left| H_{\kappa\mu} - e_i S_{\kappa\mu} \right| = 0 \qquad (1.31)$$

As we will see in Chapter 2, the coefficients $c_{\mu i}$ are determined from the secular equations 1.29 and the normalization condition

$$\langle \psi_i | \psi_i \rangle = \sum_\mu \sum_\nu c_{\mu i} S_{\mu\nu} c_{\nu i} = 1 \qquad (1.32)$$

REFERENCES

1. G. A. Segal, editor, *Semi-Empirical Methods of Electronic Structure Calculation*, Parts A, B, Plenum, New York (1977).
2. J. A. Pople and D. L. Beveridge, *Approximate Molecular Orbital Theory*, McGraw-Hill, New York (1970).
3. M. J. S. Dewar, *The Molecular Orbital Theory of Organic Chemistry*, McGraw-Hill, New York (1969).
4. S. P. McGlynn, L. G. Vanquickenborne, M. Kinoshita, and D. G. Carroll, *Introduction to Applied Quantum Chemistry*, Holt, Rinehart & Winston, New York (1972).
5. W. Kutzelnigg, *Einführung in die Theoretische Chemie*, Band 2, Verlag Chemie, Weinheim (1978).
6. J. K. Burdett, *Molecular Shapes*, Wiley, New York (1980).

CHAPTER TWO

Concepts of Bonding and Orbital Interaction

2.1. ORBITAL INTERACTION ENERGY

The derivations of Chapter 1 were very general ones. Here we look in some detail at the illustrative case of a two-center–two-orbital problem. Two atomic orbitals, χ_1 and χ_2, are centered on the two atoms A and B (2.1). (In Chapter 3 we will show

$$\chi_1 \quad \chi_2$$
$$\overset{\bullet}{A} \relbar\joinrel\relbar \overset{\bullet}{B}$$

2.1

how the results can be generalized to the case of two orbitals located on molecular fragments A and B.) The molecular orbitals resulting from the interaction between χ_1 and χ_2 can be written as

$$\psi_1 = \chi_1 c_{11} + \chi_2 c_{21}$$
$$\psi_2 = \chi_1 c_{12} + \chi_2 c_{22} \tag{2.1}$$

The overlap and interaction integrals to consider are

$$\langle \chi_1 | \chi_1 \rangle = \langle \chi_2 | \chi_2 \rangle = 1$$
$$\langle \chi_1 | \chi_2 \rangle = S_{12} \tag{2.2}$$

and

$$\langle \chi_1 | H^{\text{eff}} | \chi_1 \rangle = H_{11} = e_1^0$$
$$\langle \chi_2 | H^{\text{eff}} | \chi_2 \rangle = H_{22} = e_2^0$$
$$\langle \chi_1 | H^{\text{eff}} | \chi_2 \rangle = H_{12} = \Delta_{12} \tag{2.3}$$

where at the right-hand side of equation 2.3 we introduce a notation which will be useful. The diagonal elements H_{11} and H_{22} are sometimes called the Coulomb integrals. Recall from Section 1.3 that $S_{12} = S_{21}$ so that $\Delta_{12} = \Delta_{21}$. If χ_1 and χ_2 are so arranged that S_{12} is positive, then from equation 1.19,

$$\Delta_{12} \propto -S_{12} < 0 \qquad (2.4)$$

The molecular orbital energies in this two orbital case, e_i ($i = 1, 2$), are obtained by solving the secular determinant (equation 1.30) shown in equation 2.5 for this particular example

$$\begin{vmatrix} e_1^0 - e_i & \Delta_{12} - e_i S_{12} \\ \Delta_{12} - e_i S_{12} & e_2^0 - e_i \end{vmatrix} = 0 \qquad (2.5)$$

Expansion of equation 2.5 leads to

$$(e_1^0 - e_i)(e_2^0 - e_i) - (\Delta_{12} - e_i S_{12})^2 = 0 \qquad (2.6)$$

The form of the solutions e_i of this equation will be examined for a degenerate case $(e_1^0 = e_2^0)$ and for the general nondegenerate case $(e_1^0 \neq e_2^0)$.

A.   DEGENERATE INTERACTION

For $e_1^0 = e_2^0$, then, solution of equation 2.6 leads to two values for the e_i ($i = 1, 2$)

$$e_1 = \frac{e_1^0 + \Delta_{12}}{1 + S_{12}}$$

$$e_2 = \frac{e_1^0 - \Delta_{12}}{1 - S_{12}} \qquad (2.7)$$

When the interaction between χ_1 and χ_2 is not strong (S_{12} is small), then some very useful mathematical approximations may be used to simplify equation 2.7. Using the first two expressions in Table 2.1, the equations 2.8 and 2.9 result in

TABLE 2.1.   Some Mathematical Simplifications

Function	Approximate Expression
$\dfrac{1}{1 + x}$	$1 - x + x^2 - \cdots$
$\dfrac{1}{1 - x}$	$1 + x + x^2 + \cdots$
$\sqrt{1 + x}$	$1 + \frac{1}{2} x - \cdots$
$\dfrac{1}{\sqrt{1 + x}}$	$1 - \frac{1}{2} x + \cdots$

$$e_1 = \frac{e_1^0 + \Delta_{12}}{1 + S_{12}} = (e_1^0 + \Delta_{12})(1 - S_{12} + S_{12}^2 - \cdots)$$

$$\simeq e_1^0 + (\Delta_{12} - e_1^0 S_{12}) - S_{12}(\Delta_{12} - e_1^0 S_{12}) \tag{2.8}$$

$$e_2 = \frac{e_1^0 - \Delta_{12}}{1 - S_{12}} = (e_1^0 - \Delta_{12})(1 + S_{12} + S_{12}^2 + \cdots)$$

$$\simeq e_1^0 - (\Delta_{12} - e_1^0 S_{12}) - S_{12}(\Delta_{12} - e_1^0 S_{12}) \tag{2.9}$$

For any realistic case, e_1^0 is negative and normally $(\Delta_{12} - e_1^0 S_{12})$ is negative too (i.e., $|\Delta_{12}| > |e_1^0 S_{12}|$). So ψ_1 is stabilized by the presence of the second term in equation 2.8 but ψ_2 is destabilized by the second term in equation 2.9. Both levels are destabilized by the third term in equations 2.8 and 2.9. These results are shown pictorially in 2.2. The important result is that *with respect to the atomic orbital*

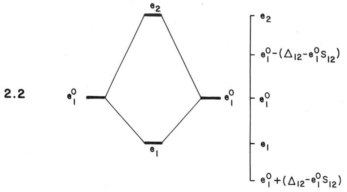

level e_1^0 the raising (destabilization) of the e_2 level is greater than the lowering (stabilization) of the e_1 level. The origin of this effect is easy to see. It arises because the orbitals χ_1 and χ_2 are not orthogonal (i.e., $S_{12} \neq 0$). Putting $S_{12} = 0$ in equation 2.7 leads to $e_i = e_i^0 \pm \Delta_{12}$ and this asymmetry disappears. Putting electrons into these resultant molecular orbitals allows calculation of the total interaction energy, ΔE, on bringing together the two atomic orbitals χ_1 and χ_2. Two important cases are shown in 2.3 and 2.4, the two-orbital–two-electron case and the two-orbital–four-

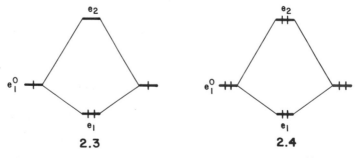

electron case, respectively. Using the results of equations 2.8 and 2.9, and weighting each orbital energy by the number of electrons in that orbital leads to

$$\Delta E^{(2)} = 2e_1 - 2e_1^0 \simeq 2(\Delta_{12} - e_1^0 S_{12})(1 - S_{12}) \qquad (2.10)$$

$$\Delta E^{(4)} = 2(e_1 + e_2) - 4e_1^0 \simeq -4S_{12}(\Delta_{12} - e_1^0 S_{12}) \qquad (2.11)$$

Since for $S_{12} > 0$ the term $(\Delta_{12} - e_1^0 S_{12})$ is negative, the two-orbital–two-electron interaction is stabilizing (i.e., $\Delta E^{(2)} < 0$) but the two-orbital–four-electron interaction is destabilizing (i.e., $\Delta E^{(4)} > 0$).

The arrangement shown in **2.3** is not the only way to put two electrons into these two molecular orbitals. **2.5** shows an alternative pattern with a total interac-

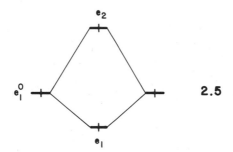

2.5

tion energy of $\Delta E^{(4)}/2$. Since the electron spins in **2.5** give rise to the lowest energy when they are parallel, this arrangement is called the high spin case, to be contrasted with the low spin arrangement of **2.3**, where they are paired. The stability of the high spin state compared to the low spin one will be examined in detail in Chapter 8 but as a general rule of thumb, when the interaction between the atomic orbitals is strong, the resultant molecular orbitals are split by a moderate amount, and the low spin situation is favored. When the two *molecular* orbitals are degenerate or close together in energy then the high spin arrangement is more stable.

B.   NONDEGENERATE INTERACTION

When $e_1^0 \neq e_2^0$, without loss of generality e_1^0 may be assumed to be lower in energy than e_2^0, that is, $e_2^0 - e_1^0 > 0$. Rearrangement of equation 2.6 leads to

$$(1 - S_{12}^2)\, e_i^2 + (2\, \Delta_{12} S_{12} - e_1^0 - e_2^0)\, e_i + (e_1^0 e_2^0 - \Delta_{12}^2) = 0 \qquad (2.12)$$

and the solutions of this quadratic equation given by

$$e_1 = \frac{-b - \sqrt{D}}{2a}$$

$$e_2 = \frac{-b + \sqrt{D}}{2a} \qquad (2.13)$$

where

$$a = 1 - S_{12}^2$$

$$b = 2\, \Delta_{12} S_{12} - e_1^0 - e_2^0$$

$$D = b^2 - 4ac \qquad (2.14)$$

with

$$c = e_1^0 e_2^0 - \Delta_{12}^2$$

Approximate expressions for e_1 and e_2 are found as follows. First D can be expanded as

$$D = (2\Delta_{12} S_{12} - e_1^0 - e_2^0)^2 - 4(1 - S_{12}^2)(e_1^0 e_2^0 - \Delta_{12}^2)$$

$$= (e_1^0 - e_2^0)^2 + 4(\Delta_{12} - e_1^0 S_{12})(\Delta_{12} - e_2^0 S_{12}) \tag{2.15}$$

From Table 2.1,

$$\sqrt{D} = -(e_1^0 - e_2^0)\left[1 + \frac{4(\Delta_{12} - e_1^0 S_{12})(\Delta_{12} - e_2^0 S_{12})}{(e_1^0 - e_2^0)^2}\right]^{1/2}$$

$$\simeq -(e_1^0 - e_2^0)\left[1 + \frac{2(\Delta_{12} - e_1^0 S_{12})(\Delta_{12} - e_2^0 S_{12})}{(e_1^0 - e_2^0)^2}\right] \tag{2.16}$$

assuming small interaction between χ_1 and χ_2 as before. [We have a negative sign in front of $e_1^0 - e_2^0$ (<0) to ensure that $\sqrt{D} > 0$.] By manipulation of equations 2.13, 2.14, and 2.16,

$$2(1 - S_{12}^2)\, e_1 = e_1^0 + e_2^0 - 2\Delta_{12} S_{12} + (e_1^0 - e_2^0)$$

$$\cdot \left[1 + \frac{2(\Delta_{12} - e_1^0 S_{12})(\Delta_{12} - e_2^0 S_{12})}{(e_1^0 - e_2^0)^2}\right]$$

$$= 2\left[e_1^0 - \Delta_{12} S_{12} + \frac{(\Delta_{12} - e_1^0 S_{12})(\Delta_{12} - e_2^0 S_{12})}{e_1^0 - e_2^0}\right]$$

$$= 2\left[e_1^0(1 - S_{12}^2) + \frac{(\Delta_{12} - e_1^0 S_{12})^2}{e_1^0 - e_2^0}\right] \tag{2.17}$$

and so

$$e_1 \simeq e_1^0 + \frac{(\Delta_{12} - e_1^0 S_{12})^2}{e_1^0 - e_2^0} \tag{2.18}$$

A similar expression is found analogously for e_2:

$$e_2 \simeq e_2^0 + \frac{(\Delta_{12} - e_2^0 S_{12})^2}{e_2^0 - e_1^0} \tag{2.19}$$

2.6 shows the orbital energies pictorially. As a result of the interaction, the lower level e_1^0 is depressed in energy and the higher level e_2^0 is raised in energy. Notice that since $e_2^0 > e_1^0$, $(\Delta_{12} - e_1^0 S_{12})^2 < (\Delta_{12} - e_2^0 S_{12})^2$. In other words, the higher energy orbital is destabilized more than the lower energy orbital is stabilized, just as found for the degenerate case above. The total interaction energies for the analogous two-orbital–two-electron and two-orbital–four-electron cases of 2.7 and 2.8 are simply obtained.

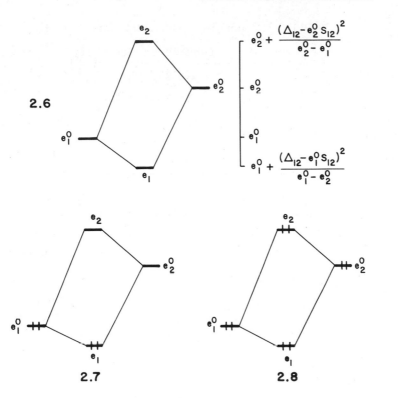

2.6

2.7   **2.8**

$$\Delta E^{(2)} = 2e_1 - 2e_1^0$$

$$\simeq 2\frac{(\Delta_{12} - e_1^0 S_{12})^2}{e_1^0 - e_2^0} \qquad (2.20)$$

$$\Delta E^{(4)} = 2(e_1 + e_2) - 2(e_1^0 + e_2^0)$$

$$\simeq -4S_{12}\left(\Delta_{12} - \frac{e_1^0 + e_2^0}{2} \cdot S_{12}\right) \qquad (2.21)$$

$\Delta E^{(2)}$ is negative since $e_1^0 - e_2^0 < 0$. We have already noted that $(\Delta_{12} - e_1^0 S_{12})$ and $(\Delta_{12} - e_2^0 S_{12})$ are negative if $S_{12} > 0$ and thus $\Delta E^{(4)}$ is positive.

The qualitative aspects of these orbital interactions may be summarized as in Table 2.2 which shows that:

1. In both degenerate and nondegenerate orbital interaction cases, a two-orbital–two-electron interaction is stabilizing while a two-orbital–four-electron interaction is destabilizing.

2. Regardless of whether the orbital picture contains two or four electrons, the magnitude of the total interaction energy increases with increasing overlap.

3. In a nondegenerate orbital interaction the magnitude of the interaction en-

TABLE 2.2.   Summary of Orbital Interactions

Case	Orbital Interaction Energy
2.3	$\Delta E^{(2)} \propto - S_{12}$
2.4	$\Delta E^{(4)} \propto S_{12}^2$
2.7	$\Delta E^{(2)} \propto - \dfrac{S_{12}^2}{\lvert e_1^0 - e_2^0 \rvert}$
2.8	$\Delta E^{(4)} \propto S_{12}^2$

ergy is inversely proportional to the energy difference between the interacting orbitals.

4. In both degenerate and nondegenerate cases the resultant upper molecular level is destabilized more than the lower one is stabilized.

These results are important ones and will be frequently used in later chapters. It is worthwhile to mention here, however, that the destabilization associated with the two-orbital–four-electron situation is behind the nonexistence of a bound molecule for He_2 or Be_2, which have this orbital pattern.

2.2.   MOLECULAR ORBITAL COEFFICIENTS

The MO coefficients c_{1i} and c_{2i} of equation 2.1 are determined from the simultaneous equations 1.29 (shown for the present case in equation 2.22) and the normalization condition equation 2.23.

$$(e_1^0 - e_i) c_{1i} + (\Delta_{12} - e_i S_{12}) c_{2i} = 0$$

$$(\Delta_{12} - e_i S_{12}) c_{1i} + (e_2^0 - e_i) c_{2i} = 0 \tag{2.22}$$

$$\langle \psi_i | \psi_i \rangle = c_{1i}^2 + 2c_{1i} c_{2i} S_{12} + c_{2i}^2 = 1 \tag{2.23}$$

The coefficients c_{1i} and c_{2i} for $i = 1, 2$ will be obtained for the degenerate and nondegenerate cases described above.

A.   DEGENERATE INTERACTION

Since $e_1^0 = e_2^0$, either of the equations 2.22 leads to

$$\frac{c_{21}}{c_{11}} = - \frac{e_1^0 - e_1}{\Delta_{12} - e_1 S_{12}} = 1 \tag{2.24}$$

and so from equation 2.23,

$$c_{11} = c_{21} = \pm \frac{1}{\sqrt{2 + 2S_{12}}} \qquad (2.25)$$

which leads to

$$\psi_1 = \pm \frac{1}{\sqrt{2 + 2S_{12}}} (\chi_1 + \chi_2) \qquad (2.26)$$

The coefficients of the MO ψ_2 are obtained in a similar manner

$$\frac{c_{12}}{c_{22}} = -\frac{e_1^0 - e_2}{\Delta_{12} - e_2 S_{12}} = -1 \qquad (2.27)$$

Use of the normalization condition leads to

$$\psi_2 = \pm \frac{1}{\sqrt{2 - 2S_{12}}} (\chi_1 - \chi_2) \qquad (2.28)$$

The nodal properties of the MOs ψ_1 and ψ_2 are shown in Figure 2.1 where the positive signs from equations 2.26 and 2.28 are arbitrarily chosen. Equation 1.11 shows that if ψ_i is an eigenfunction of H^{eff} so is $-\psi_i$. What is important, therefore, is not the overall sign of the MO ψ_i, but the relative signs of its MO coefficients. Irrespective of overall sign chosen for ψ_i, the important point is that χ_1 and χ_2 are combined in-phase in the lower lying orbital ψ_1 and out-of-phase in the higher lying orbital ψ_2. From now on we will only show one sign for our MOs.

While $c_{11} = c_{21}$ and $c_{12} = c_{22}$, it is clear from equations 2.26 and 2.28 that $c_{11} \neq c_{12}$. This is a consequence of the relationship $1 > S_{12} > 0$. The general result is that the atomic coefficients for the higher lying level in Figure 2.1 are larger than those for the lower lying level.

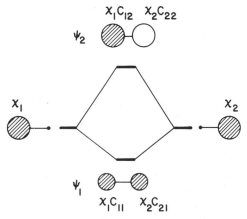

FIGURE 2.1.   Molecular orbital diagram showing details of the degenerate interaction between the two atomic s orbitals, χ_1 and χ_2.

B.   NONDEGENERATE INTERACTION

From equations 2.18 and 2.22,

$$\frac{c_{21}}{c_{11}} = -\frac{e_1^0 - e_1}{\Delta_{12} - e_1 S_{12}} = \frac{t}{1 - tS_{12}} \tag{2.29}$$

where $t = (\Delta_{12} - e_1^0 S_{12})/(e_1^0 - e_2^0)$. From Table 2.1, this ratio may be rewritten

$$\frac{c_{21}}{c_{11}} = \frac{t}{1 - tS_{12}} = t(1 + tS_{12} + \cdots) \simeq t \tag{2.30}$$

by neglecting terms greater than second order in t and S_{12}. Using the normalization condition (equation 2.23) and this result

$$\left(\frac{1}{c_{11}}\right)^2 = 1 + 2\left(\frac{c_{21}}{c_{11}}\right) S_{12} + \left(\frac{c_{21}}{c_{11}}\right)^2 = 1 + 2tS_{12} + t^2 \tag{2.31}$$

From Table 2.1, this may be rearranged and approximated as

$$c_{11} = \frac{1}{\sqrt{1 + 2tS_{12} + t^2}} \simeq 1 - tS_{12} - \tfrac{1}{2} t^2 \tag{2.32}$$

Combined with equation 2.30,

$$c_{21} = tc_{11} = t\left(1 - tS_{12} - \tfrac{1}{2} t^2\right) \simeq t \tag{2.33}$$

where, as before, terms greater than second order in t and S_{12} have been neglected. The final form of the MO ψ_1 is then

$$\psi_1 \simeq \left(1 - tS_{12} - \tfrac{1}{2} t^2\right) \chi_1 + t\chi_2 \tag{2.34}$$

with a similar expression for ψ_2

$$\psi_2 \simeq t'\chi_1 + \left(1 - t'S_{12} - \tfrac{1}{2} t'^2\right) \chi_2 \tag{2.35}$$

where $t' = (\Delta_{12} - e_2^0 S_{12})/(e_2^0 - e_1^0)$. The two functions t and t' are often called the mixing coefficients because t, for example, describes how orbital χ_2 mixes into χ_1 to give an orbital still largely χ_1 in character.
    Since invariably $\Delta_{12} - e_i^0 S_{12} < 0$,

$$t = \frac{\Delta_{12} - e_1^0 S_{12}}{e_1^0 - e_2^0} = \frac{(-)}{(-)} > 0 \tag{2.36}$$

and

$$t' = \frac{\Delta_{12} - e_2^0 S_{12}}{e_2^0 - e_1^0} = \frac{(-)}{(+)} < 0 \tag{2.37}$$

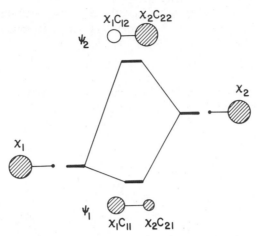

FIGURE 2.2. Molecular orbital diagram showing details of the nondegenerate interaction between two atomic s orbitals, χ_1 and χ_2.

where the symbols (+) and (−) indicate that the mathematical quantities represented by the parentheses have positive and negative signs, respectively. In a normal orbital interaction, therefore, the higher energy orbital χ_2 mixes in-phase into the lower level χ_1 to give the lower lying MO ψ_1, while the lower level χ_1 mixes out-of-phase into the higher level χ_2 to give the higher lying MO ψ_2. The magnitudes of the mixing coefficients t and t' are small when S_{12} and Δ_{12} are small. So the major orbital character of the lower lying MO, ψ_1, is given by the lower atomic orbital χ_1. Conversely, the major orbital character in ψ_2 is contributed by χ_2. As the two levels e_1^0 and e_2^0 become closer in energy, the weight of the higher atomic level χ_2 in the lower lying MO ψ_1 increases, as does the weight of χ_1 in ψ_2. As examined in the degenerate case, when $e_1^0 = e_2^0$, then χ_1 and χ_2 have equal weights in ψ_1 and ψ_2. The nodal properties of the MOs ψ_1 and ψ_2 are illustrated in Figure 2.2 where the relative magnitudes of the coefficients c_{1i} and c_{2i} ($i = 1, 2$) are represented by the relative sizes of the orbital lobes χ_1 and χ_2, respectively. Here again it may easily be shown that $t' > t$, or in other words, that the atomic coefficients for the high lying level ψ_2 in Figure 2.2 will be larger than those for the low energy combination ψ_1.

2.3. ELECTRON DENSITY DISTRIBUTION

One way that provides further insight into the energy changes that occur when χ_1 and χ_2 are allowed to interact is to use equation 1.13 along with the form of the ψ_i to calculate the new orbital energies

$$e_1 = \langle \psi_1 | H^{\text{eff}} | \psi_1 \rangle$$
$$= \langle (1 - tS_{12} - \tfrac{1}{2} t^2) \chi_1 + t\chi_2 | H^{\text{eff}} | (1 - tS_{12} - \tfrac{1}{2} t^2) \chi_1 + t\chi_2 \rangle$$
$$\simeq (1 - 2tS_{12} - t^2) e_1^0 + 2t\, \Delta_{12} + t^2 e_2^0 \qquad (2.38)$$

Here terms greater than second order in t and S_{12} have been omitted. It is easy to show that equations 2.38 and 2.18 are identical. An analogous equation holds for e_2:

$$e_2 \simeq (1 - 2t'S_{12} - t'^2) e_2^0 + 2t' \Delta_{12} + t'^2 e_1^0 \tag{2.39}$$

The origin of the various terms in these two equations is best understood by looking at the electron density distribution associated with ψ_1 and ψ_2. This is given in general by ψ_i^2. In a way analogous to the derivation of equations 2.38 and 2.39, this can be seen to be

$$\psi_1^2 \simeq (1 - 2tS_{12} - t^2) \chi_1^2 + 2t\chi_1\chi_2 + t^2\chi_2^2 \tag{2.40}$$

and

$$\psi_2^2 \simeq (1 - 2t'S_{12} - t'^2) \chi_2^2 + 2t'\chi_1\chi_2 + t'^2\chi_1^2 \tag{2.41}$$

Upon integration over space,

for ψ_1 $1 = (1 - 2tS_{12} - t^2) \langle \chi_1 | \chi_1 \rangle + 2t \langle \chi_1 | \chi_2 \rangle + t^2 \langle \chi_2 | \chi_2 \rangle$ (2.42)

for ψ_2 $1 = (1 - 2t'S_{12} - t'^2) \langle \chi_2 | \chi_2 \rangle + 2t' \langle \chi_1 | \chi_2 \rangle + t'^2 \langle \chi_1 | \chi_1 \rangle$

$$\tag{2.43}$$

These equations show that the electron density associated with ψ_1 may be decomposed into a density $(1 - 2tS_{12} - t^2)$ in the region of atom A which holds orbital χ_1, a density t^2 in the region of atom B which holds orbital χ_2, and a density $2t \langle \chi_1 | \chi_2 \rangle = 2tS_{12}$ in the region between A and B. A similar decomposition occurs for ψ_2. For positive t (the case for ψ_1) then, as shown in 2.9, there occurs a shift of electron density from the region of A to that between A and B. Energetically from equation 2.38 a stabilization results and the atoms A and B will experience an attractive contribution to their pairwise energy if ψ_1 is occupied by electrons. ψ_1 is thus a bonding molecular orbital. For the case of ψ_2, t' is negative and this results (2.10) in removal of electron density from the region between A and B (equation 2.43). A corresponding destabilization (equation 2.39) results and ψ_2 is thus an antibonding orbital.   Figure 2.3 shows this for the degenerate interaction in terms of the electron density distribution along the internuclear axis. ψ_1^2 is larger than $\chi_1^2 + \chi_2^2$ and ψ_2^2 is smaller than $\chi_1^2 + \chi_2^2$ in the bonding region.

2.9   2.10

For a polyatomic molecule with molecular orbitals described in general by equation 2.44, this analysis may be extended to give the net electron population of an atomic orbital $(P_{\mu\mu})$ and the overlap population $(P_{\mu\nu})$ between two atomic orbitals χ_μ and χ_ν located on two atoms A, B in the molecule. These two terms will cor-

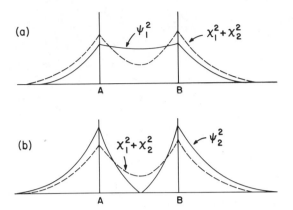

FIGURE 2.3.   (a) Buildup of electron density between the nuclei in ψ_1 compared to two super-imposed atomic densitites (dahsed curve). (b) Depletion of electron density between the nuclei in ψ_2 compared to two superimposed atomic densities (dashed curve).

$$\psi_i = \sum_\mu c_{\mu i} \chi_\mu \qquad (2.44)$$

respond, respectively, to the amount of electron density left behind in orbital χ_μ after the interaction between the constituent atomic orbitals of the molecule, and the amount transferred to the region between A and B which will contribute to AB bonding. If each molecular orbital ψ_i contains n_i electrons ($n_i = 0, 1,$ or 2), then

$$P_{\mu\mu} = \sum_i n_i c_{\mu i}^2 \qquad (2.45)$$

$$P_{\mu\nu} = \sum_i 2 n_i c_{\mu i} c_{\nu i} S_{\mu\nu} \qquad (2.46)$$

a larger overlap population, $P_{\mu\nu}$, implies a stronger bond and a larger bond order between the two atoms A and B. But the actual numbers must be used with care. A transition-metal–transition-metal single bond will invariably have a smaller overlap population than a carbon–carbon single bond. The overlap integral, $S_{\mu\nu}$, in equation 2.46 is expected to be smaller between two diffuse metal d orbitals than between $2s$ and $2p$ atomic orbitals on carbon. The gross population of χ_μ, q_μ is defined as

$$q_\mu = P_{\mu\mu} + \frac{1}{2} \sum_{\nu(\neq\mu)} P_{\mu\nu} \qquad (2.47)$$

Notice that the shared electron density, $P_{\mu\nu}$, is divided equally between the two atoms in question. The gross atomic charge on each atom is simply the sum of all the q_μ, belonging to that atom minus the nuclear charge of the atom on which orbital χ_μ is located.

We will now apply these ideas specifically to the orbital situations depicted in 2.3, 2.4, 2.7, and 2.8. Initially for the degenerate interaction of 2.2 for the orbital occupation 2.3

$$P_{11} = P_{22} = 2\left(\frac{1}{\sqrt{2 + 2S_{12}}}\right)^2 = \frac{1}{1 + S_{12}} < 1 \qquad (2.48)$$

and

$$P_{12} = 4\left(\frac{1}{\sqrt{2 + 2S_{12}}}\right)^2 S_{12} = \frac{2S_{12}}{1 + S_{12}} > 0 \qquad (2.49)$$

which shows a positive bond overlap population and a loss of electron density into the bonding region from orbitals χ_1 and χ_2. This gives rise to a stablizing situation. The converse is true for the case of **2.4** however. Here

$$P_{11} = P_{22} = 2\left(\frac{1}{\sqrt{2 + 2S_{12}}}\right)^2 + 2\left(\frac{1}{\sqrt{2 - 2S_{12}}}\right)^2 = \frac{2}{1 - S_{12}^2} > 2 \qquad (2.50)$$

and

$$P_{12} = 4\left(\frac{1}{\sqrt{2 + 2S_{12}}}\right)^2 S_{12} - 4\left(\frac{1}{\sqrt{2 - 2S_{12}}}\right)^2 S_{12} = \frac{-4S_{12}^2}{1 - S_{12}^2} < 0 \quad (2.51)$$

which shows a transfer of electron density out of the bonding region such that $P_{ii} > 2$ and the bond overlap is negative. This stems from the result, shown in Figure 2.3, that with respect to the superposed atomic orbital density $\chi_1^2 + \chi_2^2$, the electron gain resulting from occupation of ψ_2 in the region between A and B is smaller than the electron loss from this region by occupation of ψ_2. These results and the corresponding ones for the situations of **2.7** and **2.8** are summarized in Table 2.3. These are broadly similar. The two-orbital–two-electron case results in a positive bond overlap population, the two-orbital–four-electron case in a negative bond overlap population. In the former, there occurs an electron density shift from χ_1 to χ_2 as a result of the orbital interaction (i.e., $P_{22} = 2t^2 > 0$). A two-orbital–two-electron interaction is therefore often called a charge transfer interaction. As shown in **2.11**, the initially doubly occupied and empty orbitals χ_1 and χ_2 are called the donor and acceptor orbitals, respectively.

TABLE 2.3.   Summary of Population Analyses

Case	2.3	2.4	2.7	2.8
P_{11}	$\dfrac{1}{1 + S_{12}}$	$\dfrac{2}{1 - S_{12}^2}$	$2(1 - 2tS_{12} - t^2)$	$2(1 + S_{12}^2)$
P_{22}	$\dfrac{1}{1 + S_{12}}$	$\dfrac{2}{1 - S_{12}^2}$	$2t^2$	$2(1 + S_{12}^2)$
P_{12}	$\dfrac{2S_{12}}{1 + S_{12}}$	$-\dfrac{4S_{12}^2}{1 - S_{12}^2}$	$4tS_{12}$	$-4S_{12}^2$
q_1	1	2	$2(1 - tS_{12} - t^2)$	2
q_2	1	2	$2(tS_{12} + t^2)$	2

Having progressed this far with a simple two-orbital problem, how do the results change for a many-orbital system? The details of this case will be discussed in Chapter 3. However, it is not too surprising that for the particular example of **2.12** (i.e., a single "central" atom surrounded by some ligands) then the single term of equation 2.18 is replaced by an energy sum

$$e_1 \simeq e_1^0 + \sum_{j \neq 1} \frac{(\Delta_{1j} - e_1^0 S_{1j})^2}{e_1^0 - e_j^0} \tag{2.52}$$

with a corresponding summation term to describe the new wavefunction. (In this simple expression we have, of course, neglected interactions between orbitals located on different ligands.)

CHAPTER THREE

Perturbational Molecular Orbital Theory

3.1. INTRODUCTION

In principle we can perform some sort of molecular orbital calculation on molecules of almost any sort of complexity. It is, however, often extremely profitable in terms of understanding the orbital structure to relate the level arrangement in a complex system to that of a simpler one. **3.1–3.3** show three examples of different types of relationships which we will frequently use. We will be interested in seeing how the levels of the species at the left-hand side of these figures are altered electronically during these processes by using the powerful techniques of perturbation theory. We shall not derive the elements of the theory itself but will make use of its mathematical results,[1-4] which will very quickly show a striking resemblance to the orbital interaction results of Chapter 2.

electronegativity perturbation:

$$N=N=\bar{N} \longrightarrow N=N=O \qquad \textbf{3.1}$$

geometry perturbation:

3.2

intermolecular perturbation:

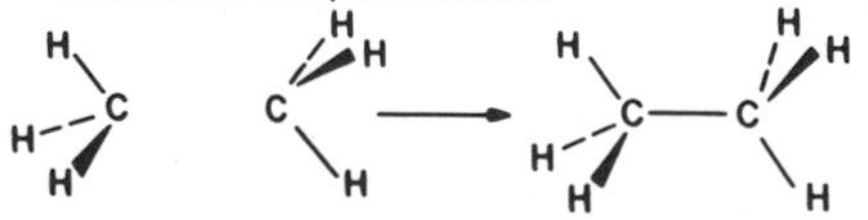

3.3

Consider a set of unperturbed (i.e., zeroth order) orbitals ψ_i^0 with energy e_i^0 corresponding to the left-hand side of **3.1** to **3.3**. In general these orbitals are given in terms of atomic orbitals as

$$\psi_i^0 = \sum_\mu c_{\mu i}^0 \chi_\mu \tag{3.1}$$

which we assume to be the eigenfunctions of H^{eff}. Thus

$$H^{\text{eff}} \psi_i^0 = e_i^0 \psi_i^0 \tag{3.2}$$

As a result of perturbation, these orbitals lead to a set of modified (i.e., perturbed) orbitals ψ_i with energy e_i corresponding to the right-hand side of **3.1** to **3.3**. In perturbation theory, it is customary to expand the perturbed energy e_i as

$$e_i = e_i^0 + e_i^{(1)} + e_i^{(2)} + \cdots \tag{3.3}$$

where $e_i^{(1)}$ and $e_i^{(2)}$ are the first- and second-order energy corrections to the unperturbed level e_i^0, respectively. As a linear combination of the unperturbed orbitals, ψ_i can be given by

$$\psi_i = t_{ii} \psi_i^0 + \sum_{j \neq i} t_{ji} \psi_j^0 \tag{3.4}$$

Here the weight of ψ_i^0 (i.e., the mixing coefficient t_{ii}) is smaller than one since ψ_i is normalized to unity like ψ_i^0. In perturbation theory, the mixing coefficients t_{ii} and t_{ji} $(j \neq i)$ are expanded as

$$t_{ii} = t_{ii}^0 + t_{ii}^{(1)} + t_{ii}^{(2)} + \cdots \tag{3.5}$$

$$t_{ji} = t_{ji}^0 + t_{ji}^{(1)} + t_{ji}^{(2)} + \cdots \tag{3.6}$$

In order to discuss the detailed expressions of the various expansion terms in equations 3.3, 3.5, and 3.6, we need to examine what happens to the molecular integrals $\langle \psi_i^0 | \psi_j^0 \rangle$ and $\langle \psi_i^0 | H^{\text{eff}} | \psi_j^0 \rangle$ as a consequence of electronegativity, geometry, or intermolecular perturbation. When there is no perturbation, the zeroth order orbitals are orthonormal, that is,

$$\langle \psi_i^0 | \psi_j^0 \rangle = \sum_\mu \sum_\nu c_{\mu i}^0 S_{\mu\nu} c_{\nu j}^0 = \delta_{ij} \tag{3.7}$$

and are eigenfunctions of H^{eff}. Thus,

$$\langle \psi_i^0 | H^{\text{eff}} | \psi_j^0 \rangle = \sum_\mu \sum_\nu c_{\mu i}^0 H_{\mu\nu} c_{\nu j}^0$$

$$= e_j^0 \langle \psi_i^0 | \psi_j^0 \rangle = e_j^0 \delta_{ij} \tag{3.8}$$

After a perturbation is switched on, some or all of the $S_{\mu\nu}$ and $H_{\mu\nu}$ elements change in size. For example, the perturbation of replacing a nitrogen with a more electronegative oxygen in **3.1** will result in a change in the $H_{\mu\mu}$ elements on this end atom and a change in all the interaction elements involving orbitals located on that atom. Recall via the Wolfsberg–Helmholz relationship (equation 1.19) $H_{\mu\nu} \propto$

$S_{\mu\nu}(H_{\mu\mu} + H_{\nu\nu})$. Since the oxygen atom orbitals have different exponents than those orbitals on the nitrogen atom they replace, the changes in $H_{\mu\nu}$ elements will arise via changes in both $S_{\mu\nu}$ and $H_{\mu\mu}$ where μ represents an orbital on the substituted atom. The geometry perturbation **3.2** will involve changing $H_{\mu\nu}$ values between some of the atomic orbitals of the basis as a result of a change in the corresponding overlap integrals $S_{\mu\nu}$ demanded by the geometry change. The intermolecular perturbation **3.3** gives rise to nonzero $S_{\mu\nu}$ and $H_{\mu\nu}$ values between the atomic orbitals of one fragment and those of the other fragment, since the two fragments are brought together in close proximity.

The changes in $S_{\mu\nu}$ and $H_{\mu\nu}$ described above make the $\langle \psi_i^0 | \psi_j^0 \rangle$ and $\langle \psi_i^0 | H^{\text{eff}} | \psi_j^0 \rangle$ integrals deviate from the values of δ_{ij} and $e_j^0 \delta_{ij}$, respectively. Let us denote the changes in $S_{\mu\nu}$ and $H_{\mu\nu}$ by $\delta S_{\mu\nu}$ and $\delta H_{\mu\nu}$, respectively. Then the deviation of $\langle \psi_i^0 | \psi_j^0 \rangle$ from δ_{ij} and that of $\langle \psi_i^0 | H^{\text{eff}} | \psi_j^0 \rangle$ from $e_j^0 \delta_{ij}$ are given by equations 3.9 and 3.10, respectively.

$$\sum_\mu \sum_\nu c_{\mu i}^0 \delta S_{\mu\nu} c_{\nu j}^0 \equiv \tilde{S}_{ij} \tag{3.9}$$

$$\sum_\mu \sum_\nu c_{\mu i}^0 \delta H_{\mu\nu} c_{\nu j}^0 \equiv \tilde{\Delta}_{ij} \tag{3.10}$$

By definition, the $\tilde{S}_{ij}$ and $\tilde{\Delta}_{ij}$ values are related to the $\langle \psi_i^0 | \psi_j^0 \rangle$ and $\langle \psi_i^0 | H^{\text{eff}} | \psi_j^0 \rangle$ integrals as shown in equations 3.11 and 3.12, respectively.

$$\langle \psi_i^0 | \psi_i^0 \rangle = 1 + \tilde{S}_{ii}$$
$$\langle \psi_i^0 | \psi_j^0 \rangle = \tilde{S}_{ij} \quad (i \neq j) \tag{3.11}$$

and

$$\langle \psi_i^0 | H^{\text{eff}} | \psi_i^0 \rangle = e_i^0 + \tilde{\Delta}_{ii}$$
$$\langle \psi_i^0 | H^{\text{eff}} | \psi_j^0 \rangle = \tilde{\Delta}_{ij} \quad (i \neq j) \tag{3.12}$$

In terms of these $\tilde{\Delta}_{ij}$ and $\tilde{S}_{ij}$ values defined above, we will examine the expressions of the various expansion terms in equations 3.3, 3.5, and 3.6 in the following. It is important to note the following qualitative relationship between $\tilde{\Delta}_{ij}$ and $\tilde{S}_{ij}$ in our discussion.

$$\tilde{\Delta}_{ij} \propto -\tilde{S}_{ij} \quad (i = j \text{ or } i \neq j) \tag{3.13}$$

In what follows we shall not derive any of the details of perturbation theory. The reader who wishes to know more is referred to the set of references at the end of the chapter. Our aim here is to show how the basic principles may be used in orbital construction.

3.2. NONDEGENERATE PERTURBATION

In this section we will consider only cases in which none of the unperturbed levels are degenerate. That is, $e_i^0 \neq e_j^0$ if $i \neq j$. For the purpose of clarity, it is important to distinguish electronegativity or geometry perturbation from intermolecular perturbation.

A. INTRAMOLECULAR PERTURBATION

Either electronegativity or geometry perturbation leads to orbital mixings among the orbitals of a single molecule under consideration and hence may be termed intramolecular perturbation. For such perturbation, the $e_i^{(1)}$ and $e_i^{(2)}$ terms of equation 3.3 are given by

$$e_i^{(1)} = \tilde{\Delta}_{ii} - e_i^0 \tilde{S}_{ii} \propto -\tilde{S}_{ii} \tag{3.14}$$

$$e_i^{(2)} = \sum_{j \neq i} \frac{(\tilde{\Delta}_{ij} - e_i^0 \tilde{S}_{ij})^2}{e_i^0 - e_j^0} \propto \sum_{j \neq i} \frac{\tilde{S}_{ij}^2}{e_i^0 - e_j^0} \tag{3.15}$$

The first-order energy, $e_i^{(1)}$, is negative (i.e., stabilizing) and positive (i.e., destabilizing), when the value of $\langle \psi_i^0 | \psi_i^0 \rangle = 1 + \tilde{S}_{ii}$ increases and decreases from one, respectively. The second-order energy $e_i^{(2)}$ results when the value of $\langle \psi_i^0 | \psi_j^0 \rangle = \tilde{S}_{ij}$ deviates from zero. The ψ_i^0 level is stabilized by a level ψ_j^0 lying higher in energy (i.e., $e_i^0 - e_j^0 < 0$) but destabilized by one lying lower (i.e., $e_i^0 - e_j^0 > 0$).

The three terms of equation 3.6 are expressed as

$$t_{ji}^0 = t_{ji}^{(2)} = 0$$

$$t_{ji}^{(1)} = \frac{\tilde{\Delta}_{ij} - e_i^0 \tilde{S}_{ij}}{e_i^0 - e_j^0} \propto \frac{-\tilde{S}_{ij}}{e_i^0 - e_j^0} \tag{3.16}$$

while those of equation 3.5 are given by

$$t_{ii}^0 = 1$$

$$t_{ii}^{(1)} = 0 \tag{3.17}$$

$$t_{ii}^{(2)} = -\sum_{j \neq i} [\tilde{S}_{ij} t_{ji}^{(1)} + \tfrac{1}{2}(t_{ji}^{(1)})^2]$$

Consequently, the perturbed orbital ψ_i can be approximated by

$$\psi_i = (1 + t_{ii}^{(2)}) \psi_i^0 + \sum_{j \neq i} t_{ji}^{(1)} \psi_j^0 \tag{3.18}$$

The weight of ψ_i^0 is reduced from one by every nondegenerate interaction ψ_i^0 makes with various orbitals ψ_j^0. In most cases of our qualitative application of perturbation theory, it is sufficient to remember that the weight of the leading term ψ_i^0 is smaller than one. With this understanding, equation 3.18 may be simplified as

$$\psi_i \simeq \psi_i^0 + \sum_{j \neq i} t_{ji}^{(1)} \psi_j^0 \tag{3.19}$$

Examples of electronegativity and geometry perturbations are discussed in Chapters 6 and 7, respectively.

B.   INTERMOLECULAR PERTURBATION

With intermolecular perturbation, we examine how the orbitals of a molecule (or fragment) are modified by those of another molecule (or fragment). A typical example of intermolecular perturbation is shown in **3.4**. The orbitals in the two

$$
\begin{array}{cc}
\text{A} & \text{B} \\
\psi^0_{kA}, e^0_{kA} \; \rule[0.5ex]{1.5em}{0.4pt} & \rule[0.5ex]{1.5em}{0.4pt} \\
 & \rule[0.5ex]{1.5em}{0.4pt} \; e^0_{jB}, \psi^0_{jB} \\
\psi^0_{iA}, e^0_{iA} \; \rule[0.5ex]{1.5em}{0.4pt} & \\
 & \rule[0.5ex]{1.5em}{0.4pt}
\end{array}
$$

3.4

stacks may be the orbitals of two fragments A, B which are brought together as in **3.3**. Let us assume for simplicity that, when two fragments are brought together, no geometry change occurs within each fragment so that there is no geometry perturbation to consider within each fragment. For those atomic orbitals χ_μ and χ_ν located on the fragments A and B, respectively, the $\delta H_{\mu\nu}$ and $\delta S_{\mu\nu}$ values are simply the $H_{\mu\nu}$ and $S_{\mu\nu}$ values of the composite system AB, respectively. Therefore, we obtain the following integrals between the fragment orbitals ψ^0_{nA} $(n = i, k)$ and ψ^0_{jB}:

$$
\tilde{S}_{nj} = \sum_{\mu \in A} \sum_{\nu \in B} c^0_{\mu n} S_{\mu\nu} c^0_{\nu j} = \langle \psi^0_{nA} | \psi^0_{jB} \rangle
$$

$$
\tilde{\Delta}_{nj} = \sum_{\mu \in A} \sum_{\nu \in B} c^0_{\mu n} H_{\mu\nu} c^0_{\nu j} = \langle \psi^0_{nA} | H^{\text{eff}} | \psi^0_{jB} \rangle \tag{3.20}
$$

The symbol $\lambda \in X$ means orbitals λ that are located on atom X. Note that $\tilde{S}_{nj}$ is the overlap integral between ψ^0_{nA} and ψ^0_{jB}, and $\tilde{\Delta}_{nj}$ is the corresponding interaction energy. Here the $\tilde{\Delta}_{nj}$ and $\tilde{S}_{nj}$ values are qualitatively related as follows:

$$
\tilde{\Delta}_{nj} \propto -\tilde{S}_{nj} \qquad (n = i, k) \tag{3.21}
$$

To examine how an orbital ψ^0_{iA} on the fragment A is modified by other orbitals, it is important to distinguish those orbitals ψ^0_{kA} on the same fragment from those orbitals ψ^0_{jB} on the other fragment B. Thus we rewrite equation 3.4 as

$$
\psi_i = t_{ii} \psi^0_{iA} + \sum_{j \in B} t_{ji} \psi^0_{jB} + \sum_{k \in A, k \neq i} t_{ki} \psi^0_{kA} \tag{3.22}
$$

and expand the coefficients t_{ii} and t_{ji} as in equations 3.5 and 3.6, respectively. The coefficient t_{ki} may also be expanded as

$$
t_{ki} = t^0_{ki} + t^{(1)}_{ki} + t^{(2)}_{ki} + \cdots \tag{3.23}
$$

Then the various terms of the mixing coefficients are given by

$$t_{ji}^0 = t_{ji}^{(2)} = 0$$

(3.24)

$$t_{ji}^{(1)} = \frac{\tilde{\Delta}_{ij} - e_{iA}^0 \tilde{S}_{ij}}{e_{iA}^0 - e_{jB}^0} \propto \frac{-\tilde{S}_{ij}}{e_{iA}^0 - e_{jB}^0}$$

and

$$t_{ki}^0 = t_{ki}^{(1)} = 0$$

$$t_{ki}^{(2)} = \sum_{j \in B} \frac{(\tilde{\Delta}_{ij} - e_{iA}^0 \tilde{S}_{ij})(\tilde{\Delta}_{jk} - e_{iA}^0 \tilde{S}_{jk})}{(e_{iA}^0 - e_{kA}^0)(e_{iA}^0 - e_{jB}^0)} \propto \sum_{j \in B} \frac{\tilde{S}_{ij}\tilde{S}_{jk}}{(e_{iA}^0 - e_{kA}^0)(e_{iA}^0 - e_{jB}^0)}$$

(3.25)

$$t_{ii}^0 = 1$$

$$t_{ii}^{(1)} = 0$$

$$t_{ii}^{(2)} = - \sum_{j \in B} [\tilde{S}_{ij} t_{ji}^{(1)} + \tfrac{1}{2} (t_{ji}^{(1)})^2]$$

(3.26)

In terms of those terms given in the above equations, ψ_i is written as

$$\psi_i = (1 + t_{ii}^{(2)}) \psi_{iA}^0 + \sum_{j \in B} t_{ji}^{(1)} \psi_{jB}^0 + \sum_{\substack{k \in A \\ k \neq i}} t_{ki}^{(2)} \psi_{kA}^0$$

(3.27)

Thus ψ_{jB}^0 mixes into ψ_{iA}^0 with the first-order mixing coefficient $t_{ji}^{(1)}$. If we arrange the orbital phases of ψ_{iA}^0 and ψ_{jB}^0 such that $\tilde{S}_{ij}$ is positive, equation 3.24 shows that ψ_{jB}^0 mixes into ψ_{iA}^0 in a bonding way (in-phase) if ψ_{jB}^0 lies higher in energy than ψ_{iA}^0 ($e_{iA}^0 - e_{jB}^0 < 0$), but in an antibonding way (out-of-phase) if ψ_{jB}^0 lies lower ($e_{iA}^0 - e_{jB}^0 > 0$). Equation 3.25 shows that, within the fragment A, ψ_{kA}^0 mixes into ψ_{iA}^0 with the second-order mixing coefficient $t_{ki}^{(2)}$ when ψ_{iA}^0 and ψ_{kA}^0 both interact with ψ_{jB}^0 of the fragment B. If the orbital phases of ψ_{iA}^0, ψ_{kA}^0, and ψ_{jB}^0 are arranged such that both $\tilde{S}_{ij}$ and $\tilde{S}_{jk}$ are positive, the sign of $t_{ki}^{(2)}$ is determined by the relative ordering of the ψ_{iA}^0, ψ_{kA}^0, and ψ_{jB}^0 levels. 3.5–3.7 show three situations

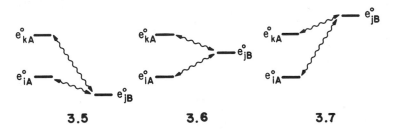

3.5   **3.6**   **3.7**

that may be encountered in practice, which differ in the ordering of these levels. For the case of **3.5**, ψ_{kA}^0 mixes into ψ_{iA}^0 in second order with a sign given by

$$t_{ki}^{(2)} \propto \frac{\tilde{S}_{ij}\tilde{S}_{jk}}{(e_{iA}^0 - e_{kA}^0)(e_{iA}^0 - e_{jB}^0)} = \frac{(+)(+)}{(-)(+)} = (-)$$

(3.28)

TABLE 3.1.   Signs of Second-Order Mixing
Coefficients

	Case		
Coefficient	3.5	3.6	3.7
$t_{ki}^{(2)}$	(−)	(+)	(+)
$t_{ik}^{(2)}$	(+)	(+)	(−)

and ψ_{iA}^0 mixes into ψ_{kA}^0 with a sign

$$t_{ik}^{(2)} \propto \frac{\tilde{S}_{kj}\tilde{S}_{ji}}{(e_{kA}^0 - e_{iA}^0)(e_{kA}^0 - e_{jB}^0)} = \frac{(+)(+)}{(+)(+)} = (+) \qquad (3.29)$$

The signs of $t_{ki}^{(2)}$ and $t_{ik}^{(2)}$ for the other cases 3.6 and 3.7 are shown in Table 3.1, which indicates that there are only two cases when the second-order mixing coefficient becomes negative: (1) when the mixing into lower lying orbital ψ_{iA}^0 occurs via a deeper, lower lying intermediate orbital ψ_{jB}^0 and (2) when the mixing into the higher lying orbital ψ_{kA}^0 occurs via an even higher lying orbital ψ_{jB}^0. These results are valid *only when* the phases of the orbitals ψ_{iA}^0, ψ_{kA}^0, and ψ_{jB}^0 are so arranged that both ψ_{iA}^0 and ψ_{kA}^0 make positive overlap with ψ_{jB}^0.

Equations 3.26 and 3.27 show that the weight of ψ_{iA}^0 is reduced from one by every nondegenerate interaction ψ_{iA}^0 makes with various orbitals ψ_{jB}^0. As already mentioned, it is sufficient to remember that the weight of ψ_{iA}^0 is less than one in most cases of our applications. With this understanding, equation 3.27 may be simplified as

$$\psi_i \simeq \psi_{iA}^0 + \sum_{j\in B} t_{ji}^{(1)}\psi_{jB}^0 + \sum_{\substack{k\in A \\ k\neq i}} t_{ki}^{(2)}\psi_{kA}^0 \qquad (3.30)$$

The energy e_i of the perturbed level ψ_i that originates from ψ_{iA}^0 is given by

$$e_i = e_{iA}^0 + e_i^{(1)} + e_i^{(2)} + \cdots \qquad (3.31)$$

where

$$e_i^{(1)} = 0$$
$$\qquad (3.32)$$
$$e_i^{(2)} = \sum_{j\in B} \frac{(\tilde{\Delta}_{ij} - e_{iA}^0 \tilde{S}_{ij})^2}{e_{iA}^0 - e_{jB}^0} \propto \sum_{j\in B} \frac{\tilde{S}_{ij}^2}{e_{iA}^0 - e_{jB}^0}$$

The first-order energy $e_i^{(1)}$ vanishes in the present case since the geometry of the fragment A, and that of the fragment B as well, is assumed to be unaffected by intermolecular perturbation. Note that the first-order orbital mixing between ψ_{iA}^0 and ψ_{jB}^0 (i.e., $t_{ji}^{(1)}$) leads to the second-order energy change $e_i^{(2)}$. The second-order orbital mixing between ψ_{iA}^0 and ψ_{kA}^0 leads to a third-order energy correction, which is not shown in equation 3.31.

As a simple example illustrating the essence of what we have just discussed, let us consider **3.8** in which s and p orbitals of A (denoted as ψ_{iA}^0 and ψ_{kA}^0, respec-

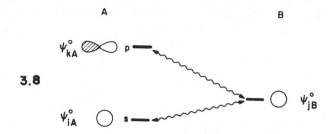

3.8

tively) interact with an s orbital (denoted as ψ_{jB}^0) of B along the internuclear axis A–B. In this case, $\tilde{\Delta}_{ij} = \Delta_{ij}$, $\tilde{\Delta}_{jk} = \Delta_{jk}$, $\tilde{S}_{ij} = S_{ij}$, and $\tilde{S}_{jk} = S_{jk}$. In **3.8** these orbitals are arranged such that both S_{ij} and S_{jk} are positive. According to equation 3.30, ψ_i is written as

$$\psi_i \simeq \psi_{iA}^0 + t_{ji}^{(1)} \psi_{jB}^0 + t_{ki}^{(2)} \psi_{kA}^0 \tag{3.33}$$

Since ψ_{jB}^0 lies higher in energy than ψ_{iA}^0, $t_{ji}^{(1)}$ is positive. ψ_{jB}^0 lies in between ψ_{iA}^0 and ψ_{kA}^0 as in **3.6** so that $t_{ki}^{(2)}$ is also positive. At this stage we will introduce a notation to indicate the first- and second-order contributions to a new orbital. In most cases of our qualitative applications, the coefficients $t_{ji}^{(1)}$ and $t_{ki}^{(2)}$ are small in magnitude compared to one so that what matters most is their signs. Therefore, we will represent the first-order term $t_{ji}^{(1)} \psi_{jB}^0$ as in equation 3.34,

$$
\begin{aligned}
t_{ji}^{(1)} \psi_{jB}^0 &= (\psi_{jB}^0), && \text{if} \quad t_{ji}^{(1)} > 0 \\
t_{ji}^{(1)} \psi_{jB}^0 &= -(\psi_{jB}^0), && \text{if} \quad t_{ji}^{(1)} < 0
\end{aligned}
\tag{3.34}
$$

and the second-order term $t_{ki}^{(2)} \psi_{kA}^0$ as in equation 3.35.

$$
\begin{aligned}
t_{ki}^{(2)} \psi_{kA}^0 &= [\psi_{kA}^0], && \text{if} \quad t_{ki}^{(2)} > 0 \\
t_{ki}^{(2)} \psi_{kA}^0 &= -[\psi_{kA}^0], && \text{if} \quad t_{ki}^{(2)} < 0
\end{aligned}
\tag{3.35}
$$

Consequently, equation 3.33 can be rewritten as

$$\psi_i \simeq \psi_{iA}^0 + (\psi_{jB}^0) + [\psi_{kA}^0] \tag{3.36}$$

A graphical representation of this equation is given in **3.9**. This shows the chemically important formation of an sp hybrid-type orbital on the left-hand atom (e.g., chlorine) as a result of interaction with a single hydrogen $1s$ orbital. Many examples of intermolecular perturbation are examined in Chapter 5.

3.9 $\psi_i \propto$

3.3. DEGENERATE PERTURBATION

A. INTRAMOLECULAR PERTURBATION

Consider for the case of intramolecular perturbation that some of the unperturbed levels ψ_i^0, ψ_j^0, ... are degenerate so that $e_i^0 = e_j^0 = \cdots$. Most common situations of importance are either doubly or triply degenerate. Under intramolecular perturbation, the symmetry of a molecule is lowered and hence the degeneracy is lifted. As will be discussed in Chapter 7 (See also Section 3.2A), lifting of the degeneracy is caused either by the first-order term $e_i^{(1)}$, which occurs when the degenerate levels do not remain normalized to one, or by the second-order term $e_i^{(2)}$, which results when the degenerate levels do not remain orthogonal to other unperturbed levels $\psi_n^0 (n \neq i, j, \ldots)$ of different energy.

B. INTERMOLECULAR PERTURBATION

Let us now consider a case of intermolecular perturbation in which an orbital ψ_{iA}^0 of the fragment A is degenerate with an orbital ψ_{jB}^0 of the fragment B. We arrange the orbital phases of ψ_{iA}^0 and ψ_{jB}^0 such that $\tilde{S}_{ij} > 0$ and thus $\tilde{\Delta}_{ij} < 0$. Then the orbitals defined in equations 3.37, which are simply linear combinations of the two

$$\frac{\psi_{iA}^0 + \psi_{jB}^0}{\sqrt{2 + 2\tilde{S}_{ij}}}$$

$$\frac{\psi_{iA}^0 - \psi_{jB}^0}{\sqrt{2 - 2\tilde{S}_{ij}}} \tag{3.37}$$

zeroth order orbitals ψ_{iA}^0 and ψ_{jB}^0, lead to the first-order energy corrections given by equations 3.38

$$e_i^{(1)} = (\tilde{\Delta}_{ij} - e_{iA}^0 \tilde{S}_{ij}) \propto -\tilde{S}_{ij}$$

$$e_j^{(1)} = -(\tilde{\Delta}_{ij} - e_{iA}^0 \tilde{S}_{ij}) \propto \tilde{S}_{ij} \tag{3.38}$$

The second-order energy corrections resulting from these orbitals are given by

$$e_i^{(2)} = e_j^{(2)} = -\tilde{S}_{ij}(\tilde{\Delta}_{ij} - e_{iA}^0 \tilde{S}_{ij}) \propto \tilde{S}_{ij}^2 \tag{3.39}$$

equation 3.39. This is analogous to our discussion in Section 2.1A concerning a two-center–two-orbital problem (See equations 2.8 and 2.9). Those two orbitals defined in equation 3.37 may further engage in nondegenerate interactions with other orbitals of the fragments A and B, thereby leading to second-order energy corrections. In a general case of intermolecular perturbation, both degenerate and nondegenerate interactions may occur between the fragments A and B. Then it is convenient to derive the orbitals of the composite system AB in two steps: (1) First, we carry out only degenerate interactions, the resulting orbitals of which are non-degenerate. (2) Second, we include these orbitals among other nondegenerate orbitals, and carry out nondegenerate interactions using the combined set of orbitals.

In intermolecular perturbation, the first-order term $e_i^{(1)}$ is zero for nondegenerate interactions but nonzero for degenerate interactions. As a consequence, degenerate and nondegenerate orbital interactions are often called first- and second-order orbital interactions, respectively. Here it is important to stress the use of some terminology to avoid confusion later on in this book. The nondegenerate interaction between ψ_{iA}^0 and ψ_{jB}^0 involves a first-order change in the orbital but a second-order change in the energy. It will always be important when using the expressions "first order" and "second order" to state whether we are referring to energy changes or orbital mixing.

Finally, we briefly comment on those cases in which neither the degenerate nor the nondegenerate perturbation treatment is quite satisfactory. To simplify our discussion, let us consider only the lower lying orbital ψ_i and its energy e_i that result from two interacting orbitals ψ_{iA}^0 and ψ_{jB}^0. For a nondegenerate interaction with $e_{iA}^0 < e_{jB}^0$, the expressions for e_i and ψ_i are given by

$$e_i = e_{iA}^0 + \frac{(\tilde{\Delta}_{ij} - e_{iA}^0 \tilde{S}_{ij})^2}{e_{iA}^0 - e_{jB}^0} \tag{3.40}$$

$$\psi_i \simeq \psi_{iA}^0 + \left(\frac{\tilde{\Delta}_{ij} - e_{iA}^0 \tilde{S}_{ij}}{e_{iA}^0 - e_{jB}^0} \right) \psi_{jB}^0 \tag{3.41}$$

These expressions are close to the exact ones if equation 3.42 is satisfied, that is,

$$(\tilde{\Delta}_{ij} - e_{iA}^0 \tilde{S}_{ij})^2 \ll |e_{iA}^0 - e_{jB}^0| \tag{3.42}$$

when the extent of perturbation is small in magnitude compared with the energy difference between the unperturbed levels. For a degenerate interaction with $e_{iA}^0 = e_{jB}^0$, the expressions of e_i and ψ_i are given by

$$e_i = e_{iA}^0 + (\tilde{\Delta}_{ij} - e_{iA}^0 \tilde{S}_{ij}) - \tilde{S}_{ij}(\tilde{\Delta}_{ij} - e_{iA}^0 \tilde{S}_{ij}) \tag{3.43}$$

$$\psi_i = \frac{\psi_{iA}^0 + \psi_{jB}^0}{\sqrt{2 + 2\tilde{S}_{ij}}} \tag{3.44}$$

From the viewpoint of numerical accuracy, a problematic situation arises when the two levels e_{iA}^0 and e_{jB}^0 are different but close in energy so that equation 3.42 is not quite satisfied. In such a case, direct solution of the appropriate 2×2 secular determinant (as described in Chapters 1 and 2) reveals the following trend: As the upper level e_{jB}^0 is lowered and becomes closer to the lower level e_{iA}^0, the weight of ψ_{iA}^0 in ψ_i decreases gradually but that of ψ_{jB}^0 in ψ_i increases gradually. Eventually, when $e_{iA}^0 = e_{jB}^0$, the weights of ψ_{iA}^0 and ψ_{jB}^0 become equal as given by equation 3.44. What is important in our qualitative applications of perturbation theory is to recognize which orbital has a greater weight. Thus equation 3.41 may be used even for the aforementioned, problematic case since it shows that ψ_{iA}^0 has a greater weight than does ψ_{jB}^0. Of course, when the two levels are very close (still $e_{iA}^0 < e_{jB}^0$), equation 3.44 may be a more accurate description for ψ_i than equation 3.41. Nevertheless, the weight of ψ_{iA}^0 would be greater, though slightly, than that of ψ_{jB}^0.

3.4. THE LINEAR H_3 MOLECULE

An example will show the application of some of the ideas introduced above. Let us start with the simple two-center–two-orbital problem described exhaustively in Chapter 2. In the language of perturbation theory these two orbitals experience a degenerate interaction for the case of H_2 where the energies of each atomic orbital are the same. The result is an in-phase (bonding) combination and an out-of-phase (antibonding) combination, between the centers A and B. A more complicated example arises when there are two orbitals on A and one on B as when the orbitals of linear H_3 are constructed from those of $H_2 + H$ (3.10). This is shown in Figure 3.1, where the relative phases of the orbitals have been chosen so that $\tilde{S}_{ij}$ and $\tilde{S}_{jk}$ are positive.

$$H_1\!-\!H_2 \ + \ H_3 \quad \longrightarrow \quad H_1\!-\!H_2\!-\!H_3 \quad \textbf{3.10}$$

The orbital ψ_i^0 will be stabilized by interaction with ψ_j^0 since the energy denominator of the second-order energy correction (equation 3.32) is negative $(e_i^0 - e_j^0 < 0)$ and, with a positive $\tilde{S}_{ij}$, the interaction integral $\tilde{\Delta}_{ij}$ is negative (equations 3.20 and 3.21). From equations 3.30, 3.34, and 3.35, the resulting orbital ψ_i is given as

$$\psi_i = \psi_i^0 + t_{ji}^{(1)}\psi_j^0 + t_{ki}^{(2)}\psi_k^0$$
$$= \psi_i^0 + (\psi_j^0) + [\psi_k^0] \tag{3.45}$$

The coefficient $t_{ji}^{(1)}$ is easily seen to be positive and from Table 3.1, $t_{ki}^{(2)}$ is also positive. (The situation in Figure 3.1 corresponds to the case **3.6**). Diagrammatically the new orbital ψ_i is then constructed as in **3.11**. It is important to recall that these mixing coefficients are smaller than one, as discussed already. The consequence of the second-order term is to diminish the atomic orbital coefficient on the left H atom and reinforce the coefficient on the middle one. If the two close

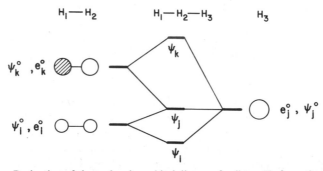

FIGURE 3.1.   Derivation of the molecular orbital diagram for linear H_3 from that of H_2 plus H.

$$\psi_i \propto \left[\; \cdots \; + \left(\cdots \right) + \left[\; \cdots \;\right]\right.$$

$$\Longrightarrow \qquad \qquad \textbf{3.11}$$

H–H distances are equal then, by symmetry the atomic orbital (AO) coefficients of the left and right H atom orbitals in the resultant ψ_i must be equal.

The new level e_j represents the sum of two interactions as shown in equation 3.46. The interaction with ψ_j^0 is destabilizing ($e_j^0 - e_i^0 > 0$), and that with ψ_k^0 is stabilizing ($e_j^0 - e_k^0 < 0$). The new orbital ψ_j is given by equation 3.47, a diagram-

$$e_j = e_j^0 + \frac{(\tilde{\Delta}_{jk} - e_j^0 \tilde{S}_{jk})^2}{e_j^0 - e_k^0} + \frac{(\tilde{\Delta}_{ji} - e_j^0 \tilde{S}_{ji})^2}{e_j^0 - e_i^0}$$

$$= e_j^0 + \frac{(-)^2}{(-)} + \frac{(-)^2}{(+)} = e_j^0 \qquad\qquad (3.46)$$

$$\psi_j = \psi_j^0 + t_{kj}^{(1)} \psi_k^0 + t_{ij}^{(1)} \psi_i^0$$

$$= \psi_j^0 + (\psi_k^0) - (\psi_i^0) \qquad\qquad (3.47)$$

matical representation of which is shown in **3.12**. Notice that the first-order mixing coefficients serve to reinforce the atomic orbital coefficients on the left-hand H

$$\psi_j \propto \left[\; \cdots \bigcirc - \left(\bigcirc\!\!-\!\!\bigcirc\right) + \left(\oslash\!\!-\!\!\bigcirc\right)\right.$$

$$\Longrightarrow \qquad\qquad \textbf{3.12}$$

atom of H_3 and cancellation occurs at the central atom. If the two H–H distances are equal then the second and third terms in **3.12** are equal in magnitude but opposite in sign. A precise cancellation occurs and a node develops at the central H atom. The astute reader will have noticed that the two energy denominators in equation 3.46, although opposite in sign, are not equal in magnitude, since we showed in Chapter 2 that the bonding combination of H_2 was stabilized less than the antibonding combination was destabilized. In particular, the denominator in the second term is larger than that for the third. However, recall the magnitudes of the coefficients in ψ_k^0 are larger than those in ψ_i^0. This leads to a larger $(\tilde{\Delta}_{jk} - e_j^0 \tilde{S}_{jk})$ term than $(\tilde{\Delta}_{ji} - e_j^0 \tilde{S}_{ji})$ in magnitude. Thus the last two terms in equation 3.46 become equal in magnitude and do in fact exactly cancel.

Finally, the ψ_k^0 level is destabilized by ψ_j^0, and the resultant molecular orbital ψ_k is given by equation 3.48. The $t_{ik}^{(2)}$ term is positive from Table 3.1, and $t_{jk}^{(1)}$ is

$$\psi_k = \psi_k^0 + t_{jk}^{(1)} \psi_j^0 + t_{ik}^{(2)} \psi_i^0$$

$$= \psi_k^0 - (\psi_j^0) + [\psi_i^0] \qquad\qquad (3.48)$$

negative since $(e_j^0 - e_k^0) < 0$. A diagrammatical representation of equation 3.48 is shown in **3.13**. So in principle the orbitals of a complex molecule (albeit H_3) may be simply derived by using the ideas of perturbation theory. In particular we have

$$\psi_k \propto \text{⬯—◯—} \cdot - \left(\text{——◯}\right) + \left[\text{◯—◯—}\right]$$

$$\Longrightarrow \text{⬯—◯—⬯} \qquad \textbf{3.13}$$

covered all of the elements of orbital interaction that we shall need for the whole book. We will construct the orbitals of much more complex molecules along similar lines by studying the first- and second-order interactions which occur as the result of a perturbation of a less complex system.

REFERENCES

1. A. Imamura, *Mol. Phys.*, **15**, 225 (1968).
2. L. Libit and R. Hoffmann, *J. Am. Chem. Soc.*, **96**, 1370 (1974).
3. M.-H. Whangbo, H. B. Schlegel, and S. Wolfe, *J. Am. Chem. Soc.*, **99**, 1296 (1977).
4. M.-H. Whangbo, in *Computational Theoretical Organic Chemistry*, I. G. Csizmadia and R. Daudel, eds., Reidel, Boston (1981), p. 233.

CHAPTER FOUR

Symmetry Considerations

4.1. INTRODUCTION

Symmetry plays an important role in chemistry. There are many everyday facts we take for granted which have a strong underlying symmetry aspect to them. One ns, three np, five nd, and so on, atomic orbitals come from the underlying symmetry of the atom; the paramagnetism of the oxygen molecule is a result of the presence of a doubly degenerate π orbital mandated by its linear structure. Jahn–Teller instabilities in molecules occur as a result of the presence of degenerate electronic states which are a consequence of a highly symmetric geometry. In this chapter we will not describe in detail the mathematics behind what chemists call group theory, but will extract, as in our use of perturbation theory in Chapter 3, the elements we shall need in our task. To the reader who wishes to know more there are several excellent books available.[1–5]

4.2. GROUPS

What is symmetry? Essentially in this book we shall be interested in two uses of this word. First we shall be interested in the "symmetry" of a molecule. When looking at objects we invariably have some feel as to whether they are highly symmetric or alternatively, not very symmetric. This needs to be quantified in some way. Second we will need to be able to classify, in terms of some symmetry description, the energy levels of molecules. Once this has been done we will find that with the use of a couple of mathematical tools we will be in a good position to be able to understand the symmetry control of the orbital structure in molecules.

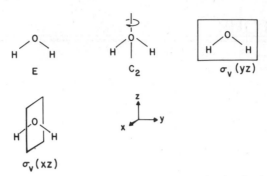

FIGURE 4.1.   Symmetry operations for the water molecule of point group C_{2v}.

Geometrical objects (including molecules, if they are regarded as being made up of balls and spokes) have an associated set of symmetry elements or operations {**R**}, rotation axes, mirror planes, and combinations of the two, such that the molecule is geometrically indistinguishable as a result of operating with any of the **R**. Figure 4.1 shows the symmetry operations associated with the water molecule. **4.1** shows the result of operating on the molecule with the C_2, or twofold rotation axis. Although the hydrogen atoms have been labeled for convenience, clearly the two geometrical arrangements, before and after, are indistinguishable. In **4.2**, reflection in a plane perpendicular to the plane of the molecule exchanges $H_{1,2}$ but in **4.3**, reflection in the plane of the molecule does not. All molecules possess the identity

$$H_1 \overset{O}{\diagup} {}_{H_2} \quad \xrightarrow{C_2} \quad H_2 \overset{O}{\diagup} {}_{H_1} \qquad \textbf{4.1}$$

$$H_1 \overset{O}{\diagup} {}_{H_2} \quad \xrightarrow{\sigma_v(xz)} \quad H_2 \overset{O}{\diagup} {}_{H_1} \qquad \textbf{4.2}$$

$$H_1 \overset{O}{\diagup} {}_{H_2} \quad \xrightarrow{\sigma_v(yz)} \quad H_1 \overset{O}{\diagup} {}_{H_2} \qquad \textbf{4.3}$$

operation, E; this is simply the null operation where we do nothing. The complete set of covering operations for the molecule together form a group. Any molecule can be assigned a point group, namely the collection of all the applicable symmetry operations which describe the geometrical relationships of its constituent atoms. The point group symbol appropriate for the collection of symmetry elements in Figure 4.1 is C_{2v}.

Having decided that the set of covering operations is a good way to describe the "symmetry" of the molecule, what about the "symmetry" of the wavefunctions describing the molecular orbital levels? Group theory tells us that, given a set of such operations for a particular molecule, only a certain type of function will be allowed. This results in powerful restrictions on the form of the wavefunctions in terms of the signs and relative magnitude of the atomic coefficients, which in their turn, fix the orbital energy. The form of these functions are determined in general

TABLE 4.1

C_{2v}	E	C_2	$\sigma_v(xz)$	$\sigma_v(yz)$	
A_1	1	1	1	1	$z^{(a)}$
A_2	1	1	-1	-1	
B_1	1	-1	1	-1	x
B_2	1	-1	-1	1	y

$^{(a)}$In the last column of the character table it is customary to list some useful functions which have the transformation properties shown. From Table 1.1 it can be seen that the angular part of a p_z orbital, for example, is simply equal to z/r. As a result, from the character table, a p_z orbital located at a point which contains all of the symmetry elements of the group will transform as a_1 symmetry.

by the character table. For the C_{2v} point group this is shown in Table 4.1. Each row consists of a set of characters $D_R(i)$ which correspond to the ith irreducible representation of the point group. These characters determine the transformation properties of a given wavefunction as a result of operating with each symmetry operation **R**. In other words

$$\mathbf{R}\psi = D_R(i)\,\psi \tag{4.1}$$

As an example take the wavefunction shown in **4.4** which we will write as

$$\psi_1 \propto \chi_1 + \chi_2 \tag{4.2}$$

$$-\text{H}_2 \qquad \textbf{4.4}$$
$$\text{H}_1$$

Using the symmetry operations defined in Figure 4.1 it is easy to see that

$$E \cdot \chi_1 = \chi_1; \qquad\qquad E \cdot \chi_2 = \chi_2$$
$$C_2 \cdot \chi_1 = \chi_2; \qquad\qquad C_2 \cdot \chi_2 = \chi_1$$
$$\sigma_v(yz) \cdot \chi_1 = \chi_1; \qquad \sigma_v(yz) \cdot \chi_2 = \chi_2$$
$$\sigma_v(xz) \cdot \chi_1 = \chi_2; \qquad \sigma_v(xz) \cdot \chi_2 = \chi_1 \tag{4.3}$$

and so

$$E \cdot \psi_1 = \psi_1; \qquad \sigma_v(yz) \cdot \psi_1 = \psi_1$$
$$C_2 \cdot \psi_1 = \psi_1; \qquad \sigma_v(xz) \cdot \psi_1 = \psi_1 \tag{4.4}$$

For the function 4.2, therefore, its transformation or symmetry properties in the C_{2v} point group are described by the characters in the top row of Table 4.1. For the C_{2v} point group there are only four different types of allowed wavefunctions or irreducible representations. They are given the Mulliken labels a_1, a_2, b_1, and b_2. The wavefunction of equation 4.2 therefore belongs to the a_1 irreducible representation of the C_{2v} point group. Since the wavefunction is symmetric with respect to all the symmetry operations of the group [i.e., $D_R(a_1) = 1$ for all R], this representation is called the totally symmetric representation. We also say that ψ_1 transforms as the a_1 irreducible representation, or is of a_1 symmetry, or belongs to the a_1 symmetry species.

For the wavefunction shown in 4.5, use of the relationships 4.3 with equation 4.5

$$\psi_2 \propto \chi_1 - \chi_2 \tag{4.5}$$

leads to

$$E \cdot \psi_2 = \psi_2; \qquad \sigma_v(yz) \cdot \psi_2 = \psi_2$$
$$C_2 \cdot \psi_2 = -\psi_2; \qquad \sigma_v(xz) \cdot \psi_2 = -\psi_2 \tag{4.6}$$

Comparison of the signs in equation 4.6 with Table 4.1 shows immediately that ψ_2 is of b_2 symmetry. A p_x orbital on the oxygen atom (4.6) may similarly be shown

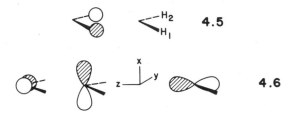

to transform as b_1, and oxygen located p_z and p_y orbitals as a_1 and b_2, respectively. The labels a, b can be seen to describe functions which are respectively symmetric and antisymmetric with respect to the twofold rotation axis. The character table dramatically restricts the types of wavefunctions that are allowed. The wavefunction in equation 4.7, for example, is not permissible, since $C_2 \cdot \psi_3 \neq (\pm 1)\,\psi_3$ as demanded by the table.

$$\psi_3 \propto \chi_1 + 2\chi_2 \tag{4.7}$$

The symmetry labels give us a useful way of classifying the molecular orbitals of molecules which we use extensively throughout the book.

4.3.   SYMMETRY-ADAPTED WAVEFUNCTIONS

In this section we will see how to build wavefunctions with given symmetry properties for collections of orbitals. Again we will not derive the group theoretical results needed but will present qualitative arguments as to their validity. First we need to introduce the concept of symmetry equivalent atoms or orbitals. Clearly the two hydrogen $1s$ orbitals of water are equivalent, specifically because any symmetry operation of the C_{2v} point group will either send χ_1 to itself or to χ_2 and likewise send χ_2 to itself or to χ_1. Thus $\{\chi_1, \chi_2\}$ form a completely equivalent set. The SF_4 molecule of 4.7 also belongs to the C_{2v} point group. Here, however, no

4.7

symmetry operation sends χ_1 or χ_2 to χ_3 or χ_4 or vice versa. The orbitals χ_1 and χ_2 are permuted amongst themselves, as are the orbitals χ_3 and χ_4, by the symmetry operations of the group. So $\{\chi_1, \chi_2\}$ and $\{\chi_3, \chi_4\}$ form two different sets of symmetry equivalent orbitals.

We now generate the orbitals of a_1 and b_2 symmetry for the hydrogen $1s$ orbitals of water with equations 4.2 and 4.5. Starting with the full complement of symmetry equivalent orbitals (χ_1 and χ_2 in this case), we force the resulting wavefunction to have particular symmetry properties by constructing it with the help of the character table and write

$$\psi(i) \propto \sum_R D_R(i) R\chi_1 \tag{4.8}$$

It does not matter whether we start off with χ_1 or χ_2 in this equation. Since they are symmetry equivalent, the same result will be found. For the C_{2v} point group the results of the operations $R\chi_1$ are given in 4.3 and the characters $D_R(i)$ in Table 2.1. For $i = a_1$, then, we may readily generate

$$\psi(a_1) \propto \chi_1 + \chi_2 + \chi_2 + \chi_1$$

$$\propto \chi_1 + \chi_2 \tag{4.9}$$

This may be normalized to give the wavefunction of equation 4.2. For $i = b_2$ then,

$$\psi(b_2) \propto \chi_1 - \chi_2 - \chi_2 + \chi_1$$

$$\propto \chi_1 - \chi_2 \tag{4.10}$$

which with the addition of a normalization constant gives equation 4.5. This technique using equation 4.8 is a very general and powerful way to construct functions of this type. The operator in this equation is called a projection operator.

One question still to be answered is how we knew to construct functions of a_1 and b_2 symmetry. Certainly if we try to construct a function of a_2 symmetry, the result is meaningless (equation 4.11)

$$\psi(a_2) \propto \chi_1 + \chi_2 - \chi_2 - \chi_1 = 0 \tag{4.11}$$

We need in general to find then how the complete set of symmetry equivalent orbitals $\{\chi_i\}$ transforms under the symmetry operations of the group. To do this we need the characters D_R. These may be generated by operating on each of the orbitals of the set $\{\chi_i\}$ with R. If $R\chi_i = d_i\chi_i$, then d_i may equal $+1$ (if R sends χ_i to itself), -1 (if R sends χ_i to minus itself), and 0 (if R sends χ_i to some other member of the set). D_R is then given by

$$D_R = \sum_i d_i \tag{4.12}$$

For the orbitals χ_1 and χ_2 the characters are simply a set of numbers which is simply

	E	C_2	$\sigma_v(xz)$	$\sigma_v(yz)$	
D_R	2	0	0	2	(4.13)

the sum of the characters for the a_1 and b_2 representations. D_R is called a reducible representation since it may be broken down into the sum of a collection of irreducible ones.

	E	C_2	$\sigma_v(xz)$	$\sigma_v(yz)$	
$D_R(a_1)$	1	1	1	1	
$D_R(b_2)$	1	-1	-1	1	
Sum	2	0	0	2	(4.14)

Very often we can decompose a sum of numbers such as those in 4.13 by inspection by using the character table. If not, then there is an equation that comes from a result called the great orthogonality theorem which does this for us (equation 4.15).

$$n_i = \frac{1}{g} \sum_R D_R D_R(i) \tag{4.15}$$

Here g is the order of the group, namely the total number of symmetry elements ($g = 4$ for C_{2v}). n_i is the number of times the ith irreducible representation occurs in the reducible one. Applying equation 4.15 to the reducible representation 4.13,

$$
\begin{aligned}
n_{a_1} &= \tfrac{1}{4} \left[2 \times 1 + 0 \times 1 + 0 \times 1 + 2 \times 1 \right] = 1 \\
n_{a_2} &= \tfrac{1}{4} \left[2 \times 1 + 0 \times 1 + 0 \times (-1) + 2 \times (-1) \right] = 0 \\
n_{b_1} &= \tfrac{1}{4} \left[2 \times 1 + 0 \times (-1) + 0 \times 1 + 2 \times (-1) \right] = 0 \\
n_{b_2} &= \tfrac{1}{4} \left[2 \times 1 + 0 \times (-1) + 0 \times (-1) + 2 \times 1 \right] = 1
\end{aligned}
\tag{4.16}
$$

And so the set of hydrogen $1s$ orbitals in water transform as $a_1 + b_2$. All of the molecular orbitals of this molecule which contain hydrogen $1s$ character will then be either of a_1 or b_2 symmetry. If of a_1 symmetry, then χ_1 and χ_2 will have the same sign and magnitude; if of b_2 symmetry, then χ_1 and χ_2 will have equal weight but their orbital coefficients will be of opposite sign.

    4.8

Another example is shown in **4.8**. Suppose that instead of OH_2, the molecule of interest was OF_2 with the same geometry and we wished to know how the two p_x orbitals on the fluorine atoms transformed. Then using equation 4.12, the set of characters for the two orbitals is

	E	C_2	$\sigma_v(xz)$	$\sigma_v(yz)$
D_R	2	0	0	-2

(4.17)

The entry under $\sigma_v(yz)$ is of different sign to the corresponding one in 4.13 since this symmetry operation sends each Fp_x orbital to $-p_x$. Reduction of 4.17 either by inspection or using equation 4.15 gives us the result that these two p_x orbitals transform as $a_2 + b_1$.

4.4.   DEGENERACIES

The case of the water molecule involved a low symmetry point group where all of the possible representations are nondegenerate. In general, this appears in the character table (Table 4.1) with a +1 entry under the identity operation E. Higher symmetry point groups contain degenerate representations. Any point group that contains a threefold (or higher) rotation axis will contain at least one degenerate representation. The entry under E reveals the order of the degeneracy. Tables 4.2 through 4.6 show doubly and triply degenerate representations, labeled with the Mulliken symbols e and t, respectively. A shorthand notation has been introduced into the character table of Tables 4.2–4.6. Here the symmetry elements have been arranged into classes. These are groupings of operations which have the same char-

TABLE 4.2

C_{3v}	E	$2C_3$	$3\sigma_v$	
A_1	1	1	1	z
A_2	1	1	-1	
E	2	-1	0	(x,y)

TABLE 4.3[a]

D_{3h}	E	$2C_3$	$3C_2$	σ_h	$2S_3$	$2\sigma_v$	
A_1'	1	1	1	1	1	1	
A_2'	1	1	-1	1	1	-1	
E'	2	-1	0	2	-1	0	(x,y)
A_1''	1	1	1	-1	-1	-1	
A_2''	1	1	-1	-1	-1	1	z
E''	2	-1	0	-2	1	0	

[a]Note the presence of primes (') and double primes (") on the Mulliken symbols. They describe representations which are, respectively, symmetric and antisymmetric with respect to reflection in the mirror plane (σ_h) perpendicular to the threefold rotation axis.

TABLE 4.4[a]

$D_{\infty h}$	E	$2C_\infty^\phi$	$\cdots$	$\infty\sigma_v$	i	$2S_\infty^\phi$	$\cdots$	$\infty C_2'$	
$A_{1g} \equiv \Sigma_g^+$	+1	+1	$\cdots$	+1	+1	+1	$\cdots$	+1	
$A_{2g} \equiv \Sigma_g^-$	+1	+1	$\cdots$	-1	+1	+1	$\cdots$	-1	
$E_{1g} \equiv \Pi_g$	+2	$2\cos\phi$	$\cdots$	0	+2	$-2\cos\phi$	$\cdots$	0	
$E_{2g} \equiv \Delta_g$	+2	$2\cos 2\phi$	$\cdots$	0	+2	$2\cos 2\phi$	$\cdots$	0	
$E_{3g} \equiv \Phi_g$	+2	$2\cos 3\phi$	$\cdots$	0	+2	$-2\cos 3\phi$	$\cdots$	0	
E_{ng}	+2	$2\cos n\phi$	$\cdots$	0	+2	$(-1)^n 2\cos n\phi$	$\cdots$	0	
$\cdots$		$\cdots$			$\cdots$	$\cdots$		$\cdots$	
$A_{1u} \equiv \Sigma_u^+$	+1	+1	$\cdots$	+1	-1	-1	$\cdots$	-1	z
$A_{2u} \equiv \Sigma_u^-$	+1	+1	$\cdots$	-1	-1	-1	$\cdots$	+1	
$E_{1u} \equiv \Pi_u$	+2	$2\cos\phi$	$\cdots$	0	-2	$2\cos\phi$	$\cdots$	0	(x,y)
$E_{2u} \equiv \Delta_u$	+2	$2\cos 2\phi$	$\cdots$	0	-2	$-2\cos 2\phi$	$\cdots$	0	
$E_{3u} \equiv \Phi_u$	+2	$2\cos 3\phi$	$\cdots$	0	-2	$2\cos 3\phi$	$\cdots$	0	
E_{nu}	+2	$2\cos n\phi$	$\cdots$	0	-2	$(-1)^{n+1} 2\cos n\phi$	$\cdots$	0	
$\cdots$		$\cdots$			$\cdots$	$\cdots$		$\cdots$	

[a]Note the special Mulliken symbols for the irreducible representations for this point group. (They apply to linear molecules such as H_2, C_2H_2, and C_3O_2.)

TABLE 4.5[a]

O_h	E	$8C_3$	$6C_2'$	$6C_4$	$3C_2(=C_4^2)$	i	$6S_4$	$8S_6$	$3\sigma_h$	$6\sigma_d$	
A_{1g}	1	1	1	1	1	1	1	1	1	1	
A_{2g}	1	1	-1	-1	1	1	-1	1	1	-1	
E_g	2	-1	0	0	2	2	0	-1	2	0	
T_{1g}	3	0	-1	1	-1	3	1	0	-1	-1	
T_{2g}	3	0	1	-1	-1	3	-1	0	-1	1	
A_{1u}	1	1	1	1	1	-1	-1	-1	-1	-1	
A_{2u}	1	1	-1	-1	1	-1	1	-1	-1	1	
E_u	2	-1	0	0	2	-2	0	1	-2	0	
T_{1u}	3	0	-1	1	-1	-3	-1	0	1	1	(x,y,z)
T_{2u}	3	0	1	-1	-1	-3	1	0	1	-1	

[a]Note the labels g, u on the Mulliken symbols. They describe irreducible representations which are, respectively, symmetric (gerade) or antisymmetric (ungerade) with respect to the inversion operation (i).

TABLE 4.6

T_d	E	$8C_3$	$3C_2$	$6S_4$	$6\sigma_d$		
A_1	1	1	1	1	1	$x^2+y^2+z^2$	
A_2	1	1	1	-1	-1		
E	2	-1	2	0	0	$(2z^2-x^2-y^2, x^2-y^2)$	
T_1	3	0	-1	1	-1		
T_2	3	0	-1	-1	1	(x,y,z)	(xy,xz,yz)

acter in each irreducible representation. Thus the C_3 and C_3^2 operations always have the same character in the C_{3v} point group and so the abbreviation $2C_3$ is used here. Similarly the three vertical mirror planes in NH_3 together make up one entry $(3\sigma_v)$ in Table 4.2. Everything we have said so far holds for degenerate representations, except we need to be a little careful in using equation 4.8 for generating the wavefunctions. An example will show the problem encountered and the way it is overcome.

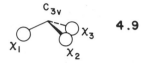

4.9

4.9 shows the three $1s$ orbitals on the hydrogen atoms in ammonia, which belongs to the C_{3v} point group. Using Table 4.2 and the approach described in the previous section it is easy to show that these three orbitals $\chi_1 - \chi_3$ transform as $a_1 + e$, that is, one nondegenerate plus one doubly degenerate representation. Generation of the a_1 wavefunction is easy. Using equation 4.8

$$\psi(a_1) \propto \chi_1 + \chi_2 + \chi_3 + \chi_1 + \chi_2 + \chi_3$$

$$\propto \chi_1 + \chi_2 + \chi_3 \tag{4.18}$$

The same answer is obtained from equation 4.8, irrespective of which of $\chi_1 - \chi_3$ is used to start with. For the doubly degenerate wavefunction however, using equation 4.8 with χ_1 gives

$$\psi_a(e) \propto 2\chi_1 - \chi_2 - \chi_3 \tag{4.19}$$

Now if instead of using χ_1 in equation 4.8, we use χ_2, the result is

$$\psi'(e) \propto 2\chi_2 - \chi_3 - \chi_1 \tag{4.20}$$

And finally, if χ_3 is used, we find

$$\psi''(e) \propto 2\chi_3 - \chi_1 - \chi_2 \tag{4.21}$$

The result is *three* different wavefunctions which apparently describe a *doubly* degenerate situation. The solution is to use a technique known as the Schmidt process. First we accept equation 4.19 as one component of the doubly degenerate pair.

The atomic orbitals χ_1, χ_2, and χ_3 are normalized, and in 4.9 the overlap between any two of them are the same. Thus,

$$\langle \chi_1 | \chi_1 \rangle = \langle \chi_2 | \chi_2 \rangle = \langle \chi_3 | \chi_3 \rangle = 1$$

$$\langle \chi_1 | \chi_2 \rangle = \langle \chi_2 | \chi_3 \rangle = \langle \chi_3 | \chi_1 \rangle = S \tag{4.22}$$

Consequently, we obtain

$$\langle 2\chi_1 - \chi_2 - \chi_3 | 2\chi_1 - \chi_2 - \chi_3 \rangle = 6(1 - S) \tag{4.23}$$

so that $\psi_a(e)$ is normalized as

$$\psi_a(e) = \frac{1}{\sqrt{6(1-S)}} (2\chi_1 - \chi_2 - \chi_3) \tag{4.24}$$

Similarly, $\psi'(e)$ and $\psi''(e)$ are normalized as follows:

$$\psi'(e) = \frac{1}{\sqrt{6(1-S)}} (2\chi_2 - \chi_3 - \chi_1)$$

$$\psi''(e) = \frac{1}{\sqrt{6(1-S)}} (2\chi_3 - \chi_1 - \chi_2) \tag{4.25}$$

The orbital $\psi'(e)$ is not orthogonal to $\psi_a(e)$, since

$$\langle \psi'(e) | \psi_a(e) \rangle = \frac{1}{6(1-S)} \langle 2\chi_1 - \chi_2 - \chi_3 | 2\chi_2 - \chi_3 - \chi_1 \rangle = -\frac{1}{2} \tag{4.26}$$

However, the orbital $\psi_b(e)$ defined in the special way of equation 4.27 is orthogonal to $\psi_a(e)$,

$$\psi_b(e) \propto \psi'(e) - \langle \psi'(e) | \psi_a(e) \rangle \psi_a(e)$$

$$\propto \chi_2 - \chi_3 \tag{4.27}$$

Since

$$\langle \chi_2 - \chi_3 | \chi_2 - \chi_3 \rangle = 2(1-S) \tag{4.28}$$

$\psi_b(e)$ is normalized as

$$\psi_b(e) = \frac{1}{\sqrt{2(1-S)}} (\chi_2 - \chi_3) \tag{4.29}$$

We note that use of the orbital $\psi''(e)$ in place of $\psi'(e)$ in equation 4.27 does not produce a new orbital that is orthogonal to both $\psi_a(e)$ and $\psi_b(e)$, since

$$\psi''(e) - \langle \psi''(e) | \psi_a(e) \rangle \psi_a(e) - \langle \psi''(e) | \psi_b(e) \rangle \psi_b(e) = 0 \tag{4.30}$$

Therefore, $\psi_a(e)$ and $\psi_b(e)$ are a set of two orthonormal orbitals belonging to the e representation. In fact, there is considerable choice as to the form of the degenerate wavefunctions. Any linear combination of $\psi_a(e)$ and $\psi_b(e)$ are also valid wavefunctions. 4.10 shows the form of the functions of equations 4.19 and 4.27. Orbitals that look like these will occur again and again throughout the book.

4.10

4.5. DIRECT PRODUCTS

Very often we need to be able to write down the symmetry species of a function which is a simple product of two functions whose symmetry we know. For example,

$$b_1 \;\; \text{---} \; \psi_2$$

4.11 C_{2v}

$$b_2 \;\; \text{---} \; \psi_1$$

we may wish to know the symmetry species of the electronic state which arises from the orbital occupancy shown in **4.11**. The symmetry labels that describe the wavefunctions ψ_1 and ψ_2 are written alongside the molecular orbital levels, using lowercase letters. The simplest function that might be expected to describe the behavior of this electronic state is just $\psi_1 \cdot \psi_2$. (We shall see in Chapter 8 how to tackle this problem properly.) The symmetry species of such a product function is obtained by multiplying the respective characters $D_R(b_1) \times D_R(b_2)$ as in 4.31. The result is clearly the set of characters corresponding to the a_2 irreducible representa-

	E	C_2	$\sigma_v(xz)$	$\sigma_v(yz)$	
$D_R(b_1)$	1	-1	1	-1	
$D_R(b_2)$	1	-1	-1	1	
$D_R(b_1) \times D_R(b_2)$	1	1	-1	-1	(4.31)

tion, that is, $b_1 \times b_2 = a_2$. The electronic state corresponding to **4.11** is thus an A_2 state, where uppercase letters are used to distinguish state symmetry from orbital symmetry. Depending upon whether the electron spins are oriented parallel or antiparallel, respectively, a triplet $(^3A_2)$ or a singlet $(^1A_2)$ state results. Using the same approach it may readily be shown that the orbital occupation situations of **4.12**

$$b_1 \;\; \text{---} \qquad\qquad b_1 \;\; \text{++}$$

4.12 C_{2v}

$$b_2 \;\; \text{++} \qquad\qquad b_2 \;\; \text{---}$$

give rise to $^1A_1$ electronic states, that is, $b_2 \times b_2 = a_1$ and $b_1 \times b_1 = a_1$. This result is found in all point groups. Any product of the two functions $\psi_a\psi_b$ which have the same symmetry species will always belong to the toally symmetric representation of the point group. This observation leads to the simplifying result that all closed shells of electrons give rise to electronic states which are totally symmetric. In determining the symmetry species corresponding to a given orbital occupation, all we need to do, as a result, is look at those electrons outside of completely filled orbitals (i.e., closed shells). Thus **4.11** may just represent the valence orbitals of a system. The symmetry species of the electronic state is not influenced by the core of completely filled levels.

Two examples involving degenerate levels are shown in **4.13** and **4.14**. For **4.13** from Table 4.2 it is readily seen that $a_2 \times e = e$, which leads to ^1E and ^3E electronic states, depending upon the orientation of the electron spins. In **4.14**, the product of the characters leads to a reducible representation

$$C_{3v}$$

a_2 ─╫─ e ═╪═

e ═╪═ e ═╪═

4.13   **4.14**

	E	$2C_3$	$3\sigma_v$	
$D_R(e)$	2	-1	0	
$D_R(e) \times D_R(e)$	4	1	0	(4.32)

and we need to use equation 4.15 to break it down into a sum of irreducible representations. In this way, or by inspection, we find $e \times e = a_1 + a_2 + e$. Depending, then, upon the electron spin orientation, the electronic states that result are $^1A_1$, $^1A_2$, ^1E, and $^3A_1$, $^3A_2$, ^3E. Note that in each of these products the total number of singlets or total number of triplets corresponded to the simple product of the dimensions of the representations involved. So $b_1 \times b_2$ gave a $(1 \times 1 = 1)$ one-dimensional product (a_2), $e \times a_2$ gave a $(2 \times 1 = 2)$ two-dimensional product (e), and $e \times e$ gave a $(2 \times 2 = 4)$ four-dimensional product $(a_1 + a_2 + e)$.

$$C_{3v}$$

e ═╪═   ─╫─ ═╪═ ─╫─   ─╫─

4.15   **4.16**   **4.17**

Problems involving degenerate levels are not always as simple. **4.15** shows double occupancy of an e orbital. **4.16** and **4.17** show that because of the restrictions of the Pauli principle, there can only be three singlets and one triplet and not (as in **4.14**) four of each. In this case we need to define a different sort of direct product. The usual simple product leads to the characters defined in equation 4.33. But the symmetric direct product which we use to generate the *singlet* levels of **4.15** is

$$D_R(i \times j) = D_R(i) \times D_R(j) \tag{4.33}$$

defined by

$$D_R^+(i \times i) = \tfrac{1}{2} [D_R(i) \times D_R(i) + D_{R^2}(i)] \tag{4.34}$$

and the antisymmetric direct product, which we use to generate the *triplet* levels of **4.15**, is defined by

$$D_R^-(i \times i) = \tfrac{1}{2} [D_R(i) \times D_R(i) - D_{R^2}(i)] \tag{4.35}$$

First, we need to identify the operations corresponding to R^2 in equations 4.34 and 4.35. For the C_{3v} point group,

$$E \times E = E; \qquad C_3^2 \times C_3^2 = C_3$$

$$C_3 \times C_3 = C_3^2; \qquad \sigma_v \times \sigma_v = E \tag{4.36}$$

and we may readily evaluate the relevant characters as

	E	$2C_3$	$3\sigma_v$
$D_R(e)$	2	-1	0
$D_R(e) \times D_R(e)$	4	1	0
$D_{R^2}(e)$	2	-1	2
$D_R^+(e \times e)$	3	0	1
$D_R^-(e \times e)$	1	1	-1

(4.37)

The set of characters for $D_R^+(e \times e)$ are seen to give rise to an $a_1 + e$ representation with the aid of Table 4.2. The set for $D_R^-(e \times e)$ corresponds to a_2. So in the C_{3v} point group, a configuration with two electrons in an e symmetry orbital gives rise to electronic states with symmetry $^1A_1$, ^1E, and $^3A_2$.

4.6.   SYMMETRY PROPERTIES AND INTEGRALS

The use of symmetry is extremely powerful in being able to simplify many orbital problems by quickly identifying those interactions which are exactly zero. In particular a rule, which we shall investigate in this section, requires that all integrals of the overlap $\langle \psi_a | \psi_b \rangle$ or interaction $\langle \psi_a | H^{\text{eff}} | \psi_b \rangle$ type are zero unless ψ_a and ψ_b transform as the *same* irreducible representation of the molecular point group. To see why this is so consider the integrals shown in **4.18** and **4.19**. Clearly the integral in equation 4.38 is nonzero

$$\int_{-a}^{a} y \, dx$$

(4.38)

for the symmetric function $y = x^2$, but is identically zero for the antisymmetric function $y = x^3$. In other words, the presence of an operation which sends y to $-y$ (in the case of **4.19** this is simply replacing x by $-x$) immediately causes the integral

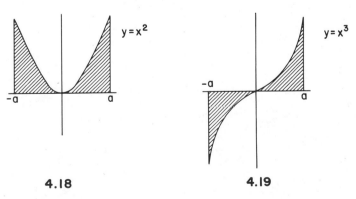

4.18   **4.19**

to become zero. Applied to our overlap and interaction integrals this means that the integrand must be totally symmetric with respect to all symmetry operations of the point group, that is, all the characters must be equal to +1 for the integral to be nonzero. In the case of the overlap integral $\langle \psi_a | \psi_b \rangle$, this means that the function $\psi_a \psi_b$ must contain the totally symmetric (a_1) representation. As we have seen above, the only way this can be true is if ψ_a and ψ_b belong to the same irreducible representation of the point group. Although we will not prove it here the Hamiltonian operator is also totally symmetric. This implies that in the interaction integral $\langle \psi_a | H^{eff} | \psi_b \rangle$, both ψ_a and ψ_b must transform as the same symmetry species of the point group for it to be nonzero. It immediately follows that only orbitals of the same symmetry may interact with each other.

4.7.   THE NONCROSSING RULE[6-8]

Assume that a system with a set of energy levels e_i is transformed in some way along a reaction coordinate, to a product. This coordinate may be a reaction in the traditional sense or perhaps a geometrical distortion. During the process some of the levels (which may correspond to molecular orbital energy levels, or those of electronic states) will go up in energy and others will drop. In general we will be able to correlate the levels before the distortion ($q = 0$) with those afterwards ($q = 1$) as in **4.20**. By making use of the point symmetry along the reaction coordinate, these levels may be given symmetry labels. The noncrossing rule forbids levels with the same symmetry to cross each other. It is easy to see why. **4.21** shows two levels of the same symmetry crossing each other in energy in the E vs. q (energy vs. reaction coordinate) diagram (dashed lines). At the crossing point they will be degenerate. If the two levels have the same symmetry then the overlap and interaction integrals $\langle \psi_a | \psi_b \rangle$ and $\langle \psi_a | H^{eff} | \psi_b \rangle$ will be nonzero and the levels given by the dashed lines will interact via a degenerate interaction as described in Chapter 3. Away from the crossing point the levels will interact by a nondegenerate process. In both cases the

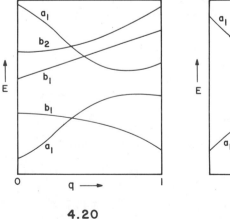

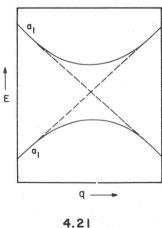

4.20   **4.21**

dashed lines will appear to "repel" each other and the solid curves will result. Although we distinguish between degenerate and nondegenerate cases here for convenience, in fact of course the strength of the interaction between ψ_a and ψ_b varies continuously as they approach each other in energy. This is called an avoided crossing. Just how much the two levels repel each other depends upon the actual magnitude of $\langle \psi_a | H^{eff} | \psi_b \rangle$ and, hence, $\langle \psi_a | \psi_b \rangle$. A strongly avoided crossing is a case where the integrals are sizable. There is then little curvature for the two levels along q. A weakly avoided crossing occurs when the two MOs have the same symmetry but their mutual overlap along q for some reason is small. The two levels will then nearly touch each other before turning around.

It is important to remember that orbitals of different symmetry can and will cross each other along a reaction coordinate. The most important case where this occurs is when the highest occupied molecular orbital (HOMO) crosses the lowest unoccupied molecular orbital (LUMO). Such HOMO–LUMO crossings are said to be symmetry forbidden and engender a high activation barrier. We shall see many examples of this behavior in later chapters and the way molecules conspire to get around the high barrier.

Since electronic states also possess symmetry, one can examine the energetic variation of several states along a reaction coordinate. The same noncrossing rule applies. This vantage point is sometimes more appropriate for cases when there is a manifold of states at close energies. This is frequently encountered in photochemically excited reactions and rearrangements.

4.8. PRINCIPLES OF ORBITAL CONSTRUCTION USING SYMMETRY PRINCIPLES[1,9]

The requirement, set out in the previous section, that two orbitals will not interact unless they are of the same symmetry species is an extremely useful one when the energy levels of molecules are constructed. We will start off by looking at a simple system, that of linear H_3.

Section 3.4 showed one way of generating the orbitals which used first- and second-order perturbation ideas. There is, as in many problems, another way that this can be done which also helps in our understanding of the level structure. In 4.22 is shown the assembly of H_3 from the pair of orbitals $\chi_{1,3}$ and the single central orbital χ_2. Generation of the orbitals of molecules by combining the atomic

$$H_1 \!-\!\!-\! H_3 \;+\; -\!\!H_2\!\!- \quad \longrightarrow \quad H_1 \!-\! H_2 \!-\! H_3 \qquad \textbf{4.22}$$
$$\chi_1 \qquad \chi_3 \qquad\quad \chi_2$$

orbitals of the central atom plus surrounding "ligands" is a very useful way of tackling this problem, and one we will use extensively in this book. First we need the symmetry properties of the pair of symmetry equivalent H $1s$ orbitals χ_1 and χ_3. The point group of linear H_3 is $D_{\infty h}$ and by using the techniques discussed above (and the characters of Table 4.4) it is very easy to show that these orbitals simply transform as $\sigma_u^+ + \sigma_g^+$. (Note the special Mulliken symbols for linear molecules). Use

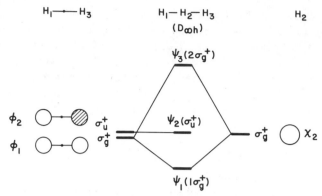

FIGURE 4.2.   Derivation of the molecular orbital diagram for linear H_3 from a central H atom orbital plus those of two "ligands."

of equation 4.8 allows generation of the proper symmetry adapted functions shown in **4.23**. The central atom orbital χ_2 is seen to be of σ_g^+ symmetry. When the interactions between χ_2 and (χ_1, χ_3) are switched on then χ_2 will only overlap and interact with the σ_g^+ combination shown in **4.23**. The σ_u^+ function will be left

$$\phi_1 \propto \chi_1 + \chi_3 = \text{O—O} \quad \sigma_g^+$$

$$\phi_2 \propto \chi_1 - \chi_3 = \text{O—⊘} \quad \sigma_u^+ \qquad \textbf{4.23}$$

$$H_1-H_2-H_3$$

$$\text{O—O—⊘} \qquad \textbf{4.24}$$
$$\chi_2$$
$$\phi_2$$

nonbonding since it is of the wrong symmetry to interact with χ_2. Their overlap integral is zero (**4.24**). Since χ_1 and χ_3 are not nearest neighbors, their overlap integral is small and so the two combinations ϕ_1, ϕ_2 will be close in energy. Figure 4.2 shows the orbital interaction diagram. The phases of the various orbitals have been drawn so that the overlap between ϕ_1 and χ_2 is positive. Note that ψ_3 is de-stabilized more than ψ_1 is stablized, a result exactly analogous to the situation in Chapter 2, where we treated the simple two-center–two-orbital problem. The details of the generation of the molecular orbitals are shown in **4.25**.

$$\psi_1 \; (1\sigma_g^+) \propto \phi_1 + \chi_2 = \text{O—O} + \text{—O—}$$
$$\Rightarrow \text{O—O—O}$$

$$\psi_2 \; (\sigma_u^+) = \phi_2 \qquad = \text{O—⊘} \qquad \textbf{4.25}$$

$$\psi_3 \; (2\sigma_g^+) \propto \phi_1 - \chi_2 = \text{O—O} - \text{—O—}$$
$$\Rightarrow \text{O—⊘—O}$$

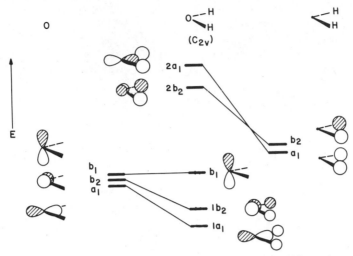

FIGURE 4.3. Assembly of the molecular orbital diagram for water (C_{2v} point group) using symmetry-adapted orbitals on the ligands and the central atom p orbitals.

As another example let us take the H_2O molecule. It may be visualized as being assembled from a central oxygen atom plus two symmetry equivalent hydrogen atoms. The hydrogen $1s$ orbitals we have already seen transform as $a_1 + b_2$. (The form of the wavefunctions are given in **4.4** and **4.5**.) The oxygen p_x, p_y, p_z orbitals (**4.6**) we have seen transform as b_1, b_2, and a_1. Thus the hydrogen a_1 combination may only interact with the oxygen p_z orbital and the hydrogen b_2 combination may only interact with the oxygen p_y orbital. In each case a bonding and antibonding pair of orbitals result (Figure 4.3). The oxygen b_1 orbital finds no ligand counterpart with the same symmetry and remains nonbonding. How the oxygen s orbital enters into this picture we will leave until Chapter 7.

This technique is a very general one for the construction of a diagram for a central atom plus ligands. Instead of working out the symmetry species of the central atom orbitals we may often use information listed in the character tables. At the right-hand side of the Tables 4.1–4.6 are shown the transformation properties of useful functions. For Table 4.1 we can see that z transforms as a_1, x as b_1, and so on, in this point group. From Table 1.1, which gives the angular dependence of atomic wavefunctions, clearly p_z and p_x will too transform as a_1 and b_1, respectively. The central atom s orbital always transforms as the totally symmetric representation. Figure 4.4 shows the assembly of the molecular orbital diagram of a planar AH_3 molecule, with point group D_{3h}. (Its character table is given in Table 4.3.) The H $1s$ orbitals transform as $a_1' + e'$, using the techniques of equations 4.12 and 4.15. Using equation 4.8, the symmetry adapted linear combinations of these basis orbitals are shown at the right-hand side of the figure. Note that the form of the degenerate pair of orbitals is just as we derived in **4.10**. Also since the hydrogen orbitals themselves are symmetric with respect to reflection in the molecular plane, their representations carry a single prime. It is interesting to see at this stage that each

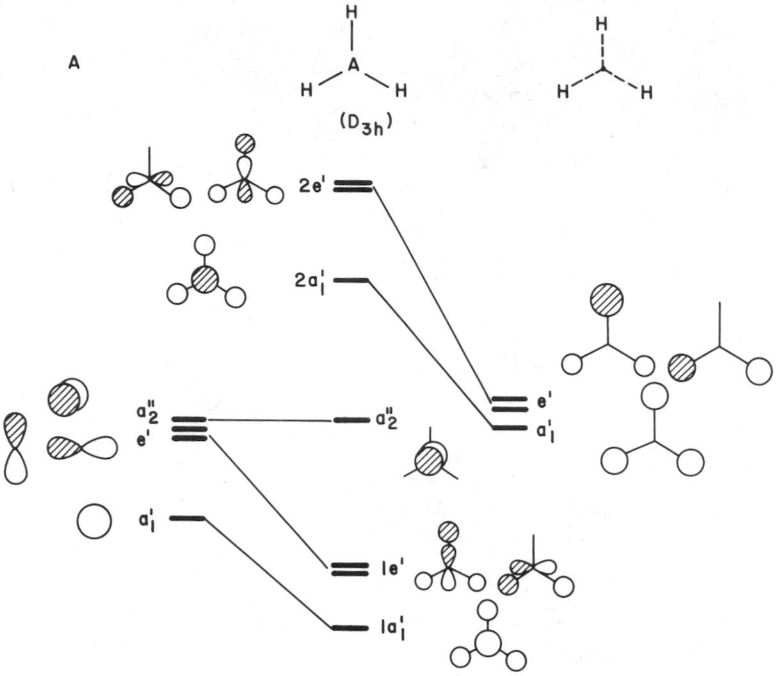

FIGURE 4.4.   Assembly of the molecular orbital diagram for the planar AH_3 molecule (D_{3h} point group) using symmetry-adapted orbitals on the ligands and the central s and p orbitals.

of the e' orbitals is nicely set up to interact with a p_x and p_y orbital on the central atom. Indeed p_x, p_y on the A atom transform as e'. p_z and s transform as a_2'' and a_1', respectively. The a_1' ligand combination has just the right symmetry to interact with the central atom s orbital. The level structure for planar AH_3 is simply produced by constructing in-phase (bonding) and out-of-phase (antibonding) molecular orbitals from orbitals at the left and right sides with the same symmetry. Of all the orbitals the p_z orbital on the central atom finds no suitable ligand combination and so remains nonbonding.

Figure 4.5 shows a diagram for tetrahedral AH_4 which belongs to the T_d point group. Here we find that the four hydrogen $1s$ orbitals transform as $a_1 + t_2$, that is, a nondegenerate and a triply degenerate representation. The central atom p orbitals transform as t_2 and the central atom s orbital as a_1. Generation of the orbital diagram is therefore quite a simple process by creating bonding and antibonding orbital pairs from central atom and ligand orbital combinations of the right symmetry. What these symmetry ideas do not give is any idea of the magnitude of the energy shifts involved. Specifically in this case, whether the upper t_2 orbital lies below or above the upper a_1 orbital is a matter that has to be resolved by numerical calculation. Shown in this diagram are the form of the triply degenerate hydrogen $1s$ orbital combinations which can be determined using the Schmidt process described earlier. Algebraically with reference to **4.26** one choice for these degenerate func-

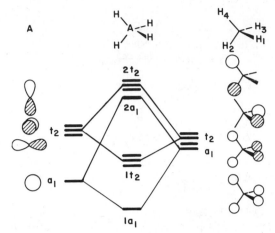

FIGURE 4.5.   Assembly of the molecular orbital diagram for the tetrahedral AH$_4$ molecule (T_d point group) using symmetry-adapted orbitals on the ligands and the central atom s and p orbitals.

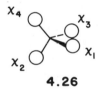

4.26

tions is

$$\psi_a(t_2) = \tfrac{1}{2}(\chi_1 - \chi_2 + \chi_3 - \chi_4)$$

$$\psi_b(t_2) = \frac{1}{\sqrt{2}}(\chi_1 - \chi_3) \qquad\qquad (4.39)$$

$$\psi_c(t_2) = \frac{1}{\sqrt{2}}(\chi_2 - \chi_4)$$

and that for the a_1 orbital is of course

$$\psi(a_1) = \tfrac{1}{2}(\chi_1 + \chi_2 + \chi_3 + \chi_4) \qquad\qquad (4.40)$$

where the factors before the parentheses are normalization constants, assuming that the overlap integrals between these orbitals are to be ignored (i.e., $\langle\chi_i|\chi_j\rangle = \delta_{ij}$).

How does the picture change however if, instead of being interested in the AH$_4$ molecule, we wanted the orbital diagram for a tetrahedral H$_4$ molecule? Now, putting the overlap integrals $\langle\chi_i|\chi_j\rangle = S_{ij}$ ($i \neq j$), the normalization constants of equations 4.39 and 4.40 change, but importantly the coefficients before the χ_i within the parentheses do not change. The new wavefunctions are simply

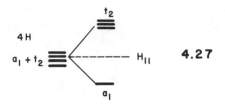

$$\psi_a(t_2) = \frac{1}{\sqrt{2(1-S)}} (\chi_1 - \chi_3) \qquad (4.41)$$

$$\psi(a_1) = \frac{1}{2\sqrt{(1+3S)}} (\chi_1 + \chi_2 + \chi_3 + \chi_4) \qquad (4.42)$$

where we have shown one t_2 component. If the H_4 problem is characterized by interaction integrals $\langle \chi_i | H^{eff} | \chi_j \rangle = H_{12}$ $(i \neq j)$, then substitution of equations 4.41 and 4.42 into equation 1.13 leads to the values of the energies,

$$e(t_2) = \frac{H_{11} - H_{12}}{(1-S)} \qquad (4.43)$$

$$e(a_1) = \frac{H_{11} + 3H_{12}}{1 + 3S} \qquad (4.44)$$

and therefore the ready synthesis of the orbital diagram of **4.27**. For simple molecules of this type then the level structure down to the relative energy of the levels involved is dominated by the symmetry of the system, and may be qualitatively understood in a few lines as we have shown here.

REFERENCES

1. F. A. Cotton, *Group Theory and Its Applications*, 2nd ed., Wiley-Interscience, New York (1971).
2. E. Heilbronner and H. Bock, *The HMO Model and its Application*, Wiley, New York (1976).
3. M. Orchin and H. H. Jaffe, *Symmetry, Orbitals and Spectra*, Wiley-Interscience, New York (1971).
4. D. M. Bishop, *Group Theory*, Oxford University Press, New York (1973).
5. A. V. Golton, in *Spectroscopy*, Vol. 2, B. P. Straughan and S. Walker, editors, Chapman and Hall, London (1970).
6. R. McWeeney, *Coulson's Valence*, Oxford University Press, New York (1979).
7. L. Salem, *Electrons in Chemical Reactions*, Wiley, New York (1982).
8. R. B. Woodward and R. Hoffmann, *The Conservation of Orbital Symmetry*, Academic Press, New York (1969).
9. K. F. Purcell and J. F. Kotz, *Inorganic Chemistry*, Saunders, New York (1977).

CHAPTER FIVE

Molecular Orbital Construction from Fragment Orbitals

5.1. INTRODUCTION

The practical aspects of the perturbation molecular orbital theory discussed in Chapter 3 will now be illustrated further by constructing the orbitals of simple molecular systems. While being illustrative in their own right, the level patterns are representative of several of the orbital patterns we will come across later. The molecules examined in this chapter have the common feature that all the atoms of the molecule are identical and each contributes one atomic orbital. At its simplest, then, these are H_n molecules (hypothetical for the most part) but all of the arguments for planar systems carry directly over to the π level structure of polyenes where one $p\pi$ orbital per carbon atom only is considered. In our discussion the first- and second-order mixing coefficients $t_{ji}^{(1)}$ and $t_{ki}^{(2)}$ are assumed to be small. Their signs are determined simply by the signs of the relevant overlap integrals $\tilde{S}_{ij}$ and $\tilde{S}_{kj}$ of equation 3.16 and by the relative energetic ordering of the orbitals involved.

5.2 TRIANGULAR H_3

This case may be constructed as in Figure 5.1 from the orbitals of the diatomic H_2 unit and the orbital of a single H atom. It is easy to see by inspection that $\langle \phi_2 | \chi_3 \rangle = 0$, a result which follows too from symmetry arguments; ϕ_2 and χ_3 are

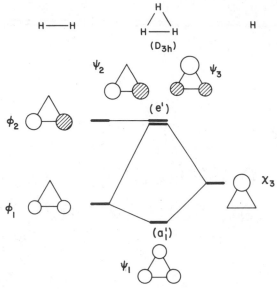

FIGURE 5.1. Assembly of the orbital diagram for equilateral triangular H_3 (D_{3h}) from those of $H_2 + H$.

of different symmetry species. We do not need to investigate this in any great detail other than to note that one is symmetric (a character of $+1$) and the other antisymmetric (a character of -1) with respect to reflection in the plane containing χ_3 and the bisector of $\chi_1 - \chi_2$. The molecular orbitals are thus constructed as in **5.1.** What is not clear in this approach is that in the orbital ψ_1, all the atomic

$$\psi_1 \propto \phi_1 + (\chi_3) = \text{△} + \left(\text{△}\right) \Rightarrow \text{△}$$

5.1

$$\psi_3 \propto \chi_3 - (\phi_1) = \text{△} - \left(\text{△}\right) \Rightarrow \text{△}$$

orbitals are of equal weight, but in its antibonding equivalent ψ_3, χ_3 carries twice the weight of χ_1 or χ_2. Neither is it clear that ψ_3 is destabilized to exactly the same energy as the orbital ψ_2, nonbonding between either $H_{1,2}$ and H_3. This result drops straight out of the group theoretical treatment of Chapter 4. Since the three H $1s$ orbitals are related by the threefold axis of the triangle *by symmetry*, they are constrained to form three molecular orbitals, one of which is nondegenerate a_1' and a pair which is degenerate (e'), just as we showed in a step preparatory to the construction of the diagram for planar AH_3 in Section 4.8. **4.10** shows the form of these orbitals. Using these wavefunctions and the expression for the energy in equation 1.13, then, since $H_{11} = H_{22} = H_{33}$ and $H_{12} = H_{13} = H_{23}$,

$$e(a_1') \propto \langle \chi_1 + \chi_2 + \chi_3 | H^{\text{eff}} | \chi_1 + \chi_2 + \chi_3 \rangle$$

$$\propto H_{11} + 2H_{12} \tag{5.1}$$

and

$$e(e') \propto \langle \chi_1 - \chi_2 | H^{eff} | \chi_1 - \chi_2 \rangle$$
$$\propto H_{11} - H_{12} \qquad \qquad (5.2)$$

or

$$e(e') \propto \langle 2\chi_3 - \chi_1 - \chi_2 | H^{eff} | 2\chi_3 - \chi_1 - \chi_2 \rangle$$
$$\propto H_{11} - H_{12} \qquad \qquad (5.3)$$

So from three isolated s orbitals, the splitting pattern appears as in **5.2**. Suppose that one H-H distance in H$_3$ is shorter than the other two. The orbitals for such an

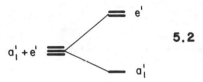

5.2

isosceles triangle are constructed in an exactly analogous way to the equilateral triangle problem of Figure 5.1. Now since the threefold axis of the latter is absent, the two orbitals ψ_2 and ψ_3 are not degenerate. The level pattern is generated in Figure 5.2. Shortening r_1 in **5.3**, relative to its value in the equilateral triangle, stabilizes ψ_1 since there is a bonding interaction between atoms 1 and 2 in this orbital. Similarly ψ_2 does not rise as high in energy as it did in Figure 5.1. Quantitatively the energy levels of the isosceles triangle may be derived from those of equilateral triangular structure by making $H_{13} = H_{23} \neq H_{12}$ in equations 5.1 and 5.3. Now equations 5.2 and 5.3 do not lead to the same energy. This is reflected

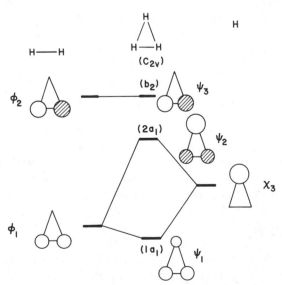

FIGURE 5.2. Assembly of the orbital diagram for an isoceles triangular H$_3$ (C_{2v}) from those of H$_2$ + H.

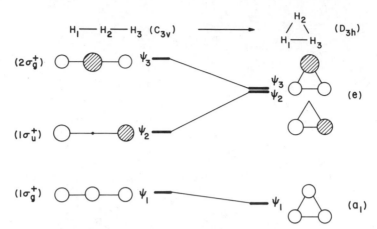

FIGURE 5.3.   Correlation diagram linking the level structures of linear and equilateral triangular H_3.

too in the symmetry species of the levels. In the new point group (C_{2v}) there are no degenerate representations and the pair of functions which used to be of e' symmetry in D_{3h} have now split into functions of $a_1 + b_2$ symmetry.

Figure 5.3 shows how the molecular orbital levels of H_3 change as the molecular geometry changes from the linear to equilateral triangular structure. As the H–H–H angle decreases from 180° to 60° the overlap integrals $\langle \chi_1 | \chi_2 \rangle$ and $\langle \chi_2 | \chi_3 \rangle$ do not change, provided the distances H_1–H_2 and H_2–H_3 remain fixed. What happens, of course, is that overlap between χ_1 and χ_3 is switched on. This stabilizes ψ_1 and ψ_3, since χ_1 and χ_3 enter with the same relative phase in both of these orbitals. During the bending process these orbitals mix together such that the atomic orbital coefficients in ψ_1 become equal at the equilateral triangular geometry. This will be covered in some depth in Chapter 7. Since in ψ_2 of the linear system, χ_1 and χ_3 are of opposite phase, as the H–H–H angle decreases, ψ_2 rises in energy and eventually becomes degenerate with ψ_3. The reader should check the resultant molecular orbitals shown in the right-hand side of Figure 5.3 and compare them with those derived in Figure 5.1. The results are, of course, the same, and we shall use both techniques, namely distortion of one geometry to another and assembly from smaller fragments to view the level structure of more complex systems.

5.3.   LINEAR H_4

The molecular orbitals of linear H_4 (5.4) may be constructed by interacting the orbitals of two H_2 fragments as shown in Figure 5.4 where the orbitals ϕ_i and ϕ_i' ($i = 1, 2$) are arranged such that $\langle \phi_i | \phi_i' \rangle > 0$. The orbitals ϕ_i and ϕ_i' enter into degenerate interactions as shown in 5.5 where each of the atomic orbital coeffi-

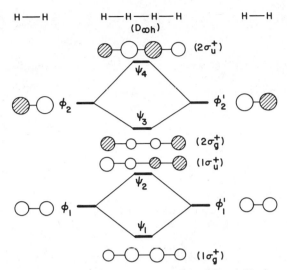

FIGURE 5.4. Generation of the molecular orbital levels of linear H$_4$ from those of two collinear H$_2$ units. The construction of this diagram occurs in two parts (see text). First-order energy shifts occur via interaction of the degenerate levels **(5.5)** followed by second-order interaction. (Figure 5.5).

$$H_1\!-\!H_2 \cdots + \cdots H_3\!-\!H_4 \longrightarrow H_1\!-\!H_2\!-\!H_3\!-\!H_4 \quad \textbf{5.4}$$
$$X_1 \quad X_2 \qquad\qquad X_3 \quad X_4$$

$\xi_1 \propto \phi_1 + \phi_1' =$

$\xi_2 \propto \phi_1 - \phi_1' =$

5.5

$\xi_3 \propto \phi_2 + \phi_2' =$

$\xi_4 \propto \phi_2 - \phi_2' =$

cients in ξ_{1-4} are equal. However ξ_1, ξ_3 have the same symmetry (they are both symmetric with respect to the mirror plane between H$_{2,3}$) and can enter into second-order interactions as shown in Figure 5.5. An analogous process is also shown for ξ_2, ξ_4. Note that the overlap between ξ_1 and ξ_3 is positive, while that between ξ_2 and ξ_4 is negative, as shown below:

$$\langle \xi_1 | \xi_3 \rangle \propto \langle \phi_1 + \phi_1' | \phi_2 + \phi_2' \rangle = \langle \phi_1 | \phi_2' \rangle + \langle \phi_1' | \phi_2 \rangle = (+) \qquad (5.4)$$

$$\langle \xi_2 | \xi_4 \rangle \propto \langle \phi_1 - \phi_1' | \phi_2 - \phi_2' \rangle = -\langle \phi_1 | \phi_2' \rangle - \langle \phi_1' | \phi_2 \rangle = (-) \qquad (5.5)$$

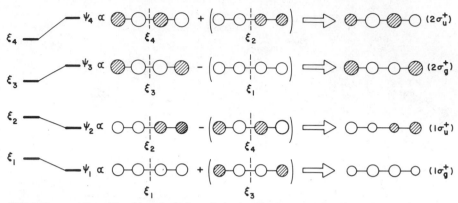

FIGURE 5.5. Second-order interaction of the levels $\xi_1 - \xi_4$ produced by degenerate interactions between two collinear H_2 units **(5.5).**

Thus the higher lying ξ_3 is mixed into the lower lying ξ_1 with a positive mixing coefficient, but the higher lying ξ_4 is mixed into the lower lying ξ_2 with a negative sign. The size of the energy shifts will be proportional to the magnitude of the overlap integrals between ξ_1 and ξ_3 and between ξ_2 and ξ_4. Note that as a result of this process the pairs of orbitals have mixed into each other so that they end up being orthogonal (i.e., have zero overlap).

5.4. RECTANGULAR AND SQUARE PLANAR H_4

The orbitals of rectangular H_4 are constructed in an exactly analogous manner in Figure 5.6. The molecular orbitals ϕ_1, ϕ_1' are degenerate as are ϕ_2' and ϕ_2 and so

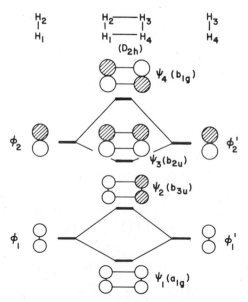

FIGURE 5.6. Assembly of the orbital diagram for rectangular H_4 from those of two H_2 units.

the wavefunctions correct to first order in the energy are given in the same manner as the case of linear H_4 in **5.5**. However, the resulting orbitals are orthogonal and so no second-order interaction between them occurs. Consider what happens when the rectangular H_4 molecule (Figure 5.6) distorts to a square, that is, r_1 becomes longer and r_2 shorter **(5.6)**. The overlap between H_1 and H_4 and between H_2 and H_3 will

$$\begin{array}{c} H_2 \overset{r_2}{-} H_3 \\ r_1| \quad | \\ H_1 - H_4 \end{array} \quad \textbf{5.6}$$

increase, and the overlap between H_1 and H_2 and between H_3 and H_4 will decrease. Increasing overlap leads to an increasing stabilization if the two atomic orbitals have the same phase in a molecular orbital and a destabilization if they have opposite phases. Thus ψ_1 and ψ_4 remain at approximately the same energy but ψ_2 rises and ψ_3 drops in energy. At the square planar geometry ψ_2 and ψ_3 have the same energy, that is, they are degenerate (Figure 5.7). As shown in Chapter 4 there is always a choice to be made in writing degenerate wavefunctions. Recall that two new functions may be generated by taking a linear combination of the old. **5.7**

$$\psi_2' \propto \psi_2 + \psi_3 = $$

$$\psi_3' \propto \psi_2 - \psi_3 = $$

5.7

shows an alternative pair of wavefunctions for this case. Note that each pair of functions are orthogonal to each other.

The energy level diagrams of Figures 5.6 and 5.7 are intimately related[1,2] to the question of the $H_2 + D_2$ exchange reaction. Shown in Figure 5.8 is an orbital

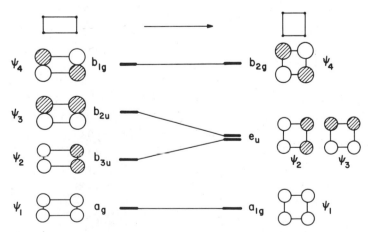

FIGURE 5.7.   Correlation diagram linking the level structures of rectangular and square H_4 molecules.

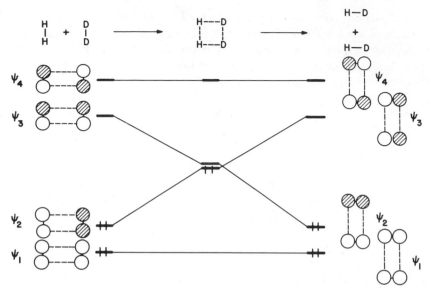

FIGURE 5.8.   Orbital correlation diagram associated with the H/D exchange for $H_2 + D_2$ proceeding via a square geometry.

correlation diagram for this process via a square geometry. This is clearly a symmetry forbidden reaction (Section 4.7) since ψ_2 and ψ_3 cross at the square geometry and are orthogonal throughout the entire reaction coordinate. A high activation energy is predicted for the process if it occurs by this pathway.

5.5.   TETRAHEDRAL H_4

In the previous section two H_2 units were linked together in a planar geometry. If they are joined together in a perpendicular manner (as in **5.8**) then a tetrahedral H_4 system results. Using the atomic orbital basis functions shown in **5.8**, the orbital

interaction diagram can be constructed as in Figure 5.9. All the overlap integrals between the two units are zero except for $\langle \phi_1 | \phi_1' \rangle$. So the molecular orbitals of the tetrahedral H_4 are simply

$$\psi_1 \propto \phi_1 + \phi_1'$$

$$\psi_2 \propto \phi_1 - \phi_1'$$

$$\psi_3 = \phi_2 \qquad\qquad (5.6)$$

$$\psi_4 = \phi_2'$$

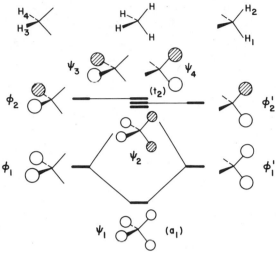

FIGURE 5.9. Assembly of the orbital diagram for tetrahcdral H_4 from those of two H_2 units. (The solid lines in this diagram which are associated with the geometrical figures are used to help the reader visualize the molecular geometry. The "bonds" between connected atoms, usually depicted by lines have been left off for clarity.)

Although it is not at all obvious from this treatment, ψ_2, ψ_3, and ψ_4 are degenerate. This was shown quite clearly, however, in Section 4.7 where we used symmetry arguments to assemble the H_4 orbital diagram from four isolated H orbitals. A single bonding (a_1) and triply degenerate antibonding (t_2) set resulted. We can simply show how in Figure 5.9 ψ_2 and ψ_3 have the same energy as ψ_4 by taking linear combinations of ψ_3 and ψ_4 as in 5.9.

5.9

Returning to the $H_2 + D_2$ exchange reaction, a tetrahedral species is not expected to serve as the transition state or intermediate either. Two electrons are placed in ψ_1 and the remaining two enter the triply degenerate t_2 level ($\psi_2 - \psi_4$). This will be an unstable situation for the same reasons the square H_4 assembly is unfavorable. Another way to look at the problem is to realize that since ψ_1 is not stabilized as much as ψ_2 is destablized, then the electronic situation that results is a simple two-orbital–four-electron destabilization.

The orbital correlation diagram for the square planar to tetrahedral transformation is shown in Figure 5.10. The hydrogen atoms at the top of the figure have been labeled for easy reference. The H_1–H_3 and H_2–H_4 distances become shorter while those for H_1–H_2 and H_3–H_4 stay roughly constant along the distortion pathway. It

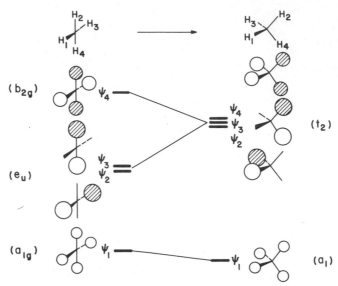

FIGURE 5.10.   Correlation diagram linking the level structures of square planar and tetrahedral H_4 molecules.

is easy to see by inspection that ψ_1 and ψ_4 are stabilized since bonding is turned on between H_1 and H_3 along with that between H_2 and H_4. ψ_2 and ψ_3 are destabilized. Notice that using the alternative descriptions for the degenerate orbitals $\psi_{2,3}$ in **5.7** for the square planar geometry leads directly to the appropriate combinations for the triply degenerate tetrahedral levels that we chose in Figure 5.9.

5.6.   PENTAGONAL H_5 AND HEXAGONAL H_6

The molecular orbitals of pentagonal H_5 may be constructed in terms of the orbitals of a bent H_3 and those of H_2 as shown in Figure 5.11. Note that $\langle \phi_1' | \phi_1 \rangle$, $\langle \phi_1' | \phi_3 \rangle$, and $\langle \phi_2' | \phi_2 \rangle$ are positive but that all other overlap integrals are zero. Using the results of Table 3.1, then, the orbital derivations of Figure 5.12 naturally follow. Again we need to resort to group theoretical considerations to show that $\psi_{4,5}$ and $\psi_{2,3}$ form two degenerate pairs of orbitals. We will return to this below.

Analogously the level structure of hexagonal H_6 may be derived as in Figure 5.13 from the orbitals of a pair of H_2 units arranged in a rectangle and an H_2 unit with a large H–H distance. For this case $\langle \phi_1 | \phi_1' \rangle$ and $\langle \phi_3 | \phi_2' \rangle$ are positive and all other overlap integrals are zero. The resulting first-order mixing of the orbitals is shown in Figure 5.14. This is not the only way this diagram can be constructed. Another obvious route is from six s orbitals using group theoretical techniques. Another is from three H_2 fragments. This interesting last approach brings us back to the H_2/D_2 exchange. **5.10** shows how two HD units may be produced from $2H_2 + D_2$ in a termolecular process. The transition state in the middle of the

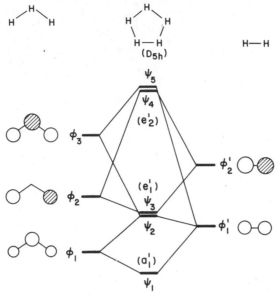

FIGURE 5.11. Assembly of the orbital diagram for pentagonal H_5 from those of bent H_3 and H_2.

$$\text{5.10}$$

diagram will be described by the orbitals of Figure 5.13. With a total of six electrons $\psi_1 - \psi_3$ are doubly occupied. The degenerate pair $\psi_{2,3}$ should lie at about the same energy as the bonding level in H_2 itself. Consequently a termolecular collision with this geometry is expected to be much more feasible than a bimolecular pathway as indeed suggested by detailed calculations. (Although such a collision is much less probable.)

The orbital structure of cyclic H_n systems or their polyene counterparts, as discussed below, form a very interesting series. We will investigate their level structure in more detail in Chapter 12, but note the emergence of a pattern in **5.11** con-

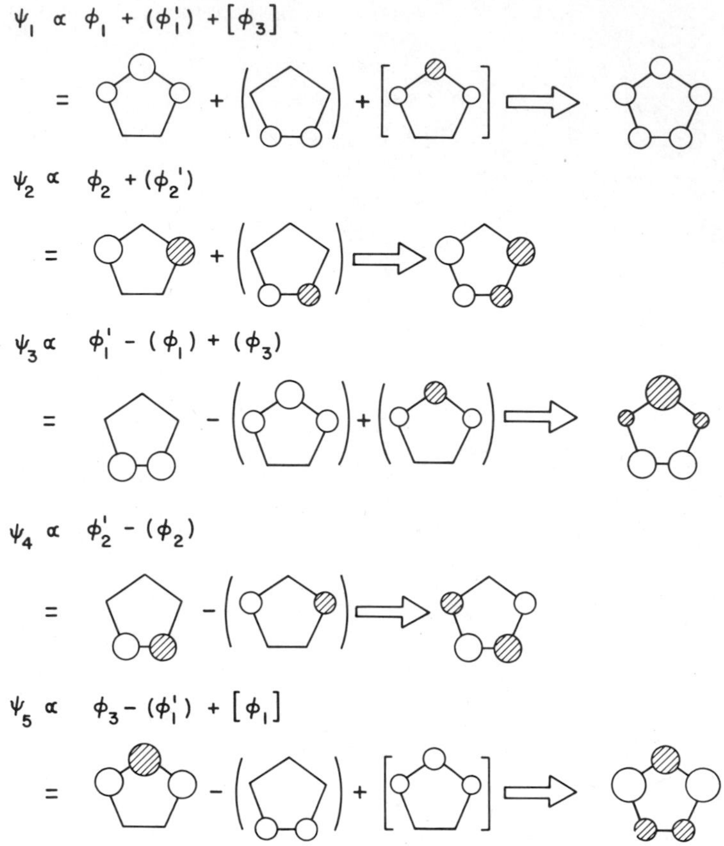

$\psi_1 \propto \phi_1 + (\phi_1') + [\phi_3]$

$\psi_2 \propto \phi_2 + (\phi_2')$

$\psi_3 \propto \phi_1' - (\phi_1) + (\phi_3)$

$\psi_4 \propto \phi_2' - (\phi_2)$

$\psi_5 \propto \phi_3 - (\phi_1') + [\phi_1]$

FIGURE 5.12. Generation of the orbital description of the levels of pentagonal H_5 from those of bent H_3 and H_2 using perturbation theory.

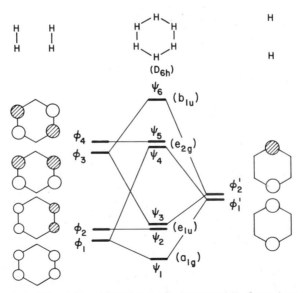

FIGURE 5.13. Assembly of the orbital diagram for hexagonal H_6 from those of two H_2 units and a pair of H atoms.

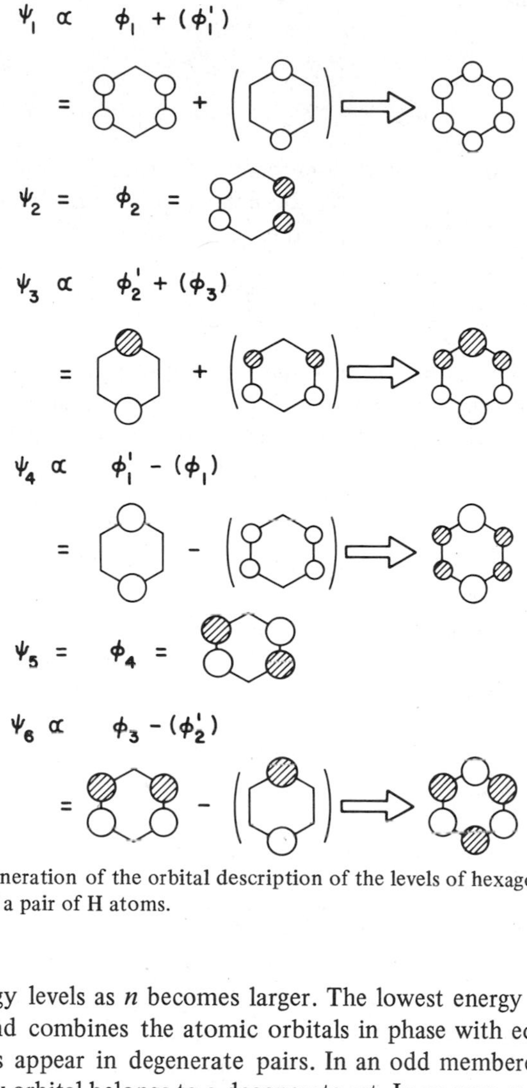

$\psi_1 \propto \phi_1 + (\phi_1')$

$\psi_2 = \phi_2 =$

$\psi_3 \propto \phi_2' + (\phi_3)$

$\psi_4 \propto \phi_1' - (\phi_1)$

$\psi_5 = \phi_4 =$

$\psi_6 \propto \phi_3 - (\phi_2')$

FIGURE 5.14.   Generation of the orbital description of the levels of hexagonal H_6 from those of two H_2 units and a pair of H atoms.

cerning the energy levels as n becomes larger. The lowest energy orbital is always nondegenerate and combines the atomic orbitals in phase with equal coefficients. Then the orbitals appear in degenerate pairs. In an odd membered ring therefore the highest energy orbital belongs to a degenerate set. In an even membered ring the highest energy orbital is nondegenerate.

5.7.   ORBITALS OF π SYSTEMS

All of the examples used so far in this chapter have employed orbitals constructed from a single s orbital on each center. However, our arguments can be carried over without change directly to the case of planar organic π systems. For example, the three $p\pi$ orbitals of cyclopropenyl (5.12) may be represented as in 5.13 which shows the view of the upper lobes of the p orbitals projected onto the molecular plane. Thus, the level structure shown in Figure 5.1 for triangular H_3 is identical

5.12   **5.13**

to that for the π orbitals of cyclopropenyl, where the orbitals are shown in perspective in **5.12**. Correspondingly, the conversion of rectangular cyclobutadiene to the square **(5.14)** follows exactly the same analysis as detailed for the H_4 prob-

5.14

lem in Section 5.4 of this chapter. With a total of four π electrons in C_4H_4, the reader can readily see from Figure 5.7 that if they are arranged in the low spin configuration (all spins paired), then the square geometry is less stable than the rectangular one.

One striking difference between the $p\pi$ orbitals of the cyclic polyene and those of the H_n molecules is that the symmetry labels are different. For example, the $p\pi$ levels of cyclopropenyl transform as $a_2'' + e''$ (double primes since they are antisymmetric with respect to reflection in the plane perpendicular to the threefold rotation axis) but the s orbitals of H_3 transform as $a_1' + e'$ (single primes since these orbitals are symmetric with respect to this symmetry operation). The breakdown into nondegenerate and degenerate orbitals (one of each) is the same in both cases.

The in-plane p orbitals of the cyclic organic systems may be derived in a similar way to their out-of-plane counterparts. For an equilateral triangle, the three in-plane tangential p orbitals **5.15** transform as $a_2' + e'$. Figure 5.15 shows how the

5.15

in-plane orbital picture may be assembled along very similar lines to the H_3 problem of Figure 5.1. Figure 5.16 shows how the form of the orbitals of the molecule are derived. Note that for the in-plane p orbital case the level picture is a two below one pattern but for the out-of-plane p orbital case, a one below two situation occurs. This arises simply because of the nodal properties of χ_3. It interacts with the higher energy orbital, ϕ_2 in Figure 5.15, but with the lower energy orbital in Figure 5.1. Another way of describing the same result is to classify these cyclic orbital problems as either of Hückel or Möbius type.[3] Hückel systems either have a zero or even number of antibonding interactions between adjacent orbitals as in **5.16**. Möbius systems have an odd number of such interactions as in **5.17**. The general result is that the energy level pattern resulting from the in-plane tangential orbitals of odd membered rings is the reverse of the pattern for the out-of-plane $p\pi$ orbitals shown in **5.11**. (For even membered rings the level patterns for in- and out-of-plane orbitals are the same.)

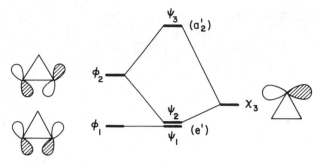

FIGURE 5.15. Assembly of the orbital diagram for the tangential in-plane p orbitals of cyclo-propenium from those of a diatomic unit and a single atom.

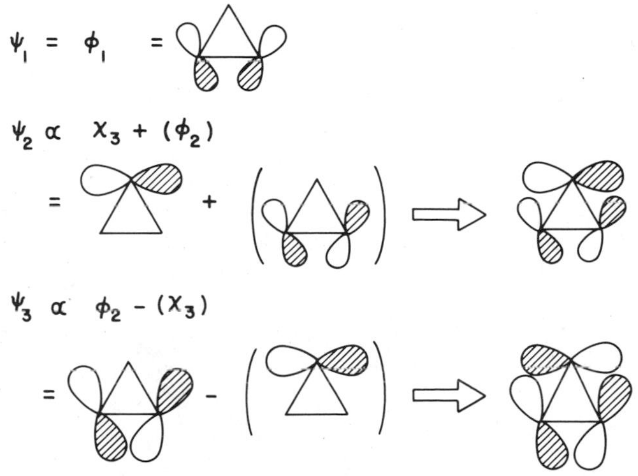

$$\psi_1 = \phi_1 =$$

$$\psi_2 \propto X_3 + (\phi_2)$$

$$= \quad + \quad \Rightarrow$$

$$\psi_3 \propto \phi_2 - (X_3)$$

$$= \quad - \quad \Rightarrow$$

FIGURE 5.16. Generation of the orbital description of the tangential in-plane p orbitals of cyclopropenium using the assembly route of Figure 15.15.

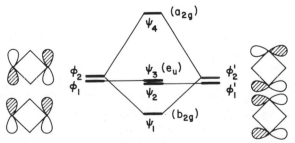

FIGURE 5.17. Assembly of the orbital diagram for the tangential in-plane orbitals of cyclo-butadiene from those of two diatomic units.

$$\psi_1 \propto \phi_2 + \phi_2'$$

$$= \quad + \quad \Rightarrow$$

$$\psi_2 = \phi_1 =$$

$$\psi_3 = \phi_1' =$$

$$\psi_4 \propto \phi_2 - \phi_2'$$

$$= \quad - \quad \Rightarrow$$

FIGURE 5.18.   Generation of the orbital description of the tangential in-plane p orbitals of cyclobutadiene using the assembly route of Figure 15.17.

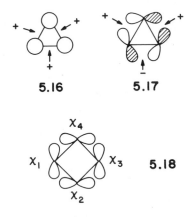

5.16   **5.17**

5.18

Finally consider the four in-plane orbitals of a square plane **(5.18)**. The orbital picture is readily assembled as in Figure 5.17 where all overlap integrals between the ϕ_i and ϕ_i' are zero except $\langle \phi_2 | \phi_2' \rangle$ which is positive. The molecular orbitals are simply constructed using a degenerate interaction as in Figure 5.18. Notice that however the phases on the orbitals $\chi_1 - \chi_4$ of **5.18** are chosen, there are either zero or an even number of changes in sign of the overlap integrals on moving round the ring. These in-plane orbitals thus form a Hückel rather than Möbius pattern, and the qualitative picture is the same as for the $p\pi$ levels of **5.11**.

REFERENCES

1. J. S. Wright, *Can. J. Chem.*, **53**, 549 (1975).
2. W. Gerhartz, R. D. Poshusta, and J. Michl, *J. Amer. Chem. Soc.*, **98**, 6427 (1976); **99**, 4263 (1977).
3. K. Yates, *Hückel Molecular Orbital Theory*, Academic Press, New York (1978).

CHAPTER SIX

Molecular Orbitals of Diatomic Molecules and Electronegativity Perturbation

6.1. INTRODUCTION

In the previous chapter we showed how the energy levels of a molecule could be derived by assembling a molecular orbital diagram from those of smaller fragments using perturbation theory. It was seen how the orbitals of a molecule, initially orthogonal, can mix together in the presence of another molecule via a second-order mixing process. There are other ways these orbitals may mix together without the presence of another fragment. One way is through an intramolecular perturbation which involves a change in the effective potential of an atomic orbital, which will be called an electronegativity perturbation (3.1). Another is via a geometry change, as typified by the example in 3.2, which will be described in more detail in Chapter 7.

In order to examine the workings of electronegativity perturbation, we will need to examine the orbitals of a molecule where the atoms are not all identical and where each atom carries more than one atomic orbital. An important feature which results is that of orbital hybridization, namely the mixing of different atomic orbitals on the same center. In this chapter we will examine the nature of such hybridization, construct the molecular orbitals of diatomic molecules from different viewpoints, and describe the essence of electronegativity perturbations.

6.2.   ORBITAL HYBRIDIZATION

When combined at a given atomic center, any two atomic orbitals add in a vectorial manner. For example, consider the orbital ϕ defined by p_x and p_y atomic orbitals as

$$\phi \propto c_1 p_x + c_2 p_y \tag{6.1}$$

The orbital addition is shown in **6.1** and **6.2** for the two cases $c_1 = c_2 > 0$ and $c_1 = -c_2 > 0$, respectively. The relative magnitudes of c_1 and c_2 control how much the orbital ϕ is tilted away from the x and y axes, respectively. The linear combination between z^2 and $x^2 - y^2$ orbitals of **6.3** leads to a $z^2 - x^2$ orbital as readily appreci-

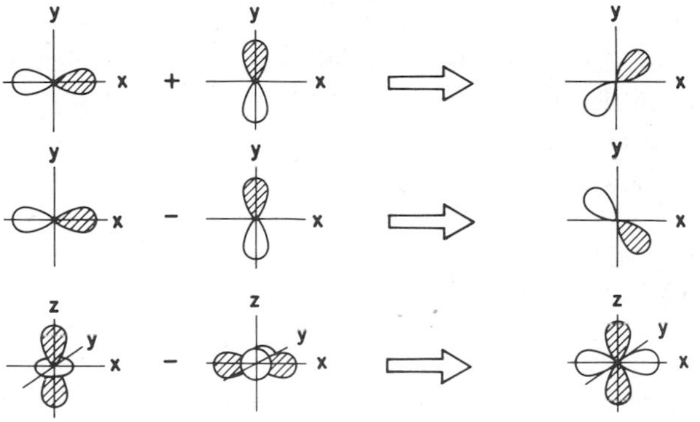

6.1

6.2

6.3

ated from Table 1.1, since

$$d_{z^2} - d_{x^2-y^2} \propto \left(\frac{3z^2 - r^2}{r^2} - \frac{x^2 - y^2}{r^2} \right) \exp(-\zeta r) \tag{6.2}$$

Because $r^2 = x^2 + y^2 + z^2$, this is proportional to

$$\left(\frac{z^2 - x^2}{r^2} \right) \exp(-\zeta r) \propto d_{z^2-x^2} \tag{6.3}$$

The mixing of atomic orbitals with different angular momentum quantum number is also controlled by a vectorial addition, and leads to various types of hybrid orbitals as shown in Figure 6.1. The variation of the overlap or interaction integrals of these functions with other orbitals as a function of geometry is given simply by a weighted sum of the contributions from each component. So if

$$\phi_{\text{hybrid}} = c_1 \chi_1 + c_2 \chi_2 \tag{6.4}$$

then

$$\langle \phi_{\text{hybrid}} | \chi_3 \rangle = c_1 \langle \chi_1 | \chi_3 \rangle + c_2 \langle \chi_2 | \chi_3 \rangle \tag{6.5}$$

(a)

(b)

(c)

ꟻIGURE 6.1.   Hybridization of atom orbitals (a) s and p, (b) d_{xz} and p_z, (c) d_z^2 and p_z.

6.3.   MOLECULAR ORBITALS OF DIATOMIC MOLECULES

We first examine the valence molecular orbitals of a homonuclear diatomic A_2 unit where each atom contributes four valence atomic orbitals (s, p_x, p_y, p_z) and we identify the internuclear axis with the z direction. The basis orbitals naturally separate into those of σ (s, p_z) and π (p_x, p_y) type. The π-type orbitals enter into a degenerate interaction and lead to π bonding and antibonding orbitals as shown in Figure 6.2. Alternatively, the $p_{x,y}$ orbitals on the two centers transform as $\pi_u + \pi_g$. Use of equation 4.8 leads to an in-phase combination of $p_{x,y}$ on the two atoms for π_u and hence a bonding pair of orbitals and an out-of-phase combination for π_g leading to an antibonding orbital. Similarly the two s orbitals transform as $\sigma_g^+ + \sigma_u^+$ and the orbitals correct to first order in the energy are shown in Figure 6.2. The two p_z orbitals also transform as $\sigma_g^+ + \sigma_u^+$ and the result of their degenerate interaction is also shown. The result is three separate first-order energetic interactions. However, the σ orbitals generated by overlap of $s-s$ and p_z-p_z functions are of

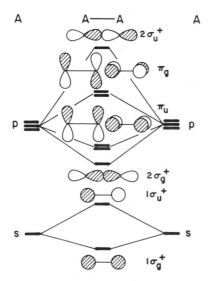

FIGURE 6.2.   Orbitals of an A_2 diatomic correct to first order in energy.

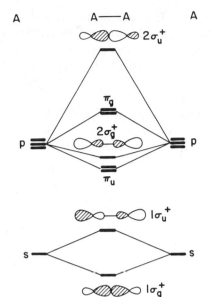

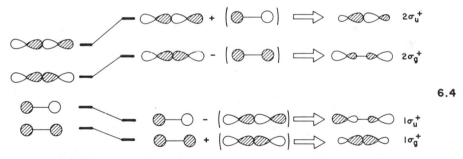

FIGURE 6.3.   Orbitals of an A_2 diatomic after second-order energetic changes (6.4) have been included. Whether $2\sigma_g^+$ or π_u lies lower in energy is system dependent.

the same symmetry, and just as in the case of the linear H_4 molecule of Section 5.3, they can interact with each other via a second-order energetic process. This is shown schematically in 6.4 and the result is the orbital diagram of Figure 6.3.

6.4

Whether the $2\sigma_g^+$ level lies below or above the $1\pi_u$ level is not a simple matter to predict. It turns out $2\sigma_g^+$ lies lower for the first row diatomics up to C_2. From then onward, $1\pi_u$ lies lower. As a result, B_2 is paramagnetic and C_2 diamagnetic. Notice that as a result of the second-order interaction, $2\sigma_g^+$ has been pushed up in energy and is now less bonding between the two A atoms than before. In fact, it has a resemblance to an orbital constructed via the in-phase addition of two lone pair orbitals. $1\sigma_u^+$ has been pushed down in this process and resembles an out-of-phase mixture of lone pair orbitals. $1\sigma_g^+$ remains the only σ orbital which is strongly A–A bonding. If we wished to establish a correspondence between the traditional Lewis structure for the 10 electron N_2 molecule (6.5) and this orbital model, the three bonds between

$$:N \equiv N: \qquad \textbf{6.5}$$

the nitrogen atoms would be identified with the double occupation of $1\sigma_g^+$ and the two components of $1\pi_u$. As we have described, the lone pair orbitals are best identified with $1\sigma_u^+$ and $2\sigma_g^+$ with the perhaps surprising result that the in-phase combination lies at higher energy than the out-of-phase combination. The explanation for this lies in the details of the orbital mixing process of **6.4**. (This "through-bond" interaction between lone pairs is discussed further in Section 7.5.)

Consider now the orbitals of the heterodiatomic AB where the atom B is more electronegative than atom A. In our language this means that the atomic levels of B lie deeper than those of A. **6.6** shows the second-order energetic interaction

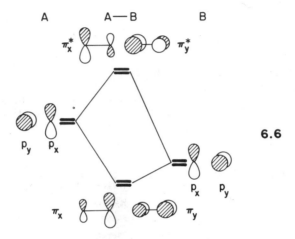

6.6

between the $p\pi$ orbitals $(p_{x,y})$ which replaces the first-order one of Figure 6.2. Note that the π bonding orbitals are weighted more heavily on atom B and the π antibonding orbitals weighted more heavily on atom A as a result. The σ orbitals may be constructed in an analogous way by first replacing the degenerate interactions of Figure 6.2 with nondegenerate ones, followed by a second-order energetic interaction as in **6.4**. The overall form of the diagram is very similar to that for the homonuclear diatomic. Alternatively the orbitals may be built up from first- and second-order mixing processes in one step as shown in Figure 6.4. The signs of the first-order mixing coefficients are easily determined by the energy differences between the relevant orbitals. The signs of the seond-order mixing are somewhat problematical. For ψ_1 there is no ambiguity. Identifying s_B and p_{zB} with ψ_{iA}^0 and ψ_{kA}^0, respectively, the second-order mixing occurs as the result of interaction with $\psi_{jB}^0 = s_A$ (case **3.6**) and with $\psi_{jB}^0 = p_{zA}$ (case **3.7**). Both lead (Table 3.1) to positive mixing coefficients $t_{ki}^{(2)}$. For ψ_2, however ψ_{iA}^0 and ψ_{kA}^0 correspond to s_A and p_{zA}, respectively, and the second-order mixing coefficients are negative for $\psi_{jB}^0 = s_B$ (case **3.5**) and positive for $\psi_{jB}^0 = p_{zB}$ (case **3.6**). Thus, the form of ψ_2 on the atomic site A (Figure 6.4c) is dependent on the relative weights of the two coefficients. A similar ambiguity arises for the form of ψ_3 on the atomic site B.

The second-order orbital mixing in ψ_2 and ψ_3 (on the sites A and B, respectively) is omitted in Figure 6.4c, since its effect would be small due to the interference of the two second-order mixing terms involved. For ψ_4 there occurs no ambiguity

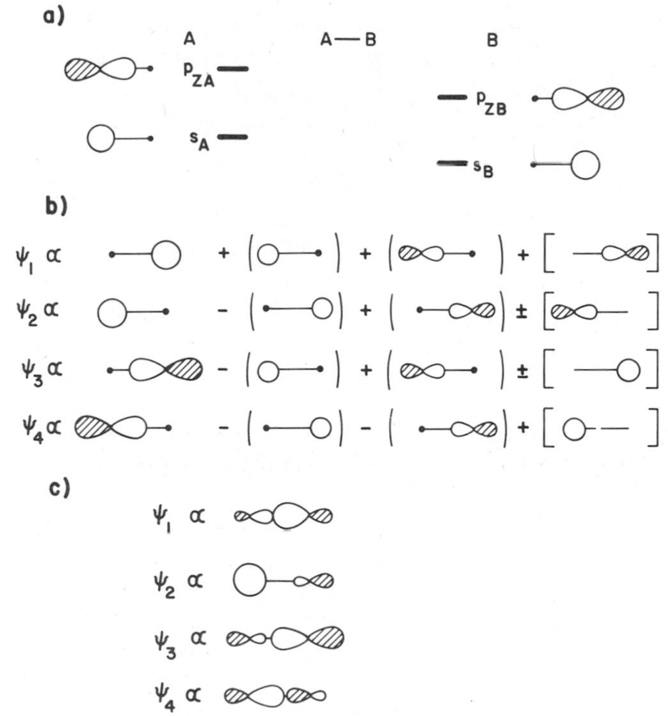

a)

A   A—B   B

p_{ZA}

p_{ZB}

s_A

s_B

b)

$\psi_1 \alpha$ — + (—) + (—) + [—]

$\psi_2 \alpha$ — - (—) + (—) ± [—]

$\psi_3 \alpha$ — - (—) + (—) ± [—]

$\psi_4 \alpha$ — - (—) - (—) + [—]

c)

$\psi_1 \alpha$

$\psi_2 \alpha$

$\psi_3 \alpha$

$\psi_4 \alpha$

FIGURE 6.4. The form of the σ orbitals of the A–B heteronuclear diatomic molecule. (a) shows the energetic disposition of the p_z and s atomic orbitals. (b) shows the signs of first- and second-order mixing coefficients obtained from perturbation theory (Table 3.1). (c) shows the final form of these orbitals. $\psi_1 - \psi_4$ are in the order of increasing energy.

concerning the sign of its second-order mixing term. It was noted in Section 3.3B that the nondegenerate perturbation treatment is not quite satisfactory when two interacting levels are close in energy so that equation 3.42 is not satisfied. This situation arises when the two levels s_A and p_{zB} become close in energy. In such a case, for example, the weight of s_A in ψ_3 approaches that of p_{zB} so that the representation of ψ_3 in Figure 6.4c is not quite correct. Direct solution of the appropriate secular determinant for such a problem shows a substantial weight on the less electronegative end A of ψ_3, as in the case of the HOMO (n_{CO}) of CO shown in **6.7**.

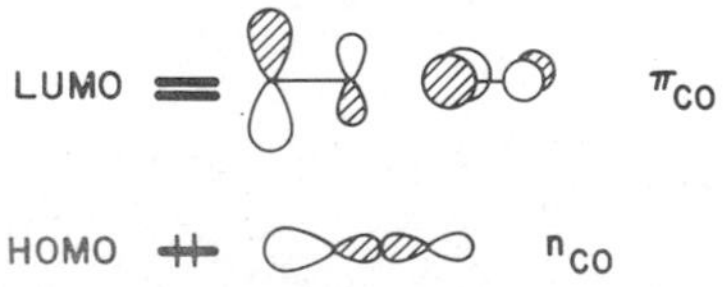

LUMO π_{CO}

6.7

HOMO n_{CO}

It is of interest to compare the nature of the HOMOs in 10-electron systems N_2 and CO. The HOMO $2\sigma_g^+$ of N_2 shown in Figure 6.3 is stabilized by degenerate orbital interaction but destabilized via nondegenerate interaction (see **6.4**). If the nondegenerate contribution is larger than the degenerate one then this orbital too can be antibonding between the atomic centers. In the AB case this would require the overlap integrals $\langle s_A | p_B \rangle$ and $\langle p_A | s_B \rangle$ to be larger than the sum of $\langle s_A | s_B \rangle$ and $\langle p_A | p_B \rangle$ in ψ_3 of Figure 6.4. In N_2 it turns out that this orbital is bonding as suggested by the vibrational frequency drop[1] on forming the positive ion: $\omega_e (N_2) = 2359.6$ cm^{-1}, $\omega_e (N_2^+) = 2207.2$ cm^{-1}. In CO, however, the HOMO, ψ_3 is antibonding; the vibrational frequency increases on forming the positive ion, $\omega_e (CO) = 2170.2$ cm^{-1}, $\omega_e (CO^+) = 2214.2$ cm^{-1}.

A result of the orbital mixing process in CO is a nicely hybridized lone pair orbital at carbon. As we have already pointed out, the π^* orbitals of this molecule contain more carbon character. These two results mean that the carbon end of this unit can act (6.7) as a σ donor and a π acceptor, a feature of this molecule that will be very important in connection with its interaction with transition metals.

6.4.   ELECTRONEGATIVITY PERTURBATION[2]

In this section we shall take the orbitals of A_2, apply an electronegativity perturbation, and see how the orbitals of AB are naturally generated. For simplicity, we shall only consider the π type orbitals and will make an approximation in the perturbation which will lead to a simple form for the relevant corrections to the energy and wavefunction. We shall simulate an increase in the electronegativity of one of the atoms of the molecule by increasing the magnitude of the Coulomb integral for an orbital (on atom B) by a small amount $\delta\alpha$ (< 0), that is, $\langle \chi_\alpha | H^{eff} | \chi_\alpha \rangle = H_{\alpha\alpha} + \delta\alpha$. In practice, such a change in $\langle \chi_\alpha | H^{eff} | \chi_\alpha \rangle$ should lead to corresponding changes in those interaction integrals $\langle \chi_\alpha | H^{eff} | \chi_\mu \rangle$ which are nonzero by symmetry via the Wolfsberg-Helmholz relationship (equation 1.19). Also, associated with a change of $H_{\mu\mu}$ values is a change in orbital exponent. As an orbital becomes more tightly bound it also becomes more contracted. These two effects we shall explicitly neglect in our discussion here.

Given that the perturbation is simply a change in $\langle \chi_\alpha | H^{eff} | \chi_\alpha \rangle$ (in our case we identify α with a p_x or p_y orbital on atom B), then using the approximations just discussed $\delta H_{\mu\nu} = 0$ (for $\mu \neq \nu$), $\delta S_{\mu\nu} = 0$ and $\delta H_{\mu\mu} = 0$ except for the case where $\mu = \alpha$. Here $\delta H_{\alpha\alpha} = \delta\alpha$. So in equation 3.9, $\tilde{S}_{ij} = 0$ for all molecular orbitals, i, j. In equation 3.10, the only term in the summation which is nonzero is that for $\mu = \nu$ ($= \alpha$) and so $\tilde{\Delta}_{ij} = c_{\alpha i}^0 \delta\alpha c_{\alpha j}^0$. As a result,

$$\tilde{\Delta}_{ij} - e_i^0 \tilde{S}_{ij} = c_{\alpha i}^0 c_{\alpha j}^0 \delta\alpha \tag{6.6}$$

which leads to

$$e_i^{(1)} = (c_{\alpha i}^0)^2 \delta\alpha$$

$$e_i^{(2)} = \sum_{j \neq i} \frac{(c_{\alpha i}^0 c_{\alpha j}^0 \delta\alpha)^2}{e_i^0 - e_j^0} \tag{6.7}$$

$$t_{ji}^{(1)} = \frac{c_{\alpha i}^0 c_{\alpha j}^0 \delta\alpha}{e_i^0 - e_j^0}$$

These results may be immediately applied to the case of the $p\pi$ orbitals of A_2 and AB. For convenience, we arrange that the coefficients of the orbital χ_α (p orbital on atom B) are positive in both π and π^* orbitals of A_2. Then identifying the π orbital with ψ_i and the π^* orbital with ψ_j of equations 6.6 and 6.7, $c_{\alpha i}^0 c_{\alpha j}^0 > 0$ and $\delta\alpha < 0$.

$$\pi_{AB} \propto \pi_{AA} + t_{ji}^{(1)}\, \pi_{AA}^*$$

$$\pi_{AB}^* \propto \pi_{AA}^* + t_{ij}^{(1)}\, \pi_{AA} \tag{6.8}$$

where from equation 6.7

$$t_{ji}^{(1)} = \frac{(+)(-)}{(-)} = (+)$$

$$t_{ij}^{(1)} = \frac{(+)(-)}{(+)} = (-) \tag{6.9}$$

which leads to **6.8**.

6.8

The new energies are given by

$$e(\pi_{AB}) = e(\pi_{AA}) + (c_{\alpha i}^0)^2\, \delta\alpha + \frac{(c_{\alpha i}^0 c_{\alpha j}^0\, \delta\alpha)^2}{e_i^0 - e_j^0}$$

$$e(\pi_{AB}^*) = e(\pi_{AA}^*) + (c_{\alpha j}^0)^2\, \delta\alpha + \frac{(c_{\alpha i}^0 c_{\alpha j}^0\, \delta\alpha)^2}{e_j^0 - e_i^0} \tag{6.10}$$

where

$$c_{\alpha i}^{02}\, \delta\alpha < 0; \qquad c_{\alpha j}^{02}\, \delta\alpha < 0$$

$$\frac{(c_{\alpha i}^0 c_{\alpha j}^0\, \delta\alpha)^2}{e_i^0 - e_j^0} < 0; \qquad \frac{(c_{\alpha i}^0 c_{\alpha j}^0\, \delta\alpha)^2}{e_j^0 - e_i^0} > 0 \tag{6.11}$$

Both first- and second-order energetic corrections lower the energy of π_{AB} relative to π_{AA}. First- and second-order corrections work in opposite directions for the π^* orbital. In most cases the first-order term dominates and both π and π^* levels lie lower in the AB molecule than they do in AA. For example, the π_{CO} and π_{CO}^* levels of a carbonyl double bond are invariably lower than the π_{CC} and π_{CC}^* levels of the comparable carbon–carbon double bond, as shown in **6.9**. Because of its lower energy and larger carbon p coefficients, nucleophiles attack the π_{CO}^* orbital of carbon with greater facility than they do the π_{CC}^* orbital in an alkene. We focus our attention

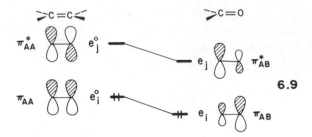

here on the HOMO–LUMO interactions of the two reactants. This is where the smallest energy gap, in terms of a perturbation expression, will be found, and, therefore, the largest two-electron–two-orbital stabilizing interaction. A nucleophile will have a high-lying HOMO and, therefore, its interaction with the LUMO of an electrophile (the carbonyl group or C–C double bond) will dominate reactivity trends in these types of problems.

Another example, shown in Figure 6.5, illustrates how the HOMOs and LUMOs of benzene may be used to construct the HOMO and LUMO of the pyridine molecule. In this treatment, we have considered only the highest unoccupied and lowest occupied orbitals of this system. The intermixing of π and π^* benzene levels in pyridine redistribute the electron density. In the perturbed π level, the electron density (via the atomic orbital coefficients) is increased on the nitrogen and two *meta*-carbon atoms. Correspondingly the coefficients at the *para* and two *ortho* positions have increased in the perturbed π^* level. Nucleophiles are therefore expected to attack *ortho* and/or *para* to this nitrogen atom and indeed they do.

As a final example we return to the linear H_3 problem and ask where a more

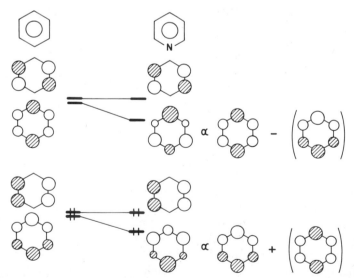

FIGURE 6.5.   Mixing of HOMO and LUMO orbitals of benzene during an electronegativity perturbation to give the pyridine molecule.

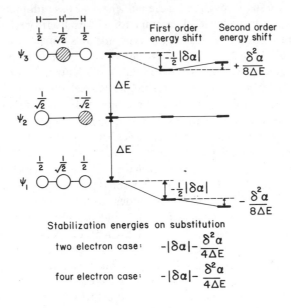

a) central atom substitution

$H—H'—H$
$\frac{1}{2}$ $-\frac{1}{\sqrt{2}}$ $\frac{1}{2}$

ψ_3

First order energy shift   Second order energy shift

$-\frac{1}{2}|\delta\alpha|$ $+\dfrac{\delta^2\alpha}{8\Delta E}$

ΔE

$\frac{1}{\sqrt{2}}$ $-\frac{1}{\sqrt{2}}$

ψ_2

ΔE

$\frac{1}{2}$ $\frac{1}{\sqrt{2}}$ $\frac{1}{2}$

ψ_1

$-\frac{1}{2}|\delta\alpha|$ $-\dfrac{\delta^2\alpha}{8\Delta E}$

Stabilization energies on substitution

two electron case: $\quad -|\delta\alpha|-\dfrac{\delta^2\alpha}{4\Delta E}$

four electron case: $\quad -|\delta\alpha|-\dfrac{\delta^2\alpha}{4\Delta E}$

b) terminal atom substitution

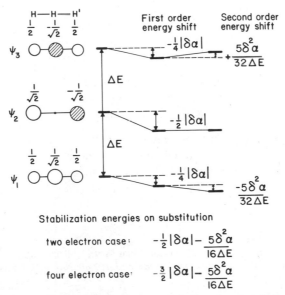

$H—H—H'$
$\frac{1}{2}$ $\frac{1}{\sqrt{2}}$ $\frac{1}{2}$

ψ_3

First order energy shift   Second order energy shift

$-\frac{1}{4}|\delta\alpha|$ $+\dfrac{5\delta^2\alpha}{32\Delta E}$

ΔE

$\frac{1}{\sqrt{2}}$ $-\frac{1}{\sqrt{2}}$

ψ_2

$-\frac{1}{2}|\delta\alpha|$

ΔE

$\frac{1}{2}$ $\frac{1}{\sqrt{2}}$ $\frac{1}{2}$

ψ_1

$-\frac{1}{4}|\delta\alpha|$ $\dfrac{-5\delta^2\alpha}{32\Delta E}$

Stabilization energies on substitution

two electron case: $\quad -\frac{1}{2}|\delta\alpha|-\dfrac{5\delta^2\alpha}{16\Delta E}$

four electron case: $\quad -\frac{3}{2}|\delta\alpha|-\dfrac{5\delta^2\alpha}{16\Delta E}$

FIGURE 6.6. Energy level shifts, obtained via perturbation theory, which result on substitution of an atom of H_3 by one of higher electronegativity. The two cases of central (a) and terminal (b) atom substitution are shown. (Within the Hückel model, described in Chapter 12, $\Delta E = \sqrt{2}\beta$.) The perturbation used is an increase in $|H_{\alpha\alpha}|$ of the relevant orbital by $|\delta\alpha|$. ($\delta\alpha, H_{\alpha\alpha} < 0$.)

electronegative atom (H') would lie in H_2H'; at the end or in the middle of the molecule. This is worked through in Figure 6.6 where we have adopted a particularly simplified form of the energy level diagram for H_3 itself. If overlap is neglected (except, of course, nonzero $\langle \chi_1 | H^{eff} | \chi_3 \rangle$ type integrals remain) then the bonding and antibonding σ_g^+ orbitals are split in energy an equal distance (ΔE) away from the exactly nonbonding σ_u^+ orbital. The orbital coefficients in this zero overlap approximation (actually no different from simple Hückel theory as we will see in Section 12.2) are also easily evaluated and are shown in the figure. For a change in Coulomb integral of $\delta\alpha$ for a hydrogen s orbital, then, the equations of 6.6 and 6.7 are readily evaluated as shown for the three levels. Replacement of the central atom by one of larger electronegativity leads to no change in the energy of the σ_u^+ orbital. This is understandable since it contains no central atom character at all. The energy change for the two-electron (H_3^+) and four-electron (H_3^-) cases are therefore identical. Substitution of an end atom leads to first-order energy changes for both orbitals, but the second-order energy correction for σ_u^+ is identically zero since it is pushed up by ψ_1 an equal amount that it is pushed down by ψ_3. Now the energy changes are different for the two- and four-electron cases. Consequently we predict that the more electronegative atom prefers to lie at a terminal position in electron-rich (four-electron) three-center systems, and in the central position for electron-poor (two-electron) three-center systems. In the case of N_2O (3.1) there are a total of eight π electrons occupying two out of the three pairs of three center π orbitals of this molecule, a situation isomorphous with that of the H_3^- problem. Correspondingly N_2O is found as NNO and not as NON.

It is interesting to correlate the electronic charge distribution for the two- and four-electron H_3 systems with the treatment above. Using the wavefunctions shown in Figure 6.6 it is easy to generate the electron densities of **6.10** and **6.11**. The details

$$
\begin{array}{cc}
\overset{+}{\text{H}-\text{H}-\text{H}} & \overset{-}{\text{H}-\text{H}-\text{H}} \\
\tfrac{1}{2} \quad | \quad \tfrac{1}{2} & |\tfrac{1}{2} \quad | \quad |\tfrac{1}{2} \\
\textbf{6.10} & \textbf{6.11}
\end{array}
$$

of the electronegativity perturbation results, therefore, in a preference for the most electronegative atom of a substituted molecule to lie at the site of highest charge density in the unsubstituted parent. This is a very important result we shall use later.

REFERENCES

1. G. Herzberg, *Spectra of Diatomic Molecules*, Van Nostrand-Rheinhold, New York (1950).
2. E. Heilbronner and H. Bock, *The HMO Model and its Application*, Wiley, New York (1976).

CHAPTER SEVEN

Molecular Orbitals and Geometrical Perturbation

7.1.   MOLECULAR ORBITALS OF AH$_2$

Let us construct the MOs of linear and bent AH$_2$ shown in **7.1** and **7.2**, respectively, where the central atom A contributes four valence atomic orbitals s, x, y, and z. We will construct the MOs based upon the perturbation method of Chapter 3, and so it is convenient to construct AH$_2$ from A and H $\cdots$ H units. The orbitals of H $\cdots$ H are the in-phase and out-of-phase combinations of hydrogen s orbitals shown in **7.3**, where the energy gap between σ_g and σ_u is small since the H $\cdots$ H distance in **7.1** and **7.2** is large in most cases of interest.

H —A— H

7.1

7.2

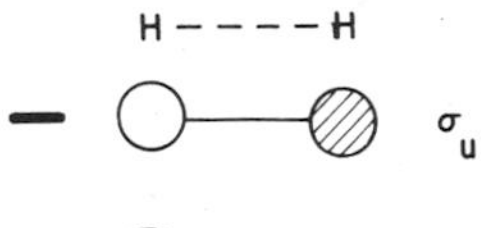

7.3

A number of relative orderings conceivable for the orbitals of A and H $\cdots$ H are shown in **7.4**. For example, the orbitals of H $\cdots$ H lie in between the s and p orbitals of A in **7.4b**, which turns out to be relevant when A is carbon. The orbitals of A shift upward or downward in energy with respect to those of carbon as A becomes more electropositive or electronegative than carbon, respectively. For simplicity, we will construct the MOs of AH$_2$ for the case of **7.4b**.

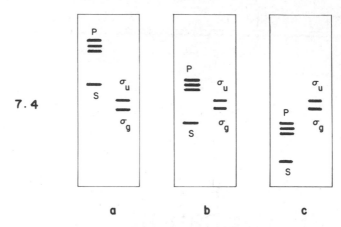

7.4

a   b   c

The orbital interaction diagram for linear AH_2 is shown in Figure 7.1, where the MOs of AH_2 are easily written down as shown in **7.5**. We have already examined in Chapter 4 how to construct the MOs of bent AH_2 based upon symmetry arguments alone. The y and z orbitals of A do not overlap with any of the two orbitals of $H \cdots H$, so they become the nonbonding orbitals of AH_2 (**7.6**). The number of nodal planes in the MOs increase upon going to orbitals of higher energy: $1\sigma_g$ has zero; $1\sigma_u$ and $1\pi_u$ each have one; $2\sigma_g$ has two; and finally there are three nodal planes in $2\sigma_u$.

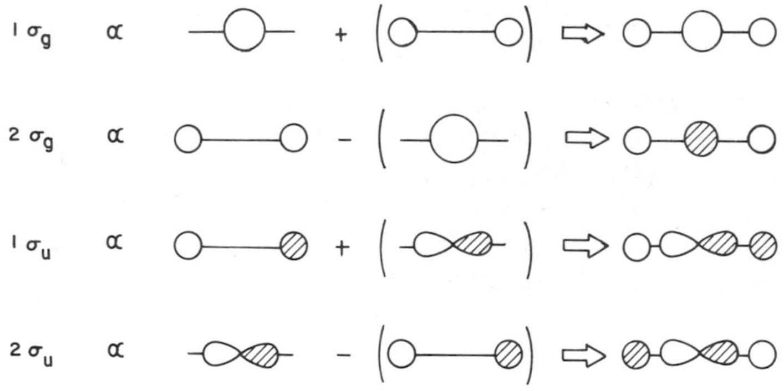

$1\sigma_g$ α

$2\sigma_g$ α

7.5

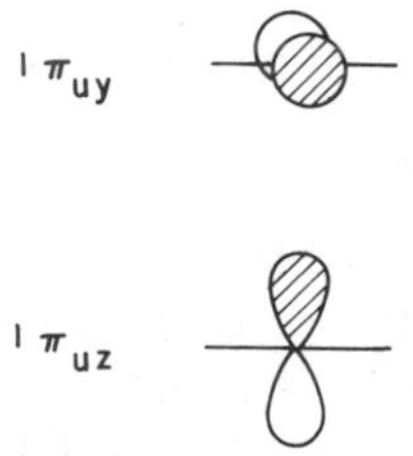

$1\sigma_u$ α

$2\sigma_u$ α

$1\pi_{uy}$

7.6

$1\pi_{uz}$

The orbital interaction diagram for bent AH_2 is shown in Figure 7.2. In a bent structure, the overlap of the s and z orbitals of A with the a_1 orbital of $H \cdots H$ is

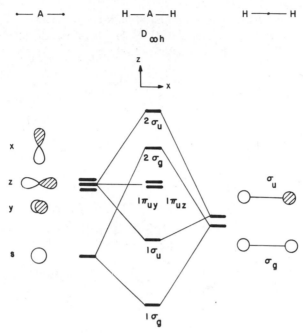

FIGURE 7.1.   An orbital interaction diagram for linear AH$_2$.

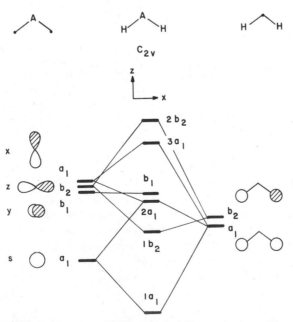

FIGURE 7.2.   An orbital interaction diagram for bent AH$_2$.

89

nonzero (7.7) so that second-order orbital mixings occur between A atom s and z orbitals via overlap with the a_1 orbital of H $\cdots$ H. Therefore, the MOs of bent AH_2 may be constructed as in 7.8. The basic orbital shapes are the same as those for

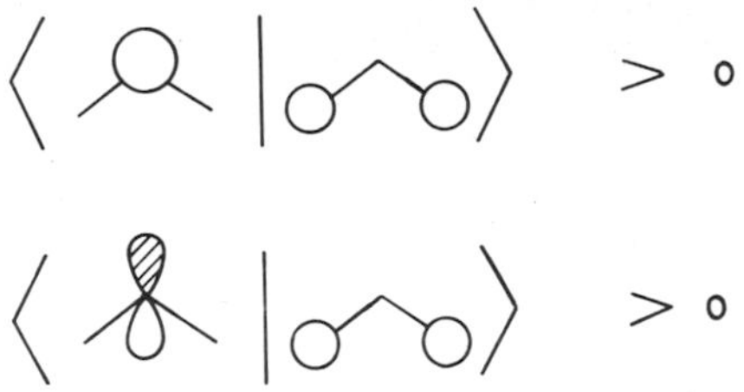

$$\langle\, \cdot \mid \cdot \cdot \,\rangle \;\; > \; 0$$

7.7

$$\langle\, \cdot \mid \cdot \cdot \,\rangle \;\; > \; 0$$

7.8

$1a_1 \qquad \alpha \quad (\text{orbital combination}) \Rightarrow$

$2a_1 \qquad \alpha \quad (\text{orbital combination}) \Rightarrow$

$3a_1 \qquad \alpha \quad (\text{orbital combination}) \Rightarrow$

$1b_2 \qquad \alpha \quad (\text{orbital combination}) \Rightarrow$

$2b_2 \qquad \alpha \quad (\text{orbital combination}) \Rightarrow$

$1b_1 \qquad = \quad (\text{orbital})$

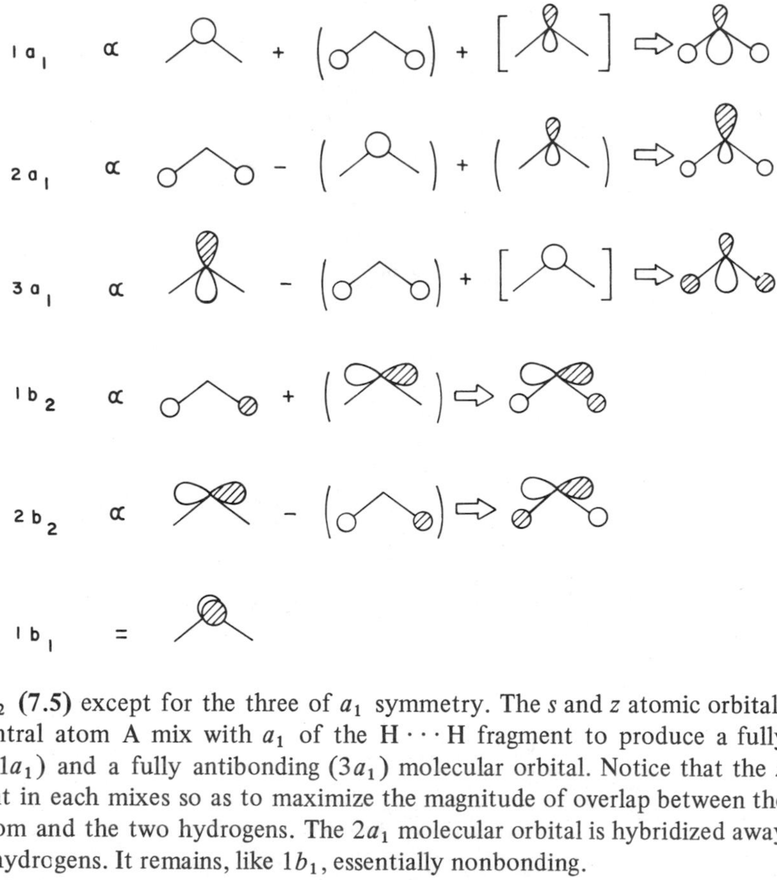

linear AH_2 (7.5) except for the three of a_1 symmetry. The s and z atomic orbitals of the central atom A mix with a_1 of the H $\cdots$ H fragment to produce a fully bonding ($1a_1$) and a fully antibonding ($3a_1$) molecular orbital. Notice that the z component in each mixes so as to maximize the magnitude of overlap between the central atom and the two hydrogens. The $2a_1$ molecular orbital is hybridized away from the hydrogens. It remains, like $1b_1$, essentially nonbonding.

With the MOs of linear and bent AH_2 derived separately, it is important to ask how the MOs of one geometry are related to those of the other. In order to find this relationship, we need to understand how the MOs of a molecule are modified under geometrical distortion.

7.2   GEOMETRICAL PERTURBATION

Suppose that a molecule undergoes a distortion. Let us examine how the MOs of one geometry are related to those of another, where the structural distortion involved is considered as a perturbation. The molecular geometries before and after distortion are said to be unperturbed and perturbed, respectively. Given any two geometries of a molecule, either one may in principle be considered as the perturbed one. In practice, however, it leads to a substantially simpler analysis if the molecular geometry of the lower symmetry is chosen as the perturbed one.

As shown in Section 3.2A, the first-order mixing coefficient and the first- and second-order energy terms are given by

$$t_{ji}^{(1)} = \frac{\tilde{\Delta}_{ij} - e_i^0 \tilde{S}_{ij}}{e_i^0 - e_j^0} \propto \frac{-\tilde{S}_{ij}}{e_i^0 - e_j^0} \tag{7.1}$$

$$e_i^{(1)} = \tilde{\Delta}_{ii} - e_i^0 \tilde{S}_{ii} \propto -\tilde{S}_{ii} \tag{7.2}$$

$$e_i^{(2)} = \sum_{j \neq i} \frac{(\tilde{\Delta}_{ij} - e_i^0 \tilde{S}_{ij})^2}{e_i^0 - e_j^0} \propto \sum_{j \neq i} \frac{\tilde{S}_{ij}^2}{e_i^0 - e_j^0} \tag{7.3}$$

where the matrix elements $\tilde{\Delta}_{ij}$ and $\tilde{S}_{ij}$ are defined as

$$\tilde{\Delta}_{ij} = \sum_\mu \sum_\nu c_{\mu i}^0 \delta H_{\mu\nu} c_{\nu j}^0 \tag{7.4}$$

$$\tilde{S}_{ij} = \sum_\mu \sum_\nu c_{\mu i}^0 \delta S_{\mu\nu} c_{\nu j}^0 \tag{7.5}$$

Note that it is the MO coefficients of the unperturbed geometry (i.e., $c_{\mu i}^0$ and $c_{\nu j}^0$) and the change in the matrix elements (i.e., $\delta H_{\mu\nu}$ and $\delta S_{\mu\nu}$) which define the $\tilde{\Delta}_{ij}$ and $\tilde{S}_{ij}$ terms. If the two geometries under consideration are identical, $\delta H_{\mu\nu} = \delta S_{\mu\nu} = 0$ for all μ and ν. Consequently, $\tilde{\Delta}_{ij} = \tilde{S}_{ij} = (\tilde{\Delta}_{ij} - e_i^0 \tilde{S}_{ij}) = 0$ for all μ and ν. Consequently, $\tilde{\Delta}_{ij} = \tilde{S}_{ij} = (\tilde{\Delta}_{ij} - e_i^0 \tilde{S}_{ij}) = 0$ in such a case. A geometrical perturbation makes most $\delta H_{\mu\nu}$ and $\delta S_{\mu\nu}$ elements different from zero, so that the values of $\tilde{\Delta}_{ij}$ and $\tilde{S}_{ij}$ deviate from zero in most cases. Thus the value of $(\tilde{\Delta}_{ij} - e_i^0 \tilde{S}_{ij})$ will often be nonzero on a geometrical perturbation.

In general the first-order energy $e_i^{(1)}$ is stabilizing (i.e., $e_i^{(1)} < 0$) if the value of $\tilde{S}_{ii}$ is positive, but destabilizing (i.e., $e_i^{(1)} > 0$) if the value of $\tilde{S}_{ii}$ is negative. The value of $\tilde{S}_{ii}$ is positive by enhancing a positive overlap, which strengthens bonding on perturbation or by diminishing a negative overlap, which weakens antibonding effects. Similarly, the value of $\tilde{S}_{ii}$ is negative by diminishing a positive overlap, which weakens bonding, or by enhancing a negative overlap, which enhances antibonding effects. For example, the $1\sigma_g$ and $1\pi_{uz}$ orbitals in linear AH_2 are normalized to unity as shown in **7.9** and **7.10**. Upon the linear to bent (i.e., $D_{\infty h} \rightarrow C_{2v}$) distortion, the $1\sigma_g$ orbital does not remain normalized to unity since the overlap between the two s orbitals increases as indicated in **7.11**. However, the $1\pi_{uz}$ orbital

7.9 $\langle \text{⬭—◯—⬭} | \text{⬭—◯—⬭} \rangle = 1$

7.10 $\langle \text{⬡} | \text{⬡} \rangle = 1$

7.11 (orbital diagram with arrow)

7.12 $\langle \text{⬭◯⬭} | \text{⬭◯⬭} \rangle > 1$

7.13 $\langle \text{⬡} | \text{⬡} \rangle = 1$

remains normalized to unity because it has no orbital contribution from hydrogen. Thus we obtain the results shown in **7.12** and **7.13**. According to **7.9** and **7.12**, the first-order energy $e_i^{(1)}$ is stabilizing for $1\sigma_g$ since the value of $\tilde{S}_{ii}$ becomes positive on the $D_{\infty h} \to C_{2v}$ distortion. **7.10** and **7.13** show that the first-order energy $e_i^{(1)}$ is zero for $1\pi_{uz}$ since the value of $\tilde{S}_{ii}$ is zero during the $D_{\infty h} \to C_{2v}$ distortion.

The first-order mixing coefficient $t_{ji}^{(1)}$ and the second-order energy $e_i^{(2)}$ are determined by the term $(\tilde{\Delta}_{ij} - e_i^0 \tilde{S}_{ij})$. This is in general negative if the value of $\tilde{S}_{ij}$ increases from zero, but positive if the value of $\tilde{S}_{ij}$ decreases from zero. This stems from the fact that $\tilde{\Delta}_{ij} \propto -\tilde{S}_{ij}$, and the magnitude of $\tilde{\Delta}_{ij}$ is generally greater than that of $e_i^0 \tilde{S}_{ij}$. Therefore, equation 7.1 shows that, if $\tilde{S}_{ij} > 0$, the coefficient $t_{ji}^{(1)}$ is positive for the mixing of an upper level ψ_j^0 into the lower level $\psi_i^0 (e_i^0 - e_j^0 < 0)$ while the coefficient $t_{ji}^{(1)}$ is negative for the mixing of a lower level ψ_j^0 into an upper level $\psi_i^0 (e_i^0 - e_j^0 > 0)$. In addition, equation 7.3 shows that a given level ψ_i^0 is lowered in energy by an upper level ψ_j^0 but raised in energy by a lower level ψ_j^0. For example, the $1\sigma_g$ and $1\pi_{uz}$ orbitals of AH$_2$ are orthogonal in a linear structure (**7.14**), but do not remain orthogonal in a bent structure (**7.15**) since the overlap between s and z is nonzero. Thus the $1a_1$ orbital of bent AH$_2$ can be approximately described in terms of $1\sigma_g$ and $1\pi_{uz}$ as **7.16**, where use is made of the fact that, in a

7.14 $\langle \text{⬭—◯—⬭} | \text{⬡} \rangle = 0$

7.15 $\langle \text{⬭◯⬭} | \text{⬡} \rangle = \langle \text{⬭◯⬭} | \text{⬡} \rangle \neq 0$

7.16 $1a_1 \quad \alpha \quad \text{⬭◯⬭} + |\text{⬡}| \Rightarrow \text{⬭◯⬭}$

linear structure, $1\sigma_g$ lies lower in energy than $1\pi_{uz}$ (see Figure 7.1). With respect to the $1\sigma_g$ level of linear AH_2, the $1a_1$ orbital of bent AH_2 is lowered in energy since the first- and second-order energy terms are both stabilizing.

7.3. WALSH DIAGRAM[1-5]

Figure 7.3 shows the MOs of linear and bent AH_2 discussed in Section 7.1. Here the MOs of linear and bent AH_2 are labeled according to their point group symmetry, and are also given the mnemonic labels such as σ, n, or σ^* to indicate their sigma-bonding, nonbonding, or sigma-antibonding character, respectively. In Section 7.2, we showed why the $1a_1$ orbital is lower in energy than the $1\sigma_g$ orbital. For the $1\pi_{uz}$ level, the first-order energy term $e_i^{(1)}$ is zero according to **7.10** and **7.13**. However, the overlap between $1\pi_{uz}$ and $2\sigma_g$ is nonzero in a bent structure **(7.15)**, thereby leading to an orbital mixing between them and hence to nonzero second-order energy terms $e_i^{(2)}$. Thus the $2a_1$ and $3a_1$ orbitals are lowered and raised with respect to $1\pi_{uz}$ and $2\sigma_g$, respectively. The nodal properties of the $2a_1$ and $3a_1$ orbitals may be constructed as shown in **7.17**, where the mixing of $1\sigma_g$ into $1\pi_{uz}$ or $2\sigma_g$ was neglected because the energy separation of $1\sigma_g$ from the $1\pi_{uz}$

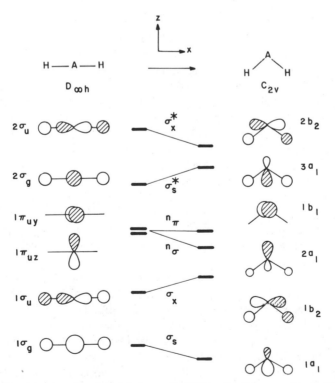

FIGURE 7.3.   Correlation between the MO levels of linear and bent AH_2.

$2a_1$ α

7.17

$3a_1$ α

or $2\sigma_g$ orbitals is large compared with the energy gap between $1\pi_{uz}$ and $2\sigma_g$. The reader can readily verify that $1\sigma_g$ will mix into $1\pi_{uz}$ with a phase opposite to that shown for $1\sigma_g$ in 7.5. Therefore, the atomic s character on A is reinforced in $2a_1$ by the first-order mixings while the atomic coefficients for the hydrogens become quite small. The nodal surfaces of $1\pi_{uz}$ and $2a_1$ are shown by the dashed lines in 7.18. Thus, the hydrogens basically follow the nodal surface in $2a_1$ on bending. This reemphasizes our previous comment that $2a_1$ is considered to be nonbonding.

7.18

Figure 1.4 shows that the overlap between the s and x atomic orbitals decreases as the HAH valence angle becomes smaller. Thus in $1\sigma_u$, the linear to bent (i.e., $D_{\infty h} \to C_{2v}$) distortion of AH_2 decreases the bonding between the s and x orbitals, and increases antibonding between the two s orbitals, thereby raising the $1\sigma_u$ level. As for $2\sigma_u$, the $D_{\infty h} \to C_{2v}$ distortion leads to a decrease in antibonding between the s and x orbitals but to an increase in the antibonding between the two s orbitals. The former predominates over the latter in magnitude due to the large $H \cdots H$ distance, so that the $2\sigma_u$ level is lowered by the $D_{\infty h} \to C_{2v}$ distortion. Finally, the $1\pi_{uy}$ level is not affected by the $D_{\infty h} \to C_{2v}$ distortion since its overlap with other orbitals vanishes at all points along the distortion coordinate.

Diagrams such as those of Figure 7.3 which show how the MO levels of a molecule vary as a function of geometrical change are known as Walsh diagrams. As we will show extensively throughout this book, a major function of a Walsh diagram is to account for the structural regularity observed for a series of closely related molecules with the same number of valence electrons. Walsh's rule for predicting molecular shapes may simply be stated as follows: A molecule adopts the structure that best stabilizes the HOMO. If the HOMO is unperturbed by the structural change under consideration, the occupied MO lying closest to it governs the geometrical preference.

Let us illustrate Walsh's rule by examining the shapes of AH_2 molecules based upon Figure 7.3. The HOMO of the two-electron AH_2 is σ_s, which is stabilized in the bent structure. Thus LiH_2^+ adopts a bent geometry. The HOMO of a three- or four-electron AH_2 is σ_x, and this orbital is destabilized on bending so that LiH_2 and BeH_2 are linear. The n_σ orbital of AH_2 lies lower in a bent structure while n_π of AH_2 is energetically unaffected by the $D_{\infty h} \to C_{2v}$ distortion. Consequently, the shape of AH_2 molecules with five to eight electrons is governed by the energetics of n_σ. Thus BH_2, CH_2, NH_2, and H_2O all adopt a bent structure (see Table 7.1).[1]

TABLE 7.1.   Typical Bond Angles in AH_2[1]

AH_2	<HAH (°)	AH_2	<HAH (°)
BH_2	131^a	AlH_2	119^a
CH_2	136^b	SiH_2	123^b
NH_2	103.3^c	PH_2	91.7^c
OH_2	104.5^d	SH_2	92.1^d

$^a(2a_1)^1$.
$^b(2a_1)^1 (1b_1)^1$.
$^c(2a_1)^2 (1b_1)^1$.
$^d(2a_1)^2 (1b_1)^2$.

7.4.   JAHN–TELLER DISTORTIONS[6-8]

In the previous section we showed an important use of a Walsh diagram in predicting molecular shapes by simply focusing upon the behavior of the HOMO (or an occupied MO lying close to it). Another important facet of a Walsh diagram lies in the ability to predict geometrical distortions by knowing how the HOMO (or an occupied MO lying close to it) is affected by the LUMO (or an unoccupied MO lying close to it) when the molecule undergoes some geometrical perturbation.

A.   SECOND-ORDER JAHN–TELLER DISTORTION

The HAH valence angles of eight-electron molecules H_2O, H_2S, and H_2Se are 104.5, 92.1, and 90.6°, respectively. This shows a steady decrease in the valence angle upon lowering the electronegativity of A. Since it is the energetic behavior of n_σ which determines the preference for a bent structure in eight-electron AH_2 systems, we will examine the behavior of the n_σ orbital in terms of the simplified Walsh diagram of **7.19**. As described by equation 7.3, the stabilization of $2a_1$ is

7.19

caused solely by the second-order term $e_i^{(2)}$, which is inversely proportional to the energy gap Δe between $1\pi_{uz}$ and $2\sigma_g$ of the linear geometry.

$$e_i^{(2)} = e(2a_1) - e^0(1\pi_{uz}) \propto -\frac{1}{\Delta e} \qquad (7.6)$$

As shown in **7.20**, the $1\pi_{uz}$ level is raised in energy upon decreasing the electronegativity of A. $2\sigma_g$, although it contains A character, behaves differently. The

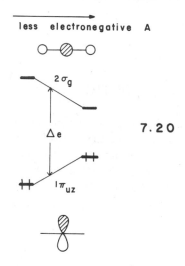

A—H bond length increases with decreasing the electronegativity of A (e.g., $r_{A-H} = 0.958$, 1.336, and 1.460 Å for H_2O, H_2S, and H_2Se, respectively).[1] This is also a reflection of the fact that, with increasing the principal quantum number n, the np atomic orbital of A becomes more diffuse and hence leads to a smaller overlap with the hydrogen $1s$ orbitals. Thus, the antibonding in $2\sigma_g$ is diminished as the electronegativity of A decreases so the net effect is to lower the energy of the $2\sigma_g$ level. The energy gap Δe between $1\pi_{uz}$ and $2\sigma_g$ becomes smaller and so the energy lowering of equation 7.6 increases upon decreasing the electronegativity of A. This provides an explanation for the decrease in the HAH valence angle of an eight-electron AH_2 system as A becomes less electronegative (See Table 7.1). It is also understandable that the inversion barrier (the amount of energy required to distort the molecule from a bent to linear geometry) increases in the order $H_2O < H_2S < H_2Se$. In other words, the downward slope of n_σ in Figure 7.3 increases in this order because of the larger mixing of $2\sigma_g$ into $1\pi_{uz}$ as A becomes less electronegative.

A structural distortion arising from a second-order energy change in the HOMO, like the $D_{\infty h} \rightarrow C_{2v}$ distortion just described, is termed a second-order Jahn–Teller distortion. This phenomenon refers to the observation that a molecule with a small energy gap between the occupied and unoccupied MOs is susceptible to a structural distortion which allows intermixing between them. Typically, it is the HOMO and LUMO that are involved in a second-order Jahn–Teller distortion. The symmetry species of the distortion which allows the mixing between the HOMO and LUMO is derived very simply from group theory. Since we need a distortion coordinate q such that the matrix element $\langle \text{HOMO} \,|\, \partial H^{\text{eff}}/\partial q \,|\, \text{LUMO}\rangle$ is nonzero, the direct product $\Gamma_{\text{HOMO}} \times \Gamma_{\text{LUMO}} \times \Gamma_q$ must contain the a_1 (i.e., totally symmetric) representation. Here Γ_{HOMO} and Γ_{LUMO} are the symmetry species of the HOMO and LUMO, respectively, while Γ_q is that of the distortion coordinate q which defines the perturbation $\partial H^{\text{eff}}/\partial q$. This implies that $\Gamma_q = \Gamma_{\text{HOMO}} \times \Gamma_{\text{LUMO}}$. In the present case for eight-electron AH_2 systems at the linear geometry, $\Gamma_{\text{HOMO}} = \pi_u$, $\Gamma_{\text{LUMO}} = \sigma_g^+$ and so $\Gamma_q = \pi_u \times \sigma_g^+ = \pi_u$, the symmetry species of the bending mode of the molecule (7.21). Eight-electron AH_2 molecules are, therefore, bent. For the four-electron case (e.g., BeH_2), $\Gamma_{\text{HOMO}} = \sigma_u^+$, $\Gamma_{\text{LUMO}} = \pi_u$ and so $\Gamma_q = \pi_g$. No normal vibrational mode exists with this symmetry in linear AH_2 and so no mixing between the HOMO and LUMO will occur on bending. Therefore four-electron AH_2 molecules are predicted to remain linear.

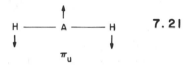

$$\text{H} \longrightarrow \text{A} \longrightarrow \text{H} \qquad 7.21$$

$$\pi_u$$

B. FIRST-ORDER JAHN–TELLER DISTORTION

Shown in 7.22 is the Walsh diagram for the equilateral triangle to linear (i.e., $D_{3h} \rightarrow C_{2v} \rightarrow D_{\infty h}$) distortion in a simple three-center system (see also Figure 5.3).

H —H—H   H—H

$D_{\infty h}$ D_{3h}

$2\sigma_g$ $1e'_z$

$1\sigma_u$ $1e'_x$ 7.22

$1\sigma_g$ $1a'_1$

It predicts that a two-electron system H_3^+ should be triangular while a four-electron system H_3^- should be linear. In a D_{3h} structure, the HOMO of H_3^- is doubly degenerate and half-filled as depicted in 7.23. The degeneracy is lifted by the $D_{3h} \rightarrow D_{\infty h}$ distortion since it stabilizes the $1e'_x$ level but destabilizes $1e'_z$. Throughout the

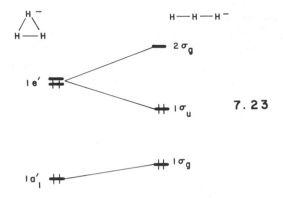

$D_{3h} \to D_{\infty h}$ distortion the overlap of $1e'_x$ with $1e'_z$ or $1a'_1$ vanishes, so that the stabilization of the $1e'_x$ level is caused by a first-order energy change that results from decreasing the extent of antibonding interactions in the $1e'_x$ orbital. This instability of triangular H_3^- is an example of a first-order Jahn–Teller distortion. In general, any nonlinear molecule with an incompletely filled, degenerate HOMO level is susceptible to a structural distortion which removes the degeneracy. In their original formulation, Jahn and Teller associated this structural instability with degenerate electronic states. These will only arise via asymmetric occupation of degenerate orbitals as in **7.23**. The symmetry species of the distortion mode which will lift this degeneracy can be determined by using group theoretical arguments. We refer the reader elsewhere[3] for a more detailed treatment. A general result is that a vibration of e symmetry will lift the degeneracy of an orbitally degenerate situation involving the occupation of e symmetry orbitals. The distortion mode which takes the triangular to the linear structure is a mixture of e' stretching and bending modes.

C. THREE-CENTER BONDING

The linear or open arrangement of a three-center–four-electron system H_3^- bears a close resemblance to the transition state geometry associated with a nucleophilic attack on a tetrahedral carbon center shown in **7.24**. Correspondingly, the triangular or closed arrangement of a three-center–two-electron system H_3^+ is related to a front–side electrophilic attack as shown in **7.25**. The relevant orbitals at the tetrahedral carbon consists of the C—X bonding and antibonding orbitals σ_{CX} and σ^*_{CX}, respectively. The electrophile or the nucleophile will possess an appropriate acceptor or a donor orbital, respectively.[9] Three MOs for the composite "supermolecule" can be readily derived which have the same local symmetry properties as the H_3 system. Of the three MOs, only the lowest is filled in electrophilic attack so that, just like H_3^+, a closed rather than open geometry is preferred in **7.25**. In a nucleophilic attack **7.24**, the donor orbital of the nucleophile is filled, and thus there are now a total of four electrons to be placed into the three MOs. Consequently, a linear geometry is the more stable one. This is a general feature which will be high-

NU: $^-$ + $\overset{\diagdown}{\underset{\diagup}{C}}$—X $\longrightarrow$ NU --- $\overset{\delta^-}{\underset{\diagup}{\overset{|}{C}}}$ --- X $\overset{\delta^-}{}$

7.24

H: $^-$ + H—H $\longrightarrow$ H --- H ---- H $\quad\overset{\delta^-}{}\qquad\overset{\delta^-}{}$

$\overset{\diagdown}{\underset{\diagup}{C}}$—X + E$^+$ $\longrightarrow$ $\overset{\displaystyle E^+}{\underset{\diagup}{C}\text{----X}}$

7.25

H—H + H$^+$ $\longrightarrow$ $\overset{\displaystyle H^+}{H\text{----}H}$

lighted in several problems throughout the book; namely, electron-deficient two-electron–three-center bonding prefers a closed arrangement while electron-rich four-electron–three-center bonding adopts an open geometry.

Recall that the degeneracy of H_3^- in a D_{3h} structure is lifted by the $D_{3h} \to D_{\infty h}$ distortion. Suppose now that we constrain the structure of H_3^- to be triangular. Then the degeneracy of H_3^- in a D_{3h} structure can be lifted, for example, by the equilateral triangle to isosceles triangle (i.e., $D_{3h} \to C_{2v}$) distortion as shown in Figure 7.4 (see also 5.3). In terms of this diagram, let us examine the energetics

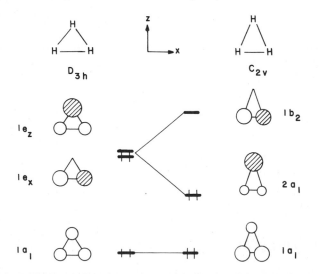

FIGURE 7.4   Correlation between the MO levels of equilateral and isosceles triangular H_3^-.

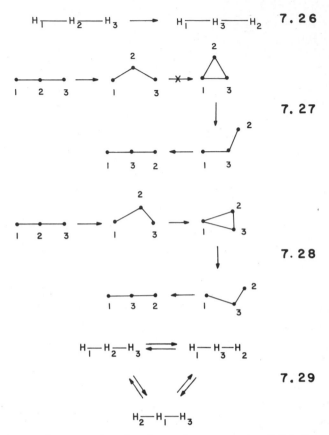

of the reaction path appropriate for the exchange reaction **7.26**. One conceivable reaction path for this exchange is **7.27** via an equilateral triangle. However, Figure 7.4 shows that the path **7.27** is less favorable than the alternative **7.28** which goes through an isosceles triangle because of the Jahn–Teller instability (and therefore local energy maximum) at the D_{3h} geometry. Figure 7.5 shows a somewhat idealized contour diagram of the potential energy surface for the interconversion **7.29**. The energy minima A, B, and C of Figure 7.5 represent the three equivalent, linear H_3^- structures. The energy maximum E lies at the equilateral triangle geometry, which is not a transition state for the interconversion since it is not a saddle point on the potential energy surface.[10] The saddle point D, located at an isosceles triangle geometry, is the transition state in the interconversion along the reaction coordinate $A \rightarrow B$. We will see that the potential energy surface of Figure 7.5 is characteristic of many chemical reactions.

7.5. BOND ORBITALS

The MOs of linear and bent AH_2 in Figure 7.3 correspond to the symmetry of the molecule, and hence are called symmetry adapted orbitals. Traditionally, molecular

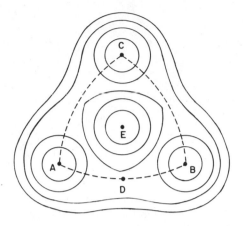

$$A = \underset{1 \quad 2 \quad 3}{\bullet\!\!-\!\!\bullet\!\!-\!\!\bullet} \qquad B = \underset{1 \quad 3 \quad 2}{\bullet\!\!-\!\!\bullet\!\!-\!\!\bullet}$$

$$C = \underset{2 \quad 1 \quad 3}{\bullet\!\!-\!\!\bullet\!\!-\!\!\bullet} \qquad D = $$

FIGURE 7.5. A contour diagram of the potential energy surface for the interconversion 7.29. The points A, B, and C represent energy minima, but the point E represents an energy maximum.

electronic structures are often described in terms of bond orbitals due to the one-to-one correspondence between a bond and a doubly occupied bonding orbital. When combined together with the valence-shell–electron-pair-repulsion (VSEPR) model,[11] the bond orbital description provides a set of simple rules for predicting molecular shapes although its applicability is limited to the consideration of ligand arrangements around a single atom. Two important rules of the VSEPR model, which are sufficient to predict the shapes of simple molecules, are as follows: (a) The best arrangement of a given number of electron pairs in the valence shell of an atom is that which maximizes the distances between them. (b) A nonbonding pair of electrons occupies more space on the surface of an atom than a bonding pair. Since bond orbital and MO descriptions of molecular electronic structures are considerably different, it is important to examine how the two approaches are related to each other.

Suppose that the HAH valence angle of AH_2 is equal to the tetrahedral angle (i.e., 109.5°). Let us construct from the valence s and p orbitals the sp^3 hybrid orbitals on A as indicated in 7.30. Two of these hybrid orbitals may be used to form bonding and antibonding orbitals with the hydrogen s along each A—H bond as shown in Figure 7.6, and the other two hybrid orbitals remain nonbonding orbitals on A. These bond orbitals (e.g., σ_{AH}, n_A, and σ^*_{AH} in Figure 7.6) do not, however, have the transformation properties with respect to the molecular geom-

7.30

etry demanded by the character table of the C_{2v} point group (Table 4.1) to which this molecule belongs. Unlike MOs, bond orbitals are therefore not eigenfunctions of the effective Hamiltonian H^{eff}.

From the vector properties of orbitals, the bond orbitals σ_{AH}, n_A and σ^*_{AH} of Figure 7.6 may be decomposed into atomic orbital contributions as shown in **7.31**. So we can easily see that the MOs of bent AH_2 are approximated by the bond orbitals as shown in Figure 7.7. Consider for instance the linear combinations of the two bonding orbitals σ_{AH}. The "positive" combination of the two leads to the σ_s MO, and the s character of A is retained. The "negative" combination leads to the σ_x, and removes the s character of A. Therefore, the two linear combinations of the degenerate bonding orbitals σ_{AH} become different in energy. Similarly, linear combinations of the two n_A orbitals or the two σ^*_{AH} orbitals lift the de-

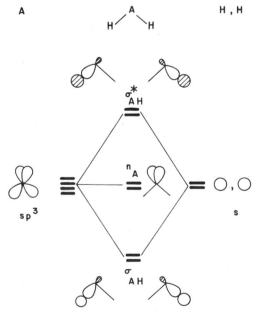

FIGURE 7.6.   An orbital interaction diagram for bent AH_2 in the bond orbital approach.

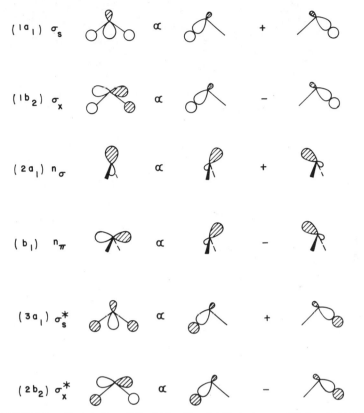

FIGURE 7.7.   Description of the MOs of bent AH_2 in terms of its bond orbitals.

7.31

generacy of the bond orbitals involved. Thus the MO levels of AH_2 are related to the bond orbitals as in **7.32**. Since these bond orbitals are not eigenfunctions of H^{eff}, the lifting of the bond orbital degeneracy is not surprising. Bond orbitals are often used as a convenient starting point in the generation of symmetry adapted MOs.

According to the bond orbital-MO correlation diagram **7.32**, eight valence electrons of H_2O are accommodated by the bond orbitals and the MOs as in **7.33** and

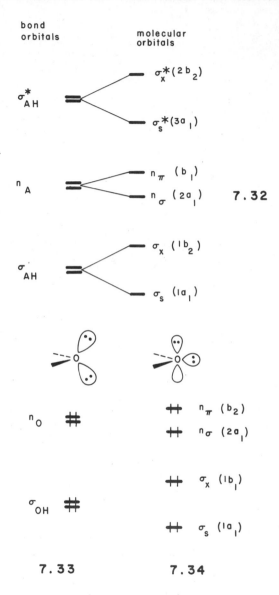

7.32

7.33   7.34

7.34, respectively. Apparently, the bond orbital and the MO descriptions of H_2O are quite different, although the decomposition described above provides links between the two. It is worthwhile, however, to comment upon the merits and limitations of the two different approaches.

The bond orbital approach predicts that H_2O is bent since the lowest energy arrangement for four electron pairs around an atom is a tetrahedral one in the VSEPR model. In addition, this approach rationalizes why the HOH angle is smaller than the tetrahedral angle since, on the surface of an atom, a nonbonding electron pair is supposed to occupy more space than does a bonding electron pair. The MO approach based upon the Walsh diagram Figure 7.3 predicts that H_2O is bent, but it does not predict how small the HOH angle would be. Nevertheless, the MO approach provides an elegant explanation for the decrease in the valence angles of eight-electron AH_2 systems in the order $H_2O > H_2S > H_2Se$. Occasionally somewhat peculiar hybridization arguments have been constructed for molecules like H_2S and H_2Se which have rather acute bond angles. Since they are close to $90°$, p atomic orbitals are used to form the A—H bonds leaving the lone pair in an unhybridized s orbital on A. It is clear from the shape of $1a_1$ and $2a_1$ that this is far removed from a realistic point of view.

An electron in an MO ψ_i is under an effective potential e_i. As we will discuss in Chapter 8, Koopmans' theorem shows that the ionization potential required to remove an electron from ψ_i is given by $-e_i$.[12,13] As schematically depicted in 7.35, a photoelectron spectrometer measures the kinetic energies of electrons

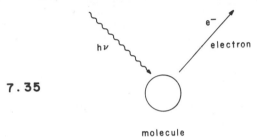

7.35

ejected from a molecule by an incident photon beam of energy $h\nu$. Such photoelectrons originate from various MO levels ψ_i, and therefore their kinetic energies (KE) are related to the orbital energies e_i by

$$-e_i = h\nu - \text{KE} \qquad (7.7)$$

Importantly, it is the MO energies e_i that are directly related to experimental ionization potentials via Koopmans' theorem. Since bond orbitals are not eigenfunctions of the effective Hamiltonian H^{eff}, their energies do not refer to the effective potentials that can be directly related to experimental ionization potentials. For example, the photoelectron spectrum of H_2O does not show ionization from two sets of degenerate levels as implied by the bond orbital picture **7.33** but four levels as expected from the MO picture **7.34** (i.e., 12.6, 13.8, 17.2, and 33.2 eV for the n_π, n_σ, σ_x, and σ_s levels, respectively). Note that the n_π and n_σ nonbonding levels are separated by 1.2 eV or about 28 kcal/mol; they are in no way close to being degenerate.

As a final example concerning the difference between the bond orbital and MO descriptions, let us consider how an electrophile E^+ might attack H_2O to form H_2O^+-E. According to the bond orbital description **7.36**, an electrophile would

7.36

approach H_2O along the axis of one nonbonding orbital of oxygen since this allows for maximum overlap between the nonbonding orbital n_O and the acceptor orbital ϕ_e of E^+. In the MO description, the interaction between H_2O and E^+ can be discussed in terms of the simplified interaction diagram **7.37**, where the

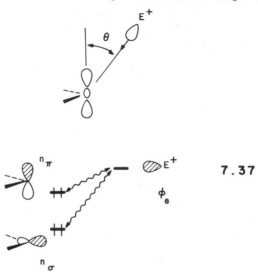

7.37

$(n_\pi - \phi_e)$ and $(n_\sigma - \phi_e)$ interactions are both stabilizing. Let us define the approach angle θ of E^+ by reference to the p orbital axis of n_π as shown in **7.37**. In terms of overlap, the magnitude of the $(n_\pi - \phi_e)$ interaction is a maximum at $\theta = 0°$ and a minimum at $\theta = 90°$ as depicted in **7.38**. The opposite situation is found for the case with the $(n_\sigma - \phi_e)$ interaction as indicated in **7.39**. Thus at $\theta \simeq 45°$, an electro-

phile can take advantage of both the $(n_\pi - \phi_e)$ and $(n_\sigma - \phi_e)$ interactions. However, n_π is closer in energy to ϕ_e than is n_σ so that, in terms of orbital energy gap, the $(n_\pi - \phi_e)$ interaction is more stabilizing than the $(n_\sigma - \phi_e)$ interaction. Thus the approach angle θ becomes smaller if the energy gap between n_π and n_σ is made greater. From our discussion using **7.19** and **7.20**, it is evident that the lowering of n_σ with respect to $1\pi_{uz}$ is greater, and hence the energy gap between n_π and

n_σ becomes greater, if the central atom of AH_2 is made less electronegative. As a consequence, the approach angle θ is predicted to be smaller for H_2S than for H_2O.[14]

7.6   THROUGH-BOND INTERACTIONS[15]

In connection with the bond orbital–MO correlation diagram **7.32**, we showed how degenerate bond orbitals centered on a single atom combine in a linear way to give symmetry-adapted MOs. Degenerate bond orbitals centered on different atoms also combine linearly to give symmetry-adapted MOs, but this occurs primarily not "through space" but "through bond." As an example of a through-bond interaction, we will return to the nonbonding orbitals of N_2 (Section 6.3). In the bond orbital picture, they are represented as in **7.40**, a description that implies the

$$\text{:}{>}\text{N} \equiv\!\!\equiv \text{N}{<}\text{:} \qquad \textbf{7.40}$$

presence of two equivalent nonbonding electron pairs. The bond orbitals of N_2 may be constructed by employing two sp hybrid orbitals and two p orbitals on each nitrogen atom as shown in Figure 7.8, where the bond orbitals σ, n_+, n_-, and σ^* are given in **7.41**. The π and π^* orbitals of Figure 7.8 are identical to the

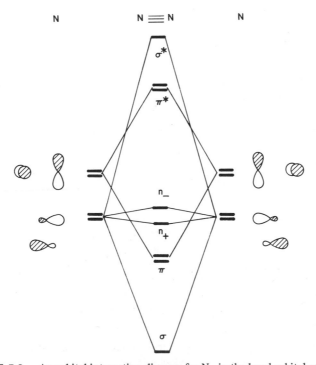

FIGURE 7.8.   An orbital interaction diagram for N_2 in the bond orbital approach.

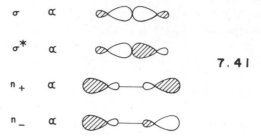

7.41

corresponding MOs of a homodiatomic molecular examined in Chapter 6. In Figure 7.8 the in-phase nonbonding orbital n_+ lies slightly lower than the out-of-phase nonbonding orbital n_- because of the small overlap between the two sp hybrid orbitals through space. Let us compare this result with the MO description of a homonuclear diatomic molecule in Chapter 6. This showed the in-phase nonbonding orbital $2\sigma_g$ to lie at higher energy than the out-of-phase nonbonding orbital $1\sigma_u$. This apparent discrepancy originates from the complication that, in the bond orbital description the n_+ orbital has nonzero overlap with σ and the n_- orbital a nonzero overlap with σ^*. Therefore, we need to include interaction between n_+ and σ and between n_- and σ^*. It is this through-bond interaction that leads to the ordering of the nonbonding orbitals shown in Figure 7.9. That is, n_- is lowered in energy by σ^* to become $1\sigma_u$, while n_+ is raised in energy to become $2\sigma_g$. According to the photoelectron spectrum of N_2, the ionization potentials from the $1\sigma_u$ and $2\sigma_g$ levels are 15.5 and 18.8 eV, respectively. The 3.3 eV energy

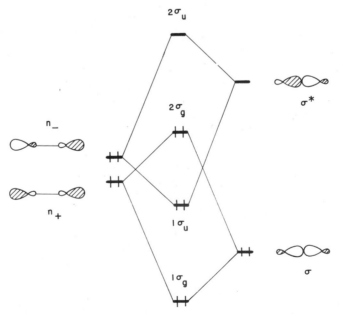

FIGURE 7.9.   Change in the relative ordering of the nonbonding levels of N_2 derived from the bond orbital approach by through-bond interaction.

difference between the two lone pairs and the fact that the in-phase combination lies at a higher energy than the out-of-phase combination may seem disconcerting but it is an experimental fact. Further details on this topic are presented in Section 11.3.

REFERENCES

1. B. M. Gimarc, *Molecular Structure and Bonding*, Academic, New York (1980).
2. A. D. Walsh, *J. Chem. Soc.*, 2260, 2266, 2288, 2296, 2301, 2306, 2318, 2321, 2325, 2330 (1953).
3. J. K. Burdett, *Molecular Shapes*, Wiley, New York (1980).
4. R. J. Buenker and S. D. Peyerimhoff, *Chem. Rev.*, 74, 127 (1974).
5. L. C. Allen, *Theor. Chim. Acta*, 24, 117 (1972).
6. L. Salem, *The Molecular Orbital Theory of Conjugated Systems*, Benjamin, New York (1966).
7. L. S. Bartell, *J. Chem. Ed.*, 45, 754 (1968).
8. H. A. Jahn and E. Teller, *Proc. Roy. Soc.*, A161, 220 (1937).
9. N. T. Anh and C. Minot, *J. Am. Chem. Soc.*, 102, 103 (1980); A. Veillard and A. Dedieu, *ibid*, 94, 6730 (1972); C. Minot, *Nouv, J. Chim.*, 5, 319 (1981).
10. J. W. McIver and R. E. Stanton, *J. Am. Chem. Soc.*, 94, 8618 (1972).
11. R. J. Gillespie, *Molecular Geometry*, Van Nostrand-Reinhold, London (1972).
12. D. W. Turner, A. P. Baker, and C. R. Brundle, *Molecular Photoelectron Spectroscopy*, Interscience, New York (1970).
13. J. W. Rabalais, *Principles of Ultraviolet Photoelectron Spectroscopy*, Wiley, New York (1977).
14. P. A. Kollman, *J. Am. Chem. Soc.*, 94, 1838 (1972); P. A. Kollman, J. McKelvy, A. Johansson, and S. Rothenberg, *ibid*, 97, 955 (1975).
15. R. Hoffmann, *Acc. Chem. Res.*, 4, 1 (1971).

CHAPTER EIGHT

State Wavefunctions and State Energies

8.1. INTRODUCTION

Up until now we have avoided discussion of the nature of the effective Hamiltonian H^{eff} that has figured prominently in the expressions for the interaction integrals. We have also left until this chapter consideration of a related problem, shown in **2.3** and **2.5**. Given two orbitals of different energies, and two electrons, what factors influence the relative stabilities of the possible singlet and triplet states? In the case of atoms where electrons enter degenerate p or d orbitals, one of Hund's rules tells us that the state of highest spin multiplicity will be most stable. To put this in perspective, **8.1** shows the strategy which we will use in understanding molecular orbital calculations. Much of what we have to say elsewhere is accessible by considering the steps A–C and describing problems in terms of one-electron energies. But as mentioned above, there are several situations which do not make sense until we take the next step and switch on electron–electron interactions. Finally we shall also find some material that requires a higher level of sophistication still and forces us to include the last step in **8.1**. Discussion of these problems requires the study of the molecular Hamiltonian $\hat{H}$ and the state wavefunction Φ of a molecule. Since $\hat{H}$ and Φ should describe all the electrons, $1, 2, 3, \ldots, N$ present in a molecule, they are functions of the electron coordinates

$$\hat{H} = \hat{H}(1, 2, 3, \ldots, N) \tag{8.1}$$

$$\Phi = \Phi(1, 2, 3, \ldots, N) \tag{8.2}$$

where each electron number μ ($= 1, 2, 3, \ldots, N$) refers to the spatial coordinates x_μ, y_μ, and z_μ as well as the spin coordinate s_μ of the electron μ. The importance of

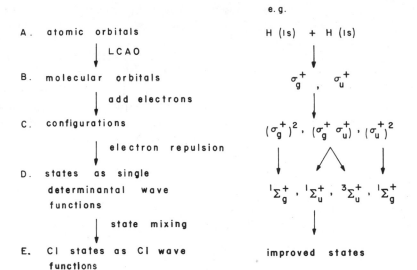

$$8.1$$

$\hat{H}$ and Φ originates from the fact that the total energy of a molecule in the state Φ is given by the expectation value

$$E = \langle \Phi | \hat{H} | \Phi \rangle \tag{8.3}$$

if Φ is normalized to unity

$$\langle \Phi | \Phi \rangle = 1 \tag{8.4}$$

8.2.   THE MOLECULAR HAMILTONIAN AND STATE WAVEFUNCTIONS[1]

In terms of atomic units, in which the electron mass m, charge e, and the constant $h/2\pi$ are taken to be unity, the molecular Hamiltonian $\hat{H}$ may be written as

$$\hat{H} = \sum_{\mu=1}^{N} \left\{ \hat{h}(\mu) + \sum_{\nu < \mu} \hat{g}(\mu, \nu) \right\} + \sum_{A>B} \sum \frac{Z_A Z_B}{r_{AB}} \tag{8.5}$$

where

$$\hat{h}(\mu) = -\frac{1}{2} \nabla_\mu^2 - \sum_A \frac{Z_A}{r_{\mu A}} \tag{8.6}$$

$$\hat{g}(\mu, \nu) = \frac{1}{r_{\mu\nu}} \tag{8.7}$$

The first term of the core-Hamiltonian $\hat{h}(\mu)$ in the equation 8.6 is the kinetic energy of the electron μ which can be expressed as given by the equation 8.8. The second term of $\hat{h}(\mu)$ is the energy of attraction between the electron μ and all the nuclei of

$$-\frac{1}{2}\nabla_\mu^2 = -\frac{1}{2}\left(\frac{\partial^2}{\partial x_\mu^2} + \frac{\partial^2}{\partial y_\mu^2} + \frac{\partial^2}{\partial z_\mu^2}\right) \tag{8.8}$$

the molecule. $r_{\mu A}$ is the distance between the electron μ and the nucleus of atom A with charge Z_A. The electron–electron repulsion between electrons μ and ν is given by equation 8.7, where $r_{\mu\nu}$ is the distance between electrons μ and ν. The total nuclear–nuclear repulsion V_{nn} is represented by the last term of equation 8.5

$$V_{nn} = \sum_{A>B}\sum \frac{Z_A Z_B}{r_{AB}} \tag{8.9}$$

where r_{AB} is the distance between the nuclei of atoms A and B.

One way of phrasing the Pauli exclusion principle is the requirement that the electronic state wavefunction Φ be antisymmetric with respect to the interchange of any two electron coordinates. For example,

$$\Phi(2,1,3,\ldots,N) = -\Phi(1,2,3,\ldots,N) \tag{8.10}$$

Within the framework of molecular orbital theory, the simplest wavefunction satisfying this antisymmetric property is the determinant constructed from all the occupied MOs known as the Slater determinant. As an example, we will consider a $2n$ electron closed-shell molecule that has doubly occupied levels $\psi_1, \psi_2, \psi_3, \ldots, \psi_n$ as in **8.2**. The functions describing up-spin and down-spin character of an elec-

8.2

$$
\begin{array}{ll}
— & \psi_{n+2} \\
— & \psi_{n+1} \\
⊣⊢ & \psi_n \\
⊣⊢ & \psi_{n-1} \\
\vdots & \\
⊣⊢ & \psi_2 \\
⊣⊢ & \psi_1
\end{array}
$$

tron μ are written as $\alpha(\mu)$ and $\beta(\mu)$, respectively, and satisfy the orthonormality relationship

$$\int \alpha(\mu)\,\alpha(\mu)\,ds_\mu \equiv \langle \alpha(\mu)|\alpha(\mu)\rangle = 1$$

$$\int \beta(\mu)\,\beta(\mu)\,ds_\mu \equiv \langle \beta(\mu)|\beta(\mu)\rangle = 1 \tag{8.11}$$

$$\int \alpha(\mu)\,\beta(\mu)\,ds_\mu \equiv \langle \alpha(\mu)|\beta(\mu)\rangle = 0$$

The up-spin and down-spin wavefunctions for electron μ in an MO ψ_i are given by the products $\psi_i(\mu)\alpha(\mu)$ and $\psi_i(\mu)\beta(\mu)$, respectively. For the purpose of simplicity, we write these as

$$\psi_i(\mu)\alpha(\mu) \equiv \psi_i(\mu)$$
$$\psi_i(\mu)\beta(\mu) \equiv \bar{\psi}_i(\mu) \tag{8.12}$$

The Slater determinant Φ for the electron configuration **8.2** is then given by

$$\Phi(1,2,3,\ldots,N) = \frac{1}{\sqrt{(2n)!}} \begin{vmatrix} \psi_1(1) & \psi_1(2) & \psi_1(3) & \cdots & \psi_1(2n) \\ \bar{\psi}_1(1) & \bar{\psi}_1(2) & \bar{\psi}_1(3) & \cdots & \bar{\psi}_1(2n) \\ \psi_2(1) & \psi_2(2) & \psi_2(3) & \cdots & \psi_2(2n) \\ \bar{\psi}_2(1) & \bar{\psi}_2(2) & \bar{\psi}_2(3) & \cdots & \bar{\psi}_2(2n) \\ \vdots & \vdots & \vdots & & \vdots \\ \psi_n(1) & \psi_n(2) & \psi_n(3) & \cdots & \psi_n(2n) \\ \bar{\psi}_n(1) & \bar{\psi}_n(2) & \bar{\psi}_n(3) & \cdots & \bar{\psi}_n(2n) \end{vmatrix} \tag{8.13}$$

$$\equiv \mathcal{A}\psi_1(1)\bar{\psi}_1(2)\psi_2(3)\bar{\psi}_2(4)\cdots\psi_n(2n-1)\bar{\psi}_n(2n) \tag{8.14}$$

where $\mathcal{A}$ is an antisymmetrizing operator.

8.3. THE FOCK OPERATOR[1]

We quote the following result without proof or further discussion. When applied to the state wavefunction Φ, the variation principle leads to the Fock equation

$$\hat{F}\psi_i = e_i\psi_i \tag{8.15}$$

where $\hat{F}$, the Fock operator, controls the form of the MOs ψ_i and their "orbital energies" e_i and is the effective one-electron Hamiltonian H^{eff} for the configuration **8.2**. Since $\hat{F}$ and ψ_i depend only upon the coordinate of a single electron, equation 8.15 may be written as

$$\hat{F}(\mu)\,\psi_i(\mu) = e_i\psi_i(\mu) \tag{8.16}$$

where $\mu = 1, 2, 3, \ldots, 2n$.

The Fock operator is made up of three terms. The first is a one-electron term describing the core potential and the other two are two-electron terms which contain the electron–electron repulsion energies. For an electron μ located in the MO ψ_i, the core potential, namely the kinetic plus nuclear–electron attraction energies, is given by the expectation value of the core-Hamiltonian $\hat{h}(\mu)$

$$h_i = \int \psi_i(\mu)\,\hat{h}(\mu)\,\psi_i(\mu)\,d\tau_\mu$$

$$\equiv \langle \psi_i(\mu)|\hat{h}(\mu)|\psi_i(\mu)\rangle \equiv \langle \psi_i|\hat{h}|\psi_i\rangle \tag{8.17}$$

For two electrons μ and ν ($\mu \neq \nu$) accommodated in the MOs ψ_i and ψ_j, respectively, the two-electron terms are the Coulomb repulsion J_{ij} and the exchange repulsion K_{ij} energies.

$$J_{ij} = \iint \frac{\psi_i(\mu) \, \psi_i(\mu) \, \psi_j(\nu) \, \psi_j(\nu)}{r_{\mu\nu}} \, d\tau_\mu \, d\tau_\nu$$

$$= \int \psi_i(\mu) \left(\int \frac{\psi_j(\nu) \, \psi_j(\nu)}{r_{\mu\nu}} \, d\tau_\nu \right) \psi_i(\mu) \, d\tau_\mu$$

$$\equiv [\psi_i(\mu) \, \psi_i(\mu) | \psi_j(\nu) \, \psi_j(\nu)] \equiv (\psi_i \psi_i | \psi_j \psi_j) \tag{8.18}$$

$$K_{ij} = \iint \frac{\psi_i(\mu) \, \psi_j(\mu) \, \psi_i(\nu) \, \psi_j(\nu)}{r_{\mu\nu}} \, d\tau_\mu \, d\tau_\nu$$

$$= \int \psi_i(\mu) \left(\int \frac{\psi_i(\nu) \, \psi_j(\nu)}{r_{\mu\nu}} \, d\tau_\nu \right) \psi_j(\mu) \, d\tau_\mu$$

$$\equiv [\psi_i(\mu) \, \psi_j(\mu) | \psi_i(\nu) \, \psi_j(\nu)] \equiv (\psi_i \psi_j | \psi_i \psi_j) \tag{8.19}$$

From these definitions, it follows that $K_{ii} = J_{ii}$. Also note that the Coulomb repulsion between two electrons is independent of their spins while the exchange repulsion vanishes unless their spins are the same. J_{ij} is repulsive (i.e., positive) and represents the electrostatic repulsion between electron μ in orbital ψ_i and electron ν in orbital ψ_j. It increases as the overlap between ψ_i and ψ_j increases. In other words, the closer the electrons are the larger their mutual repulsion. The exchange integral arises purely as a result of the expansion of equation 8.13, that is, the requirement that the state wavefunction be antisymmetric with respect to electron exchange. Notice that the integrand of equation 8.19 involves the exchange of two electrons compared to equation 8.18; hence its name. K_{ij} represents, in a sense, a correction to the Coulomb repulsion term J_{ij} for the case of two electrons with parallel spins. When the electron spins are parallel then, in another phrasing of the Pauli principle, they cannot occupy the same region of space. The exchange repulsion, therefore, has no classical analog. Although intrinsically positive, it is subtracted from Coulomb repulsion to give the total electron–electron energy. It is important to realize that with the inclusion of K_{ij} we are not correlating the motions of the electrons. Electrons of opposite spins are still allowed to move completely independent of one another.

In order to specify the Fock operator $\hat{F}$, itself, we introduce the Coulomb operator $\hat{J}_j$ and the exchange operator $\hat{K}_j$ for an electron in the MO ψ_j.

$$\hat{J}_j(\mu) \, \psi_i(\mu) \equiv \left(\int \frac{\psi_j(\nu) \, \psi_j(\nu)}{r_{\mu\nu}} \, d\tau_\nu \right) \psi_i(\mu) \tag{8.20}$$

$$\hat{K}_j(\mu) \, \psi_i(\mu) \equiv \left(\int \frac{\psi_j(\nu) \, \psi_i(\nu)}{r_{\mu\nu}} \, d\tau_\nu \right) \psi_j(\mu) \tag{8.21}$$

Then the integrals J_{ij} and K_{ij} are simply the expectation values of the operators $\hat{J}_j(\mu)$ and $\hat{K}_j(\mu)$ (see equations 8.18 and 8.19),

$$J_{ij} = \int \psi_i(\mu)\, \hat{J}_j(\mu)\, \psi_i(\mu)\, d\tau_\mu$$

$$\equiv \langle \psi_i(\mu) | \hat{J}_j(\mu) | \psi_i(\mu) \rangle \equiv \langle \psi_i | \hat{J}_j | \psi_i \rangle \tag{8.22}$$

$$K_{ij} = \int \psi_i(\mu)\, \hat{K}_j(\mu)\, \psi_i(\mu)\, d\tau_\mu$$

$$\equiv \langle \psi_i(\mu) | \hat{K}_j(\mu) | \psi_i(\mu) \rangle \equiv \langle \psi_i | \hat{K}_j | \psi_i \rangle \tag{8.23}$$

In terms of $\hat{h}(\mu)$, $\hat{J}(\mu)$, and $\hat{K}_j(\mu)$, the Fock operator $\hat{F}(\mu)$ for 8.2 is written as

$$\hat{F}(\mu) = \hat{h}(\mu) + \sum_{j=1}^{n} [2\hat{J}_j(\mu) - \hat{K}_j(\mu)] \tag{8.24}$$

or, simply,

$$\hat{F} = \hat{h} + \sum_{j=1}^{n} (2\hat{J}_j - \hat{K}_j) \tag{8.25}$$

8.4. STATE ENERGY

The orbital energy e_i for the ith level of the electron configuration 8.2 is the expectation value of the Fock operator

$$e_i = \langle \psi_i | \hat{F} | \psi_i \rangle \tag{8.26}$$

Using equation 8.25,

$$e_i = \langle \psi_i | \hat{h} | \psi_i \rangle + \left\langle \psi_i \left| \sum_{j=1}^{n} (2\hat{J}_j - \hat{K}_j) \right| \psi_i \right\rangle$$

$$= h_i + \sum_{j=1}^{n} (2J_{ij} - K_{ij}) \tag{8.27}$$

Thus an electron in one of the occupied MOs ψ_i ($i = 1, 2, 3, \ldots, n$) feels the core potential h_i as well as the electron–electron repulsion arising from the presence of other electrons. The Fock operator $\hat{F}$, though determined only in terms of the occupied MOs, determines the energy of both the occupied and the unoccupied levels. As will be shown in the next section, the orbital energy e_i of an unoccupied MO ψ_i ($i = n + 1, n + 2, \ldots$) refers to the potential that an extra electron feels if it were placed in that orbital.

The total electron–electron repulsion V_{ee} in 8.2 is given by

$$V_{ee} = \sum_{i=1}^{n} \sum_{j=1}^{n} (2J_{ij} - K_{ij}) \tag{8.28}$$

Thus the total energy E of the configuration **8.2** is simply

$$E = \sum_{i=1}^{n} 2h_i + V_{ee} + V_{nn} \qquad (8.29)$$

Combining equations 8.27 and 8.29,

$$E = \sum_{i=1}^{n} 2e_i - V_{ee} + V_{nn} \qquad (8.30)$$

The sum of the first two terms in equation 8.30 is the electronic energy. Note that the total energy is not equal to the sum of all the occupied orbital energies. However, importantly it may be shown that[2]

$$-V_{ee} + V_{nn} \simeq \tfrac{1}{3} E \qquad (8.31)$$

and so

$$E \simeq \tfrac{3}{2} \sum_{i=1}^{n} 2e_i \qquad (8.32)$$

This relationship, though approximate, justifies in part the use of orbital energy changes alone in discussing molecular structure and reactivity problems.

8.5. EXCITATION ENERGY

The electron configuration **8.2** is a typical closed shell, in which all the occupied MOs are doubly filled. Let us examine the stability of such a state with respect to those states in which some of the high lying occupied MOs are singly filled. To simplify our discussion, consider the various electronic configurations shown in **8.3**

8.3

which result from a simple two-orbital–two-electron case. In the following, the MOs ψ_1 and ψ_2 are to be determined from the eigenvalue equation associated with the singlet ground state configuration Φ_G

$$\hat{H}\Phi_G = E_G \Phi_G \qquad (8.33)$$

where

$$\Phi_G = \frac{1}{\sqrt{2}} \begin{vmatrix} \psi_1(1) & \psi_1(2) \\ \bar{\psi}_1(1) & \bar{\psi}_1(2) \end{vmatrix} = \frac{1}{\sqrt{2}} [\psi_1(1)\,\bar{\psi}_1(2) - \bar{\psi}_1(1)\,\psi_1(2)] \qquad (8.34)$$

and the energy E_G is given by

$$E_G = \langle \Phi_G | \hat{H} | \Phi_G \rangle \tag{8.35}$$

Application of the variation principle to equation 8.35 leads to the Fock equation

$$\hat{F}\psi_i = e_i \psi_i \quad (i = 1, 2) \tag{8.36}$$

where

$$\hat{F} = \hat{h} + 2\hat{J}_1 - \hat{K}_1 \tag{8.37}$$

Note that the MOs ψ_i $(i = 1, 2)$ are determined if $\hat{F}$ is known, but $\hat{F}$ is defined in terms of the occupied MO ψ_1 (via $\hat{J}_1$ and $\hat{K}_1$) that is yet to be determined. This problem is solved by the method of self-consistent field (SCF) iteration: In the first cycle of iteration, a trial MO ψ_1 is assumed to obtain $E_G^{(1)}$ and $\hat{F}^{(1)}$. In the second cycle of iteration, we solve equation 8.36 for $\hat{F}^{(1)}$ to find new MOs $\psi_i^{(2)}$ $(i = 1, 2)$ and use $\psi_1^{(2)}$ to generate $E_G^{(2)}$ and $\hat{F}^{(2)}$. In the third cycle of iteration, equation 8.36 is solved for $\hat{F}^{(2)}$ to obtain new MOs $\psi_i^{(3)}$ $(i = 1, 2)$ and hence $E_G^{(3)}$ and $\hat{F}^{(3)}$. If such an iteration is repeated n times, the state energies at various cycles of iteration satisfy the following relationship[3]

$$E_G^{(1)} \geqslant E_G^{(2)} \geqslant \cdots \geqslant E_G^{(n-1)} \geqslant E_G^{(n)} \tag{8.38}$$

due to the variation principle. When the energy difference between the last two iterations is negligibly small, the SCF iteration is said to be converged after n iterations. In such a case no further iteration could improve the wavefunction. In our discussion, the MOs ψ_i and the orbital energies e_i $(i = 1, 2)$ are assumed to be those determined from a converged SCF iteration for the state Φ_G.

From equation 8.27 the orbital energies e_1 and e_2 are given by

$$e_1 = \langle \psi_1 | \hat{h} + 2\hat{J}_1 - \hat{K}_1 | \psi_1 \rangle = h_1 + 2J_{11} - K_{11} = h_1 + J_{11} \tag{8.39}$$

$$e_2 = \langle \psi_2 | \hat{h} + 2\hat{J}_1 - \hat{K}_1 | \psi_2 \rangle = h_2 + 2J_{12} - K_{12} \tag{8.40}$$

Thus e_1 is the effective potential exerted on an electron in the MO ψ_1 of Φ_G. If an extra electron is placed in the MO ψ_2 of Φ_G, that electron would feel the effective potential given by e_2. In all the electronic states of **8.3**, the molecular geometry is assumed to be the same so that the relative stability of those states can be examined by simply comparing their electronic energies.

The electronic energies of the singlet ground state Φ_G and the triplet state Φ_T are determined by the form of the wavefunctions. The expressions of Φ_G and Φ_T are given by equations 8.34 and 8.41, respectively.

$$\Phi_T = \frac{1}{\sqrt{2}} \begin{vmatrix} \psi_1(1) & \psi_1(2) \\ \psi_2(1) & \psi_2(2) \end{vmatrix} = \frac{1}{\sqrt{2}} [\psi_1(1)\psi_2(2) - \psi_2(1)\psi_1(2)] \tag{8.41}$$

A bit of arithmetic leads to the form of the energies

$$E_G = \langle \Phi_G | \hat{H} | \Phi_G \rangle$$

$$= 2h_1 + J_{11} = 2e_1 - J_{11} \tag{8.42}$$

$$E_T = \langle \Phi_T | \hat{H} | \Phi_T \rangle$$

$$= h_1 + h_2 + J_{12} - K_{12} = e_1 + e_2 - J_{11} - J_{12} \tag{8.43}$$

The electronic energies of the configurations Φ_1 and Φ_2 are the same and are evaluated in an analogous way via a knowledge of the state functions

$$\Phi_1 = \frac{1}{\sqrt{2}} \begin{vmatrix} \psi_1(1) & \psi_1(2) \\ \bar{\psi}_2(1) & \bar{\psi}_2(2) \end{vmatrix} = \frac{1}{\sqrt{2}} [\psi_1(1)\bar{\psi}_2(2) - \bar{\psi}_2(1)\psi_1(2)]$$

$$\Phi_2 = \frac{1}{\sqrt{2}} \begin{vmatrix} \bar{\psi}_1(1) & \bar{\psi}_1(2) \\ \psi_2(1) & \psi_2(2) \end{vmatrix} = \frac{1}{\sqrt{2}} [\bar{\psi}_1(1)\psi_2(2) - \psi_2(1)\bar{\psi}_1(2)]$$

(8.44)

which leads to

$$E_1 = \langle \Phi_1 | \hat{H} | \Phi_1 \rangle$$
$$= E_2 = \langle \Phi_2 | \hat{H} | \Phi_2 \rangle$$
$$= h_1 + h_2 + J_{12} = e_1 + e_2 - J_{11} - J_{12} + K_{12}$$

(8.45)

Unlike the states Φ_G and Φ_T, however, the configurations Φ_1 and Φ_2 are not eigenfunctions of the so-called total spin angular momentum operator $\hat{S}^2$.[4] The singlet excited state Φ_S is given by the linear combination

$$\Phi_S = \frac{1}{\sqrt{2}} (\Phi_1 - \Phi_2)$$

(8.46)

a function which does satisfy this requirement. From equation 8.44 it is easy to show the relationship

$$\langle \Phi_1 | \hat{H} | \Phi_2 \rangle = -(\psi_1 \psi_2 | \psi_1 \psi_2) = -K_{12}$$

(8.47)

which leads to the electronic energy of the singlet excited state Φ_S

$$E_S = \langle \Phi_S | H | \Phi_S \rangle$$
$$= h_1 + h_2 + J_{12} + K_{12} = e_1 + e_2 - J_{11} - J_{12} + 2K_{12}$$

(8.48)

One state wavefunction describing the triply degenerate triplet state is given by equation 8.41, and the other two are as follows:

$$\Phi_T' = \frac{1}{\sqrt{2}} \begin{vmatrix} \bar{\psi}_1(1) & \bar{\psi}_1(2) \\ \bar{\psi}_2(1) & \bar{\psi}_2(2) \end{vmatrix} = \frac{1}{\sqrt{2}} [\bar{\psi}_1(1)\bar{\psi}_2(2) - \bar{\psi}_2(1)\bar{\psi}_1(2)] \quad (8.49)$$

$$\Phi_T'' = \frac{1}{\sqrt{2}} (\Phi_1 + \Phi_2)$$

(8.50)

The MOs ψ_1 and ψ_2 are occupied by up-spin electrons in Φ_T, but by down-spin electrons in Φ_T'. It can be easily shown that the electronic energy of Φ_T' or Φ_T'' is the same as that of Φ_T.

$$E_T = \langle \Phi_T | \hat{H} | \Phi_T \rangle = \langle \Phi_T' | \hat{H} | \Phi_T' \rangle = \langle \Phi_T'' | \hat{H} | \Phi_T'' \rangle$$

(8.51)

Collecting the above results together,

$$E_T - E_G = (e_2 - e_1) - J_{12}$$
$$E_S - E_T = 2K_{12}$$

(8.52)

Consequently, if $(e_2 - e_1) - J_{12} > 0$, the relative stability of the ground, the triplet, and the singlet excited states is given as in **8.4**. Thus, with electron–electron repulsion explicitly taken into consideration, the excitation energy is not simply given by the orbital energy difference $(e_2 - e_1)$. Since $K_{12} > 0$, the triplet state is always more stable than the singlet excited state.

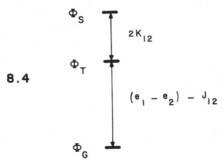

8.4

8.6.   IONIZATION POTENTIAL AND ELECTRON AFFINITY

The ionization potential (IP) and the electron affinity (EA) of a molecule M are defined as the energies required for the ionization processes

$$M \longrightarrow M^+ + e^-, \quad \text{IP}$$
$$M^- \longrightarrow M + e^-, \quad \text{EA}$$
(8.53)

As indicated in **8.5**. we may construct the cation state Φ_+ and the anion state Φ_- by

8.5

using the MOs ψ_1 and ψ_2 obtained from the ground state Φ_G. The electronic energies of those ionic states are evaluated via construction of the state wavefunctions using equation 8.13. $\Phi_+ = \psi_1(1)$, but to get Φ_- requires evaluation of the determinant made up of the MOs ψ_1, $\bar{\psi}_1$, and ψ_2. We find

$$E_+ = \langle \Phi_+ | \hat{H} | \Phi_+ \rangle = h_1 = e_1 - J_{11} \tag{8.54}$$

$$E_- = \langle \Phi_- | \hat{H} | \Phi_- \rangle$$
$$= 2h_1 + h_2 + J_{11} + 2J_{12} - K_{12}$$
$$= 2e_1 + e_2 - J_{11} \tag{8.55}$$

and, combined with equation 8.42,

$$E_+ - E_G = -e_1 = \text{IP} \tag{8.56}$$

$$E_G - E_- = -e_2 = \text{EA} \tag{8.57}$$

In general, for the electron configuration **8.2**, the ionization potential associated with an electron removal from an occupied MO ψ_i $(i = 1, 2, 3, \ldots, n)$ is given by $-e_i$. This is known as Koopmans' theorem.[5] The energy of an unoccupied MO ψ_i $(i = n + 1, n + 2, \ldots)$ refers to the potential exerted on an extra electron placed in that orbital. Thus the electron affinity of removing such an electron in an unoccupied MO ψ_i $(i = n + 1, n + 2, \ldots)$ is given by $-e_i$. These simple results are obtained because of the implicit assumption that the electrons not involved in the ionization process are not perturbed. Neither does the molecular geometry relax in the ionic states. Photoelectron spectroscopy (Section 7.5) yields experimental values of the ionization potentials associated with each occupied MO. Thus Koopmans' theorem provides a direct comparison between theory and experiment, although its applicability is limited because of the assumption employed to obtain it.

8.7. ELECTRON DENSITY DISTRIBUTION AND THE MAGNITUDES OF COULOMB AND EXCHANGE REPULSIONS

We noted in Section 8.3 that, for two MOs ψ_i and ψ_j, there are two kinds of electron–electron repulsions to consider, namely the Coulomb repulsion J_{ij} and the exchange repulsion K_{ij}. The electron density distributions associated with ψ_i and ψ_j are given by $\psi_i\psi_i$ and $\psi_j\psi_j$, respectively. As noted above these electron densities lead to J_{ij} (equation 8.18). On the other hand, it is the overlap density distribution $\psi_i\psi_j$ that defines K_{ij} (equation 8.19).

The magnitude of K_{ij} is therefore small unless the overlap density $\psi_i\psi_j$ is large, in some region of space. Consider for example the two p orbitals ψ_a and ψ_b arranged perpendicular to each other as shown in **8.6**. Since the large amplitude re-

8.6

gion of ψ_a does not coincide with that of ψ_b, the overlap density $\psi_a\psi_b$ is small in all regions compared with the "diagonal" density $\psi_a\psi_a$ or $\psi_b\psi_b$. Thus in **8.6**, the exchange repulsion K_{12} is substantially smaller than the Coulomb repulsion J_{12}. Given two MOs ψ_i and ψ_j, the magnitudes of J_{ij} and K_{ij} will satisfy the relationship[1]

$$J_{ij} \geqslant K_{ij} \geqslant 0 \tag{8.58}$$

where the equality $J_{ij} = K_{ij}$ arises when ψ_i and ψ_j are identical.

Another example that illustrates the effect of the overlap density $\psi_i\psi_j$ upon the magnitude of K_{ij} involves two conjugated hydrocarbons. Listed in Table 8.1 are the experimental values of the first ionization potential (IP), the electron affinity (EA), the first singlet excitation energy $(E_S - E_G)$, and the first triplet excitation energy $(E_T - E_G)$ for azulene **8.7** and anthracene **8.8**.[6] For simplicity of notation, the

TABLE 8.1.   The Ionization Potentials,
Electron Affinities, and Excitation Energies
of Azulene and Anthracene

Quantity[a]	Molecule	
	Azulene	Anthracene
IP	7.4	7.4
EA	0.7	0.6
$E_S - E_G$	1.8	3.3
$E_T - E_G$	1.3	1.8
J_{12}	5.4	5.0
K_{12}	0.25	0.75

[a]All the quantities are given in electron volts.

HOMO and LUMO of **8.7** and **8.8** are denoted by the subscripts 1 and 2, respectively. According to Koopmans' theorem, the IP and EA values of **8.7** and **8.8** are related to their HOMO and LUMO energies as

$$e_2 = -\text{EA}$$
$$e_1 = -\text{IP}$$
(8.59)

Despite the fact that the HOMO and LUMO energies of the two molecules are virtually identical, azulene is blue but anthracene is colorless (i.e., $E_S - E_G$ = 1.8 and 3.3 eV for **8.7** and **8.8**, respectively). To explore the cause of this difference we

8.7   **8.8**

note from equations 8.42 and 8.48 that the singlet excitation energy is given by

$$E_S - E_G = e_2 - e_1 - J_{12} + 2K_{12}$$
(8.60)

In addition, the magnitudes of J_{12} and K_{12} are estimated as follows:

$$J_{12} = e_2 - e_1 - (E_T - E_G) = -\text{EA} + \text{IP} - (E_T - E_G)$$
(8.61)

$$K_{12} = (E_S - E_T)/2 = (E_S - E_G)/2 - (E_T - E_G)/2$$
(8.62)

The J_{12} and K_{12} values derived in this way are listed in Table 8.1, which reveals that the J_{12} values of azulene and anthracene are nearly the same. Therefore, the difference in the $(E_S - E_G)$ values of the two molecules originates largely from the fact that the K_{12} value of azulene is substantially smaller than that of anthracene. Schematically depicted in **8.9** and **8.10** are the nodal properties of the HOMO and LUMO of the two molecules. They show that the large amplitude regions of the HOMO coincide with those of the LUMO in anthracene, while this is not the case

8.9

ψ_1 ψ_2

8.10

ψ_1 ψ_2

for azulene. Therefore, the overlap density $\psi_1\psi_2$ of azulene is small in most regions, compared with that of anthracene. This leads to an exchange repulsion K_{12} smaller for azulene than for anthracene.[6]

8.8. LOW VS. HIGH SPIN STATES

According to **8.4**, the triplet state Φ_T may become more stable than the singlet ground state Φ_G, if $(e_2 - e_1) - J_{12} < 0$, that is, when the orbital energy difference $(e_2 - e_1)$ is small compared with the Coulomb repulsion J_{12}. Using equations 8.39 and 8.40, this may be expressed in terms of the difference in the core potentials

$$E_T - E_G = (h_2 - h_1) - J_{11} + J_{12} - K_{12}$$

$$\simeq h_2 - h_1 - J_{11} \tag{8.63}$$

where we have approximated the energy difference by making use of the relationship $J_{12} \gg K_{12}$. This shows that the high spin state Φ_T becomes more stable than the low spin state Φ_G, when the core-potential difference $(h_2 - h_1)$ is smaller than the electron–electron repulsion J_{11} resulting from the orbital double occupancy (i.e., electron pairing) in ψ_1. When the MOs ψ_1 and ψ_2 are degenerate, $(h_2 - h_1)$ vanishes, so that the high spin is more stable than the low spin state. This is a special case of Hund's first rule, that out of a collection of atomic states, the one with the highest spin multiplicity lies lowest in energy. It means in the molecular case, for example, that the lowest energy (or ground) electronic state of the oxygen molecule with the configuration $(1\pi_g)^2$ of Figure 6.3 will be a triplet, paramagnetic molecule.

In Sections 7.1 and 7.2, the MOs of linear and bent AH_2 were developed. There are six valence electrons in CH_2 and so the π_u set is half-filled in a linear geometry. At a bent geometry $2a_1$ is the HOMO and b_1 is the LUMO for the ground singlet state. The energy gap between the $2a_1$ and b_1 levels is strongly dependent on the HCH angle, as can be seen from Figure 7.1. For reasons discussed in Section 7.3 the $2a_1$ orbital is stabilized as the HCH angle decreases and b_1 remains at constant energy. Therefore, in the triplet state for CH_2 where b_1 and $2a_1$ are singly occupied, some of the driving force for bending is lost. Based upon this orbital rationale, it is expected that the triplet state of CH_2 should be less bent than its singlet analog.

The result of an *ab initio* calculation on several electronic states of CH_2 is discussed in Section 8.10. The relative stability of various electronic states as a function of molecular geometry is a general problem approachable only by direct calculation. Given a pair of degenerate levels and two electrons at the level of discussion here, the triplet state is always more stable than the singlet. Whether there is some other lower energy structure where this degeneracy is removed, and a singlet state lies lowest in energy **(8.11)**, is difficult to probe qualitatively.

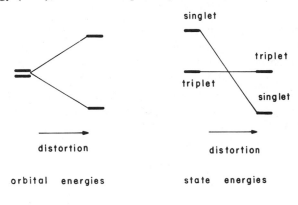

singlet

triplet

triplet

singlet

distortion

distortion

orbital   energies

state   energies

8.11

8.9.   ELECTRON–ELECTRON REPULSION AND CHARGED SPECIES

As already noted, an important part of the orbital energy arises from electron–electron repulsion. The effective potential for an electron in the MO ψ_1 of Φ_+ **(8.5)**, e_1^+, is given by

$$e_1^+ = h_1 = e_1 - J_{11} < e_1 \qquad (8.64)$$

where e_1 is the effective potential for an electron in the MO ψ_1 of Φ_G **(8.3)**. Thus, removal of an electron lowers the orbital energy, or equivalently, addition of an electron raises the orbital energy.

As an indication of the importance of charge on the energy levels of molecules, we look at the variation in carbonyl stretching vibrational frequencies in the series $Mn(CO)_6^+$, $Cr(CO)_6$, and $V(CO)_6^-$. All are low spin d^6 isoelectronic molecules. The orbital details of the attachment of the carbon monoxide ligand to a transition metal are reserved for Section 15.1, but an important part is played by the acceptor behavior of the carbonyl π_{CO}^* level **(8.12)**. The more important donation from metal d to carbonyl π_{CO}^*, the lower the carbonyl vibrational frequency. The mixing between these two sets of orbitals is intimately linked to their energy separation, Δe. If we imagine assembling one of these carbonyls from $M^{\pm}$ (or M) and six CO, then the carbonyl levels will remain fixed in energy for all members of this series, but the location of the metal levels will depend upon the charge. From what

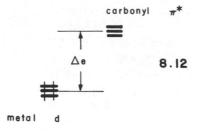

we have said above, their energy will decrease in the order $|H_{\mu\mu}(M^+)| > |H_{\mu\mu}(M)| > |H_{\mu\mu}(M^-)|$, leading to Δe values which decrease in the same order. Metal $d \rightarrow \pi^*_{CO}$ donation will then increase in the order $M^+ < M < M^-$ and the negative ion should have the lowest vibrational frequency. Experimentally $Mn(CO)_6^+$, $Cr(CO)_6$, and $V(CO)_6^-$ have vibrational frequencies (t_{1u}) of 2094, 1984, and 1843 cm^{-1}, respectively.[7] A similar effect is seen upon reduction of the organometallic molecule, $Co_2(\eta^5 - C_5Me_5)_2(\mu - CO)_2$.[8] One electron enters the b_2 orbital of **8.13**, a level

which is metal–metal antibonding but which contains no CO character. On reduction the Co–Co distance accordingly increases by 0.034 Å. However, this is accompanied by a decrease in the Co–CO bond length of 0.024 Å. Clearly, a simple one-electron orbital picture will not rationalize this result. What happens in fact is that the metal-located levels rise in energy on addition of the extra electron, and interaction with the π^*_{CO} levels increases as the energy separation Δe (of **8.12**) decreases. This leads to a contraction of the Co–CO distance and a decrease in the CO stretching vibrational frequencies of about 80 cm^{-1}.

Orbital energy also depends upon core potential. The HOMOs of H_2O and HO^- are oxygen $2p$ orbitals as shown in **8.14**. The HOMO of HO^- is raised with respect to that of H_2O, due to the loss of an atom that provides attractive potential. As a result, OH^- is more basic and nucleophilic.

8.10.   CONFIGURATION INTERACTION[9]

So far it has been implicitly assumed that each electronic state of a molecule is represented by a single configuration (i.e., Slater determinant). In general, an electronic state of a molecule is better described if a linear combination of many configurations is used as its state wavefunction. Physically this process provides the dynamic correlation of electrons, not allowed for in a single determinant wavefunction. Because electrons repel each other electrostatically, there will be a tendency for them to stay apart even though their spins are different. In other words, we need to correct for the assumption, employed so far, that each electron moves independently of all the electrons in the average field provided by them. As an example, consider two electron configurations Φ_μ and Φ_ν which are orthonormal, that is,

$$\langle \Phi_\mu | \Phi_\mu \rangle = \langle \Phi_\nu | \Phi_\nu \rangle = 1$$
$$\langle \Phi_\mu | \Phi_\nu \rangle = 0 \tag{8.65}$$

so that their expectation values are given by

$$E_\mu = \langle \Phi_\mu | \hat{H} | \Phi_\mu \rangle$$
$$E_\nu = \langle \Phi_\nu | \hat{H} | \Phi_\nu \rangle \tag{8.66}$$

Assuming that the interaction energy $\langle \Phi_\mu | \hat{H} | \Phi_\nu \rangle$ between Φ_μ and Φ_ν is nonzero, configuration interaction (CI) wavefunctions Ψ_i^{CI} ($i = 1, 2$) may be written as

$$\Psi_i^{CI} = c_{1i} \Phi_\mu + c_{2i} \Phi_\nu \tag{8.67}$$

We demand these CI wavefunctions to be normalized to unity (i.e., $\langle \Psi_i^{CI} | \Psi_i^{CI} \rangle = 1$) and also to be eigenfunctions of the molecular Hamiltonian $\hat{H}$:

$$\hat{H} \Psi_i^{CI} = E_i^{CI} \Psi_i^{CI} \tag{8.68}$$

Then the state energies E_i^{CI} ($i = 1, 2$) are determined by solving the secular equation

$$\begin{vmatrix} E_\mu - E_i^{CI} & \langle \Phi_\mu | \hat{H} | \Phi_\nu \rangle \\ \langle \Phi_\mu | \hat{H} | \Phi_\nu \rangle & E_\nu - E_i^{CI} \end{vmatrix} = 0 \tag{8.69}$$

in an exactly analogous way to the generation of MOs from atomic orbitals in Section 1.3.

As a practical example of CI wavefunctions, let us consider how to improve the singlet ground state Φ_G of 8.3. The obvious question is to determine what configurations can mix into Φ_G. First, we examine the interaction between Φ_G and Φ_1. Note that the MOs used in constructing Φ_1 are obtained from a closed-shell configuration Φ_G, and Φ_1 differs from Φ_G only in one MO. In such a case the interaction $\langle \Phi_G | \hat{H} | \Phi_1 \rangle$ vanishes, a result known as Brillouin's theorem.[10] This can be shown as follows:

$$\langle \Phi_G | \hat{H} | \Phi_1 \rangle = \langle \psi_1 | \hat{h} | \psi_2 \rangle + (\psi_1 \psi_1 | \psi_1 \psi_2)$$
$$= \langle \psi_1 | \hat{h} | \psi_2 \rangle + \langle \psi_1 | \hat{J}_1 | \psi_2 \rangle \tag{8.70}$$

Because of the relationship,

$$\langle \psi_1 | \hat{J}_1 | \psi_2 \rangle = \langle \psi_1 | \hat{K}_1 | \psi_2 \rangle = (\psi_1 \psi_1 | \psi_1 \psi_2) \tag{8.71}$$

equation 8.70 can be rewritten as

$$\langle \Phi_G | \hat{H} | \Phi_1 \rangle = \langle \psi_1 | \hat{h} + 2\hat{J}_1 - \hat{K}_1 | \psi_2 \rangle$$

$$= \langle \psi_1 | \hat{F} | \psi_2 \rangle = e_2 \langle \psi_1 | \psi_2 \rangle = 0 \tag{8.72}$$

Similarly, $\langle \Phi_G | \hat{H} | \Phi_2 \rangle = 0$. Thus the singlet excited configuration Φ_S (equation 8.46) cannot mix into Φ_G. The only configuration of **8.3** that can mix with Φ_G is the doubly excited configuration Φ_E, since

$$\langle \Phi_G | \hat{H} | \Phi_E \rangle = (\psi_1 \psi_2 | \psi_1 \psi_2) = K_{12} > 0 \tag{8.73}$$

Therefore, the CI wavefunction Ψ_i^{CI} ($i = 1, 2$) may be written as

$$\Psi_i^{CI} = c_{1i} \Phi_G + c_{2i} \Phi_E \tag{8.74}$$

and the state energies E_i^{CI} ($i = 1, 2$) of these CI wavefunctions Ψ_i^{CI} are obtained from

$$\begin{vmatrix} E_G - E_i^{CI} & K_{12} \\ K_{12} & E_E - E_i^{CI} \end{vmatrix} = 0 \tag{8.75}$$

Without loss of generality, it may be assumed that $E_1^{CI} < E_2^{CI}$. Then if the energy difference between Φ_G and Φ_E is substantially greater than the interaction energy K_{12} between them, we obtain the following results

$$\Psi_1^{CI} \simeq \Phi_G + \frac{K_{12}}{E_G - E_E} \Phi_E$$

$$\tag{8.76}$$

$$E_1^{CI} \simeq E_G + \frac{K_{12}^2}{E_G - E_E}$$

Thus E_1^{CI} is lower in energy than E_G, and a better description of the singlet ground state is given by Ψ_1^{CI}, the leading configuration of which is Φ_G. In Ψ_1^{CI}, Φ_E mixes into Φ_G with a negative mixing coefficient, since the interaction energy $\langle \Phi_G | \hat{H} | \Phi_E \rangle = K_{12}$ is positive.

For qualitative discussions of chemical problems, one of the most important uses of CI wavefunctions arises when we deal with potential energy surfaces for chemical reactions. Let us suppose that the MOs ψ_1 and ψ_2 in **8.3** are functions of a reaction coordinate q as shown in Figure 8.1a, and the symmetry properties of these MOs are different throughout the reaction coordinate. Thus Figure 8.1a is a typical example of a symmetry-forbidden thermal reaction. To make our example more concrete, ψ_1 and ψ_2 of Figure 8.1a might be considered as, for example, ψ_2 and ψ_3 of Figure 5.8 which describes one geometrical possibility for the H_2/D_2 exchange reaction. The energies of the states resulting from the configurations $(\psi_1)^2$ and $(\psi_2)^2$ (i.e., Φ_G and Φ_E, respectively) vary as shown by the dashed lines in Figure

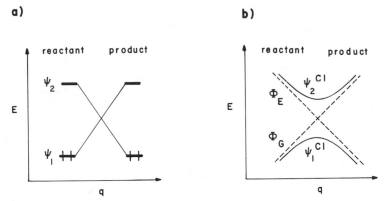

FIGURE 8.1.   Orbital and state energy correlation diagrams of a typical symmetry-forbidden thermal reaction, where q refers to an appropriate reaction coordinate: (a) orbital energy and (b) state energy.

8.1b. The energies of the CI wavefunctions Ψ_1^{CI} and Ψ_2^{CI}, which are obtained by solving equation 8.75, behave as shown by the solid lines. This is another example of the noncrossing rule discussed in Section 4.7. The symmetry-forbidden nature of the reaction is indicated by the presence of a barrier between reactant and product in the potential energy surface for the state Ψ_1^{CI}.

We note that near the reactant site where $E_G < E_E$, Ψ_1^{CI} is given by

$$\Psi_1^{CI} \simeq \Phi_G + \frac{K_{12}}{E_G - E_E}\ \Phi_E \tag{8.77}$$

At the transition state where $E_G = E_E$, Ψ_1^{CI} is expressed as

$$\Psi_1^{CI} = \frac{\Phi_G - \Phi_E}{\sqrt{2}} \tag{8.78}$$

Finally, near the product site where $E_E < E_G$, Ψ_1^{CI} is written as

$$\Psi_1^{CI} = \Phi_E + \frac{K_{12}}{E_E - E_G}\ \Phi_G \tag{8.79}$$

Consequently, the new wavefunction Ψ_1^{CI} provides a continuous transformation from a wavefunction predominantly Φ_G to one predominantly Φ_E in character as the reaction proceeds. Another place where configuration interaction is important is in looking at processes where bonds are made or broken, or just even stretched as in the transition states of many chemical reactions. **8.15** shows how the bonding and antibonding levels of H_2 become closer in energy as the internuclear distance is increased. **8.16** shows how the two electronic states $\Phi_G = (\psi_1)^2$ and $\Phi_E = (\psi_2)^2$ become closer in energy as a result and hence how configuration interaction becomes increasingly important.

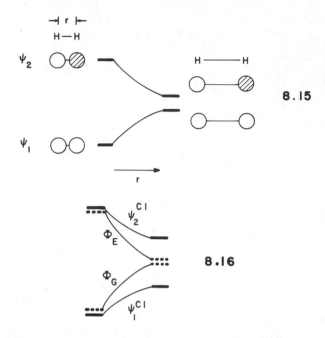

8.15

8.16

As a practical example of configuration interaction, let us examine the result of an *ab initio* calculation on carbene CH_2 summarized in Figure 8.2.[11] An orbital description of the configurations leading to those states of Figure 8.2 is given in **8.17** for bent CH_2 and in **8.18** for linear CH_2. For simplicity, other low-lying filled levels are not shown, and the Slater determinants resulting from the various electron configurations are denoted by parentheses. At any nonlinear geometry there are two states of $^1A_1$ symmetry. The upper state mixes into the lower one stabiliz-

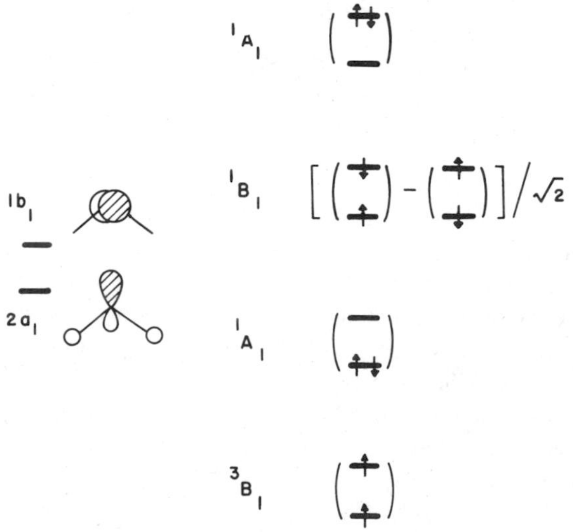

8.17

$$^1\Sigma_g^+ \quad \left[\left(\uparrow\downarrow \ -\right) + \left(- \ \uparrow\downarrow\right)\right]/\sqrt{2}$$

$$^1\Delta_g \quad \left[\left(\uparrow\downarrow \ -\right) - \left(- \ \uparrow\downarrow\right)\right]/\sqrt{2}$$

8.18

$$^1\Delta_g \quad \left[\left(\uparrow \ \downarrow\right) - \left(\downarrow \ \uparrow\right)\right]/\sqrt{2}$$

$$^3\Sigma_g^- \quad \left(\uparrow \ \uparrow\right)$$

ing the latter, and this in turn complicates the task of evaluating the singlet–triplet energy difference. For linear CH_2, each of the $^1\Sigma_g^+$ and $^1\Delta_g$ states is represented by a linear combination of two configurations of identical energy. It is not obvious from the orbital representations of 8.18 why the two components of the $^1\Delta_g$ state should be the same in energy. Let us represent the two atomic p orbitals of 8.18 by ϕ_a and ϕ_b. Then it can be easily shown that the electronic energy of the upper $^1\Delta_g$ is given by

$$E(^1\Delta_g) = 2h_a + J_{aa} - K_{ab} \tag{8.80}$$

while that of the lower $^1\Delta_g$ state is given by

$$E(^1\Delta_g) = 2h_a + J_{ab} + K_{ab} \tag{8.81}$$

In deriving equations 8.80 and 8.81, we employed the relationships that $h_b = h_a$,

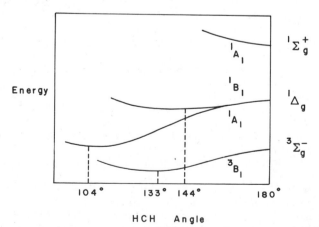

FIGURE 8.2.   Calculated state energies of carbene CH_2 as a function of the HCH valence angle.

and $J_{bb} = J_{aa}$. Consequently, the degeneracy of the two $^1\Delta_g$ states requires that

$$J_{aa} - K_{ab} = J_{ab} + K_{ab} \qquad (8.82)$$

This is the case,[12] although we will not show the proof.

Figure 8.2 shows that the $^1B_1$ state has an optimum HCH angle in the same region as the $^3B_1$ state, while the molecule in these two states is significantly less bent than that in the lower $^1A_1$ state. It is easy to understand this observation based upon the Walsh diagram for AH_2 shown in Figure 7.3. In both $^1B_1$ and $^3B_1$ states, the $2a_1$ and b_1 orbitals are singly occupied. On the other hand, the lower $^1A_1$ state is dominated by the configuration that has the $2a_1$ level doubly occupied but the b_1 level empty. As noted in Figure 7.3 the $2a_1$ level is lowered upon bending but the b_1 level is energetically invariant to bending. In general, those systems which have two closely spaced orbitals of different symmetry but only two electrons in them are referred to as diradicals.[13] Without elaborate calculations it is not possible to predict whether the ground electronic state of a diradical will be a singlet or triplet, but we can establish *trends* using the singlet–triplet energy differences from Walsh diagrams and a detailed knowledge of how orbitals are perturbed during geometrical distortion. In Section 7.4A we discussed why the HAH valence angle of H_2S is smaller than that of H_2O. As the HAH angle decreases, the stabilization of the $2a_1$ level is greater for H_2S than H_2O. Similarly, with respect to a linear geometry, the $^1A_1$ state is stabilized more for SiH_2 than CH_2. In fact, for SiH_2, the $^1A_1$ state is calculated to be approximately 4.6 kcal/mol more stable than the $^3B_1$ state.[14] For a quantitative or even semiquantitative understanding of diradical systems, configuration interaction is of vital importance. The H_2/D_2 exchange, rectangular vs. square planar cyclobutadiene, $Cr(CO)_5$, and $Fe(CO)_4$ dynamics are all diradical situations. What we have learned from CH_2 is applicable to all of them.

REFERENCES

1. C. C. J. Roothaan, *Rev. Mod. Phys.*, **23**, 69 (1951); A. Szabo and N. S. Ostlund, *Modern Quantum Chemistry*, Macmillan, New York (1982).
2. K. Ruedenberg, *J. Chem. Phys.*, **66**, 375 (1977).
3. I. G. Csizmadia, *Theory and Practice of MO Calculations on Organic Molecules*, Elsevier, New York (1976).
4. R. L. Liboff, *Introductory Quantum Mechanics*, Holden-Day, San Francisco (1980).
5. T. Koopmans, *Physica*, **1**, 104 (1933).
6. J. Michl and E. W. Thulstrup, *Tetrahedron*, **32**, 205 (1976).
7. D. M. Adams, *Metal Ligand and Related Vibrations*, St. Martins Press, New York (1968).
8. L. M. Cirjak, R. E. Ginsberg, and L. S. Dahl, *Inorg. Chem.*, **21**, 940 (1982).
9. I. Shavitt, in *Methods of Electronic Structure Theory*, H. F. Schaefer, editor, Plenum, New York (1977), Chapter 6.
10. L. Salem, *The Molecular Orbital Theory of Conjugated Systems*, Benjamin, New York (1966).
11. S. V. O'Neil, H. F. Schaefer, and C. F. Bender, *J. Chem. Phys.*, **55**, 162 (1971).
12. W. T. Borden, in *Diradicals*, W. T. Borden, editor, Wiley, New York (1982), Chapter 1.
13. L. Salem and C. Rowland, *Angew. Chem., Int. Ed. Engl.*, **11**, 92 (1972).
14. B. Wirsam, *Chem. Phys. Lett.*, **14**, 214 (1972).

CHAPTER NINE

Molecular Orbitals of Small Building Blocks

9.1. MO'S FROM BOND ORBITALS

In Section 7.5 we found it convenient to construct the MOs of AH_2 by symmetry adaptation of bond orbitals. This analysis can easily be extended to other systems. Some bond orbitals may already transform as an irreducible representation of the molecular point group. It is usually only degenerate bond orbitals that require symmetry adaptation in constructing MOs. In the following we will first derive the MOs of AH, pyramidal AH_3, and tetrahedral AH_4 based upon the appropriate orbital interaction diagrams, and then analyze those MOs in terms of bond orbitals. As in 7.4b, the hydrogen s orbital may be assumed to lie energetically in between the s and p orbitals of A. For convenience, the bond orbitals of AH, AH_3, and AH_4 may be constructed by assuming sp^3 hybridization on A.

A. AH

The interaction diagram for the orbitals of A and H is shown in Figure 9.1. The p_x and p_y orbitals of A do not overlap with the hydrogen s orbital, and so become the n_x and n_y nonbonding orbitals (9.1) of AH, respectively. According to the perturbation treatment of Chapter 3, σ_{AH}, n_σ, and σ_{AH}^* are derived as shown in 9.2. In this MO description, the second-order orbital mixing represented by brackets is essential for creating the orbital hybridization on A.

The bond orbitals of AH which result from combining the sp^3 hybrid orbitals of A with a hydrogen s orbital are shown in Figure 9.2. One sp^3 hybrid orbital interacts with hydrogen s to form the σ_{AH} and σ_{AH}^* levels, and the remaining three

131

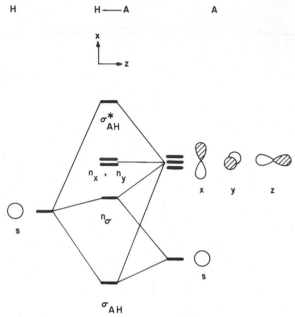

FIGURE 9.1.   An orbital interaction diagram for AH.

hybrid orbitals become nonbonding orbitals on A. Note that the bond orbitals σ_{AH} and σ_{AH}^* are already symmetry adapted (i.e., transform as one of the irreducible representations of the $C_{\infty v}$ point group), and thus are similar in character to the corresponding MOs. The three sp^3 nonbonding orbitals of A are not symmetry adapted in this sense, but a linear combination of them may be constructed as described in Section 5.2 to produce a set of symmetry-adapted MOs. Let us recall the vector decomposition of a hybrid orbital as exemplified in **9.3**, and, for convenience of a

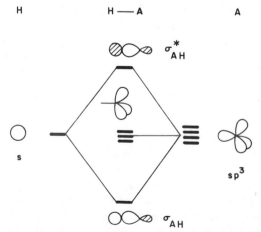

FIGURE 9.2.   An orbital interaction diagram for AH in the bond orbital approach.

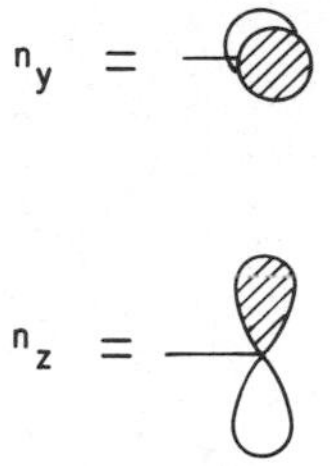

$$n_y =$$

9.1

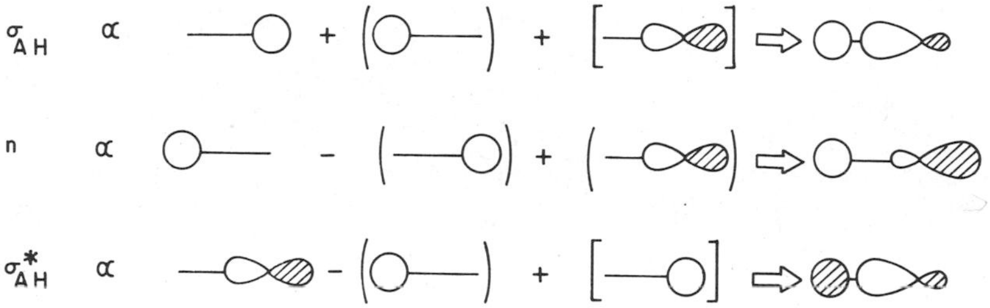

$$n_z =$$

$$\sigma_{AH} \quad \alpha \qquad$$

$$n \quad \alpha \qquad \qquad \textbf{9.2}$$

$$\sigma_{AH}^{*} \quad \alpha \qquad$$

9.3

$$=$$

$$=$$

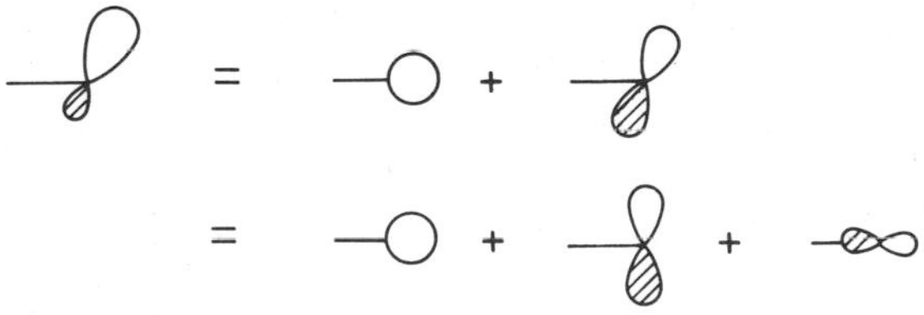

9.4 $$=$$

graphical presentation, let us represent a hybrid orbital by showing only the large front lobe (**9.4**). Then the MOs n_x, n_y, and n_σ may be approximated by the three sp^3 nonbonding orbitals as shown in **9.5**. Consequently, the bond orbitals and MOs of AH are correlated as indicated in **9.6**. Notice the splitting of the three sp^3 hybrid orbitals into a degenerate pair and a single nondegenerate orbital, a result demanded by group theory.

B. PYRAMIDAL AH₃

The interaction between the orbitals of H_3 and Λ shown in Figure 9.3 gives rise to the MOs of pyramidal AH_3 shown in **9.7**. The $1e_x$ and $2e_x$ orbitals are the in-phase and out-of-phase combinations of e_x and p_x, respectively. Likewise, $1e_y$ and $2e_y$ are the analogous in-phase and out-of-phase combinations of e_y and p_y, respectively. The $1a_1$, $2a_1$, and $3a_1$ orbitals are slightly more complicated and can be derived from Figure 9.3 as shown in **9.8**. Notice that an ideal pattern exists for the three a_1 MOs we saw for AH_2 (**7.5**), AH (**9.2**), and AH_3 (**9.8**). The lowest and highest MOs

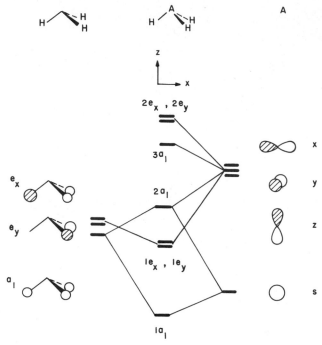

FIGURE 9.3.   An orbital interaction diagram for pyramidal AH_3.

of the three are always hybridized toward the hydrogen(s). The middle, nonbonding MO is concentrated on A and hybridized away from the hydrogen(s). This is a characteristic pattern that we will see again for the hypervalent molecules in Chapter 14 and the transition metal building blocks in Chapters 17–19. Whenever two different atomic orbitals on one center combine with an ensemble of hydrogen s or ligand σ donor orbitals, a fully bonding combination at low energy and a fully antibonding combination at high energy is produced. A nonbonding orbital at moderate energy is left behind which is hybridized away from the surrounding hydrogens (or ligands).

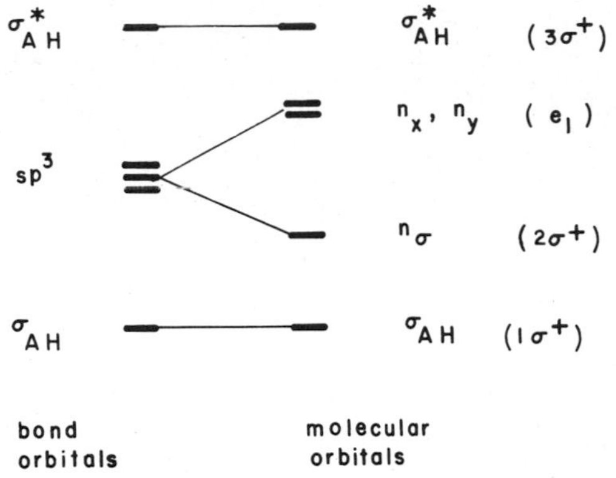

9.6

σ^*_{AH} σ^*_{AH} $(3\sigma^+)$

sp^3 n_x, n_y (e_1)

n_σ $(2\sigma^+)$

σ_{AH} σ_{AH} $(1\sigma^+)$

bond orbitals   molecular orbitals

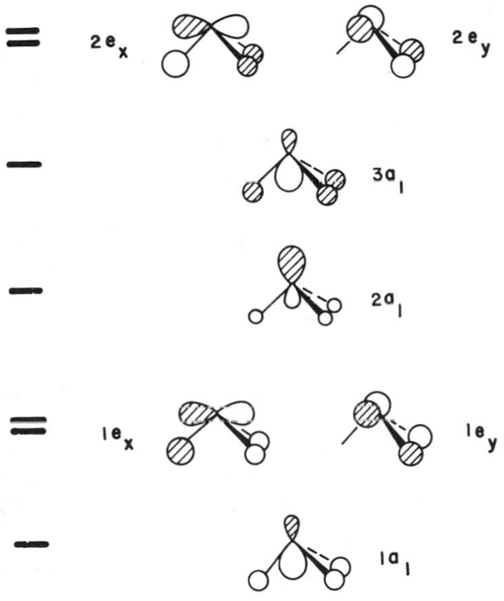

$2e_x$ $2e_y$

$3a_1$

$2a_1$ 9.7

$1e_x$ $1e_y$

$1a_1$

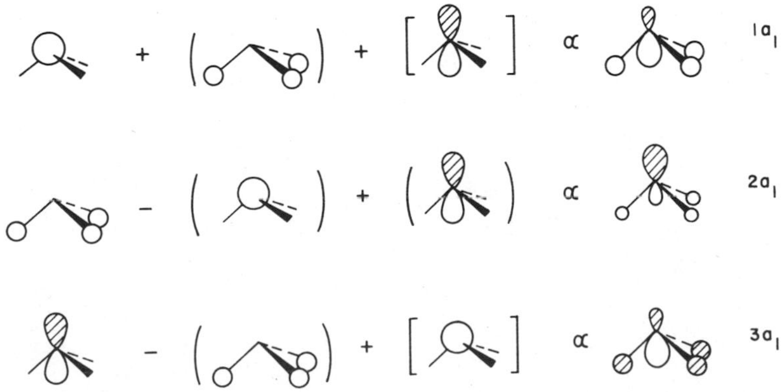

α $1a_1$

9.8 α $2a_1$

α $3a_1$

135

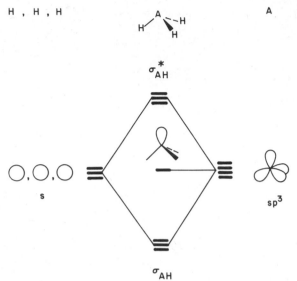

σ_{AH}

FIGURE 9.4.   An orbital interaction diagram for pyramidal AH_3 in the bond orbital approach.

Formation of the bond orbitals of pyramidal AH_3 is shown in Figure 9.4. The bond orbitals σ_{AH} and σ_{AH}^* are not symmetry adapted since triply degenerate species are not allowed in the C_{3v} point group. The main features of the MOs, $1a_1$, $1e_x$, and $1e_y$, are well approximated by suitable linear combinations of the three σ_{AH} orbitals as in **9.9**. Similarly, the MOs $3a_1$, $2e_x$, and $2e_y$ are approximated in terms of the three σ_{AH}^* orbitals as in **9.10**. The sp^3 nonbonding orbital is already symmetry adapted and similar in description to the $2a_1$ MO except for the absence of hydrogen s character. We could build this in by allowing it to interact with $1a_1$ and $3a_1$ (**9.9** and **9.10**, respectively) in a fashion that is similar to the way the lone pair and σ orbitals interacted after symmetry adaptation in the N_2 molecule (Section 7.6). The energy of the nonbonding orbital does not change much by this extra orbital mixing and remains concentrated on A. The bond and molecular orbitals are correlated as shown in **9.11**.

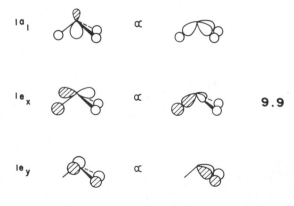

9.9

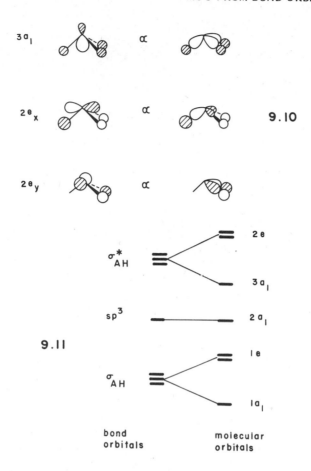

C.   TETRAHEDRAL AH₄

The orbital interaction diagram between H_4 and A is shown in Figure 9.5. The triply degenerate MOs of tetrahedral H_4 are an alternative representation of those shown in Figure 4.5 (see also 5.9). The orbital interaction diagram of Figure 9.5 leads to the MOs of tetrahedral AH₄ shown in 9.12. The a_1, t_{2x}, t_{2y}, and t_{2z} orbitals of H_4 combine in-phase (out-of-phase) with the s, p_x, p_y, and p_z orbitals of A to form the MOs $1a_1$, $1t_{2x}$, $1t_{2y}$, and $1t_{2z}$ ($2a_1$, $2t_{2x}$, $2t_{2y}$, and $2t_{2z}$), respectively. As for the bond orbitals, 9.13 shows that the four sp^3 hybrid orbitals of A interact with the hydrogen s orbitals to form four σ_{AH} and four σ_{AH}^* orbitals. Symmetry adaptation of the four σ_{AH} bond orbitals directly leads to the $1a_1$, $1t_{2x}$, $1t_{2y}$, and $1t_{2z}$ MOs as indicated in 9.14. Similarly, the $2a_1$, $2t_{2x}$, $2t_{2y}$, and $2t_{2z}$ MOs can be well approximated by suitable linear combinations of the four σ_{AH}^* orbitals. Thus the bond orbitals and MOs of tetrahedral AH₄ are correlated as shown in 9.15.

With the MOs of tetrahedral H_4 chosen as in Figure 4.5, the MOs of AH₄ are given by 9.16. We note that the triply degenerate MOs $1t_2$ or $2t_2$ of 9.16 can be

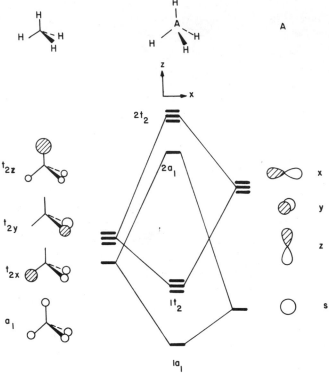

FIGURE 9.5. An orbital interaction diagram for tetrahedral AH_4.

9.12

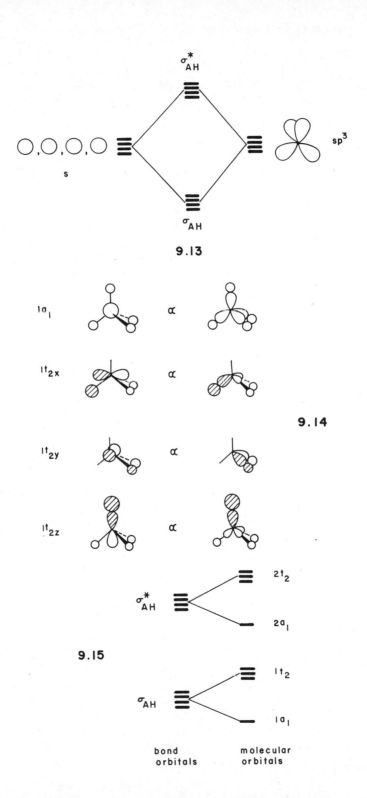

9.13

9.14

9.15

bond
orbitals

molecular
orbitals

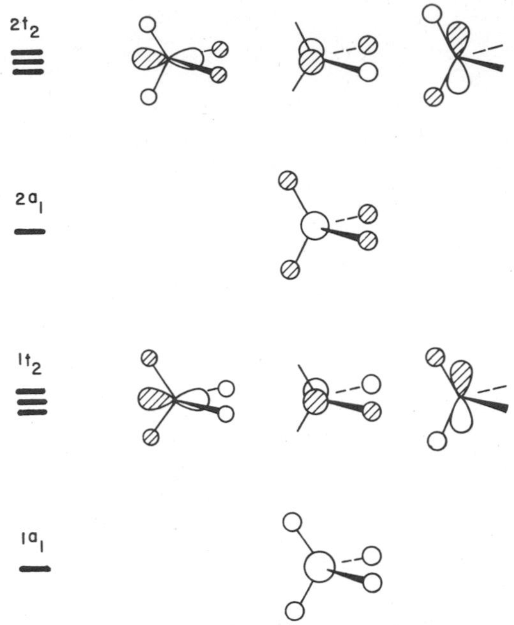

$2t_2$

$2a_1$

9.16

$1t_2$

$1a_1$

expressed as linear combinations of the corresponding MOs of **9.12**. Such linear combinations among a set of degenerate MOs do not lift the degeneracy, as already pointed out in Section 7.5. There are many connections between the molecular orbitals of the AH_n series. For example, in tetrahedral AH_4 the four atomic orbitals of A find a one-to-one match with the four orbitals of tetrahedral H_4. Consequently four bonding and four antibonding MOs are produced. One hydrogen atom can be removed from AH_4 to produce pyramidal AH_3. Thus a nonbonding orbital localized on A is produced. The three bonding and three antibonding MOs of AH_3 find their counterparts that have an essentially identical composition in AH_4. Likewise, in AH_2 two-hydrogen atoms have been removed which creates two A-centered nonbonding orbitals. In AH there are three A-centered nonbonding orbitals.

9.2.   SHAPES OF AH_3 SYSTEMS[1-4]

The MOs of trigonal planar AH_3 were already constructed by reference to the orbitals of H_3 and A in Figure 4.4. During the course of the trigonal planar to pyramidal (i.e., $D_{3h} \rightarrow C_{3v}$) distortion, the MO levels of AH_3 vary as shown in Figure 9.6. Here we have correlated the levels already derived for planar and pyramidal AH_3 units. Several trends in Figure 9.6 are immediately obvious. The $1a_1'$ level is stabilized slightly upon pyramidalization, since overlap between the hydrogen s orbitals increases. In the $1e'$ and $2e'$ levels, the magnitude of the overlap integral between the hydrogen s and p orbitals of A decreases upon pyramidalization. Recall that a similar situation occurred for the b_2 orbitals in AH_2 on bending (see Section 7.3). Thus the bonding $1e'$ set rises but the antibonding $2e'$ set falls in energy. From

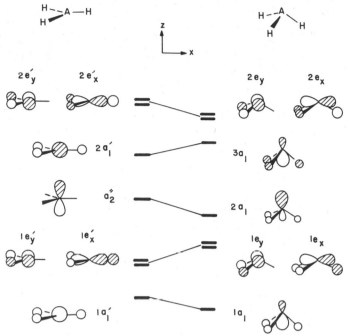

FIGURE 9.6.   An orbital interaction diagram for planar AH$_3$.

first-order energy considerations, we expect the $2a_1'$ level to go down in energy and a_2'' to remain unchanged in energy. This clearly does not occur. Upon pyramidalization, both $2a_1'$ and a_2'' become orbitals of a_1 symmetry so that the two may mix together. As a result $2a_1'$ is destabilized on pyramidalization and a_2'' is stabilized. This orbital mixing will be examined in more detail below. We will now study the shapes and properties of AH$_3$ systems using the Walsh diagram of Figure 9.6.

A.   SIX-ELECTRON SYSTEMS

Figure 9.6 shows that a six-electron AH$_3$ system such as CH$_3^+$ and BH$_3$ should be planar. The HOMO, $1e'$, of such a species is destabilized upon pyramidalization. The LUMO is the p orbital of A which lies perpendicular to the molecular plane. Such an empty orbital, concentrated on one atomic center, is a good electron acceptor and is responsible for the Lewis acid character in boranes and carbonium ions. Let us consider a nucleophilic addition to a carbonium ion CH$_3^+$ by representing a nucleophile Nu: by a filled nonbonding hybrid orbital n_σ. Figure 9.7 shows how n_σ is stabilized, yielding the C–Nu bonding orbital σ. The resultant σ level is derived as in **9.17**. If the hydrogen atoms distort away from the incoming nucleophile, antibonding between s components in the a_1' orbitals and n_σ is decreased. Furthermore, bonding between s components in the a_1' orbitals and a_2'' is increased. So as shown in **9.18**, the σ orbital is further stabilized if the carbon configuration becomes tetrahedral. In a sense, the orbital interaction shown in Figure 9.7 serves to transfer

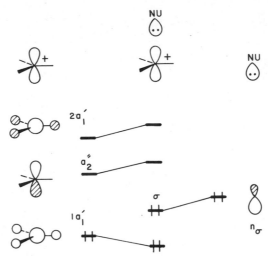

FIGURE 9.7. An orbital interaction diagram for the addition of a nucleophile to CH_3^+.

electron density from n_σ to a_2'', which induces the geometrical response (see Figure 9.6) of pyramidalization of the CH_3 moiety.

It is of interest to consider the formation of diborane $(BH_3)_2$, **9.19**, in terms of the HOMO and LUMO of two BH_3 fragments. The most stable arrangement of two BH_3 units in the incipient stage of dimerization is given by **9.20**, which maximizes the extent of the two HOMO–LUMO interactions **9.21**. Note again that on dimeriza-

HOMO — LUMO   LUMO — HOMO

9.21

tion a pyramidalization occurs at each BH_3 unit. This takes place via a mechanism exactly analogous to the one considered in Figure 9.7. Here electron transfer takes place from one component of the $1e'$ orbital (**9.21**).

B.   EIGHT-ELECTRON SYSTEMS[1]

Figure 9.6 shows that an eight-electron AH_3 system such as CH_3^- or NH_3 should be pyramidal since the a_2'' HOMO is markedly stabilized upon pyramidalization. The resultant HOMO in the new geometry $(2a_1)$ resembles an sp^3 nonbonding orbital. Figure 9.8 shows the potential energy diagram for pyramidal inversion in an eight-electron AH_3 system. Upon decreasing the electronegativity of A, the pyramidal inversion barrier E_{inv} increases and the HAH valence angle decreases. For instance, calculations show that the inversion barriers are 5.8, 30-36, and 46 kcal/mol while the HAH angles are 106.7, 93.3, and 92.1° for NH_3, PH_3, and AsH_3, respectively.[5-7] These observations are analogous to the corresponding one in eight electron AH_2 systems, and may be rationalized in terms of **9.22**. The geometry perturbation argument of Chapter 7 shows that the HOMO $2a_1$ of pyramidal AH_3 is given by **9.23**. This orbital mixing stabilizes the $2a_1$ level with respect to a_2''. The magnitude of the stabilization is inversely proportional to the energy gap, Δe, between the a_2'' and $2a_1'$ levels. Thus,

$$E_{inv} \propto e(a_2'') - e(2a_1) \propto \frac{1}{\Delta e} \tag{9.1}$$

For exactly the same reasons discussed in Section 7.4A, the a_2'' and $2a_1'$ levels of

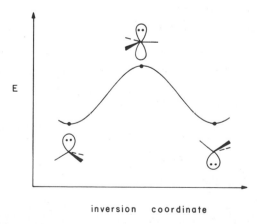

E

inversion   coordinate

FIGURE 9.8.   A potential energy diagram for pyramidal inversion in an eight-electron AH_3.

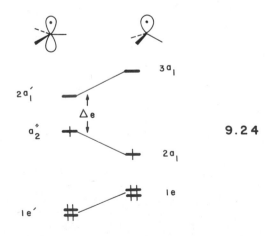

planar AH_3 are raised and lowered, respectively, as A becomes less electronegative. Therefore a smaller energy gap Δe, a larger inversion barrier, and a smaller HAH angle are found in eight-electron AH_3 systems when A becomes less electronegative.

The above discussion also provides a rationale for the computational results that seven-electron AH_3 radicals with a first-row atom A (e.g., CH_3 and NH_3^+) are planar,[8] but those with a second-row atom (e.g., SiH_3 and PH_3^+) are pyramidal.[9] The orbital occupancy of a seven-electron AH_3 system is shown in **9.24**. The a_2'' level, singly occupied, is of lower energy in the pyramidal geometry. However the subjacent level, doubly degenerate and completely filled, is of lower energy at the planar geometry. A pyramidal structure is expected only if the energy lowering associated with a_2'' is substantial. This occurs for AH_3 systems containing a second-row atom A with a smaller electronegativity, and hence a smaller Δe value than their first-row counterparts. Overlap is probably just as important as the energy gap in influencing problems such as these.

9.3. π-BONDING EFFECTS OF LIGANDS

Let us consider AL_n systems in which L refers to a ligand (i.e., an atom or a group) that contains more than just a single s orbital and $n = 1, 2, 3, \ldots$. In most cases L contributes one electron in a sigma fashion to A as shown in **9.25** for planar AL_3. The hybrid orbital used by L may be the sp^3 hybrid of an alkyl group in **9.26** or the sp hybrid of a halogen atom in **9.27**. The local symmetry of the hybrid orbitals in **9.26** or **9.27** is the same as that for a hydrogen s orbital. Thus the way the atomic orbitals of A overlap with the ligand orbitals is the same whether the ligand carries a single s orbital or a directed orbital of this type. As a result, the overall shape and symmetry of the valence, A-centered orbitals of AL_n are the same as those of AH_n. Therefore, the MOs of AH_n developed so far can be used to discuss the structures of AL_n systems. For example, the orbital of $N(CH_3)_3$ corresponding to the $2a_1$ orbital of pyramidal AH_3 is that shown in **9.28**.

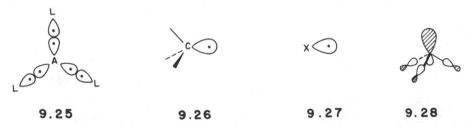

| **9.25** | **9.26** | **9.27** | **9.28** |

A very electronegative atom substituted at A causes all orbitals to decrease in energy (see Section 6.4), especially those that have appreciable orbital character associated with the electronegative substituent. There is a direct relationship between the energy lowering and the coefficients on the atoms that are made more electronegative. Consider the level ordering of planar AH_3 in Figure 9.6. Replacement of the hydrogens by fluorine atoms would cause all of the MO levels to shift downward except for a_2''. The effect of π bonding between the central atom and the ligands will be to raise a_2'' in energy. Consequently, the energy gap Δe in **9.22** will be smaller for NF_3 than for NH_3. This suggests an explanation for the computational result that the NF_3 inversion barrier is larger (78.5 kcal/mol) and the FNF valence angle is smaller (102.4°) than the corresponding value for NH_3 (106.7°).[10] The NF_3 inversion barrier of 78.5 kcal/mol is larger than the first bond dissociation energy of 57 kcal/mol so that bond breaking is energetically favored over inversion.

The external ligand L may be an atom or a group that possesses π-bonding capabilities. In such a case it is convenient to separate the effects due to σ and π bonding. To do this, we first develop the valence orbitals around the central atom A, neglecting the π-bonding capabilities of L, and then introduce the π-bonding effects of L as a perturbation. As an example, consider $H_2\ddot{A}$-L in which L stands for a group with a π-acceptor orbital ϕ_a (e.g., an empty p orbital of BH_2, that of an electropositive metal such as Li, or a π^*_{CO} orbital of a carbonyl group), which can engage in π interaction with the nonbonding orbital (analogous to a_2'') of A. The π interactions $(n_\pi - \phi_a)$ and $(n_\sigma - \phi_a)$ that occur in planar and pyramidal $H_2\ddot{A}$-L are shown in **9.29** and **9.30**, respectively. The stabilization of the nonbonding orbital on A is governed by the overlap between the n and ϕ_a orbitals and by the energy gap Δe between them. As can be seen from **9.31**, the overlap $\langle n_\pi | \phi_a \rangle$ in **9.29** is greater

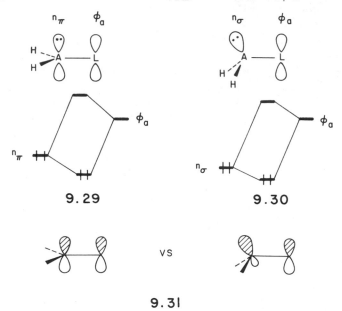

9.31

than the overlap $\langle n_\sigma | \phi_a \rangle$ in **9.30**. Since the n_π level of AH_2 is higher in energy than the n_σ level, the energy gap Δe is smaller in **9.29**. Therefore, in terms of both overlap and energy gap considerations, the stabilization of n_π in **9.29** is greater than that of n_σ in **9.30**. Such a π interaction can stabilize the planar structure over the pyramidal one if it is larger than the competing σ-effect, described above, responsible for pyramidal eight-electron AH_3 molecules. As shown for the ground state geometries in **9.32** and **9.33**. This is the case for $H_2\ddot{A}$-L with a first-row A atom but not for $H_2\ddot{A}$-L when a second-row A atom is present. For the case of the heavier A atom, although the energy gap Δe between n_π and ϕ_a is small, the $\langle n_\pi | \phi_a \rangle$ overlap is small too because of the longer A-L bond and more diffuse valence orbitals. In addition, recall that the inversion barrier of $\ddot{A}H_3$ is larger for a second-row compared to first-row atom. As a result the $(n_\pi - \phi_a)$ interaction in **9.29** is not large enough to make $H_2\ddot{A}$-L planar when A comes from the second row.[11,12] It should be noted that the nitrogen center of **9.32** becomes pyramidal when the π interaction is cut off by rotation around the N-L bond as depicted in **9.34**. Thus, rotation about the C-N bond in amides is intimately coupled with inversion at the amide nitrogen.[13] Likewise, rotation about the C-C bond in the enolate ion :$\bar{C}H_2CH{=}O$

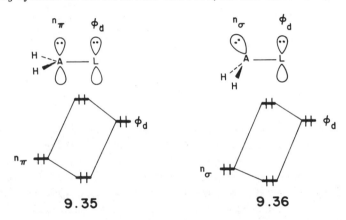

9.32   **9.33**   **9.34**

and that about the C–N bond in :$\overline{C}H_2NO_2$ are calculated to be complicated by inversion at the carbanion center.[14] A similar situation occurs for the species $N(SiH_3)_3$ which has a planar NSi_3 skeleton, and $P(SiH_3)_3$ which has a pyramidal PSi_3 skeleton.[11] Here the acceptor orbital ϕ_a is one component of the $2e$ orbitals of SiH_3 (see Figure 9.6 and Section 10.5).

Consider now the species $H_2\ddot{A}$–L, where L is a group with π-donor orbital ϕ_d (e.g., halogen atoms, NH_2 group, etc.). The result is the opposite to that described above for the acceptor case. Here the effects of π and σ bonding work in the same direction, and the barrier to inversion around the atom A is greater than that in the parent AH_3 system. As shown in **9.35** and **9.36**, the interaction between the non-

9.35   **9.36**

bonding orbital of A and the donor orbital of L is a destabilizing two-orbital–four-electron one. The destabilization is greater at the planar structure **9.35** since the overlap integral $\langle n_\pi | \phi_d \rangle$ is greater than $\langle n_\sigma | \phi_d \rangle$. Therefore the π interaction between the A and L centers increases the inversion barrier. The effect of π acceptors and donors on bending in AH_2 can be understood in an identical fashion. This has a dramatic impact on not only the reactivity of carbenes but also the singlet–triplet energy difference (Section 8.10).[15] Notice in the case of the singlet configuration that there is an acceptor orbital (b_2) and a donor orbital $(2a_1)$ on the central atom. There is an interesting result, analogous to the observation concerning the planar skeleton of $N(SiH_3)_3$, in silicon–oxygen chemistry.[16] The average value of the Si–O–Si angle in silicates (~149°) lies between the linear and tetrahedral extremes. There is also evidence from the variety of angles known in such species that the bending around the oxygen is rather soft. Both of these observations may be interpreted in terms of Si–O π bonding, which best stabilizes the linear geometry while σ effects best stabilize a considerably nonlinear geometry.

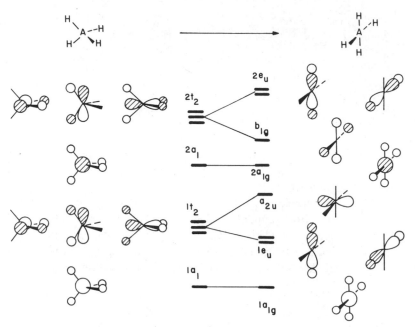

FIGURE 9.9.   Correlation between the MO levels of tetrahedral and planar AH_4.

9.4.   SHAPES OF AH_4 SYSTEMS

Figure 9.9 shows how the orbitals of tetrahedral and planar AH_4 molecules are correlated. During the $T_d \to D_{4h}$ distortion, two members of the $1t_2$ set are stabilized since overlap between the p orbital on A and the two hydrogen s orbitals is increased. But the third member is destabilized since all bonding between the p orbital of A and the hydrogen s orbitals is lost. This leads to the $1e_u$ and a_{2u} levels. For identical reasons, two members of the antibonding $2t_2$ set are destabilized and the third member is lowered in energy, yielding the $2e_u$ and b_{1g} levels. For the moment, we show the a_{2u} level to lie below b_{1g} in Figure 9.9, but we note that the relative ordering of the two orbitals depends upon the electronegativity difference between A and H (or L) as will be discussed later.

In eight-electron AH_4 systems such as CH_4 and NH_4^+, the HOMO of a square planar structure is the a_{2u} level. According to Figure 9.9, an eight-electron AH_4 molecule prefers to be tetrahedral. The tetrahedral configuration of AH_4 may in principle be inverted as shown in **9.37**. However, the barrier for this inversion is

$$\text{\Large $\times$} - \text{\Large $\times$} - \text{\Large $\times$} \qquad \textbf{9.37}$$

exceedingly large compared with that for the pyramidal inversion in eight-electron AH_3 systems. For example, the barrier for the configuration inversion in CH_4 is estimated to require about 160 kcal/mol,[17] in sharp contrast to the pyramidal inversion barrier of about 6 kcal/mol in NH_3.[5] This observation may be understood by reference to **9.38** and **9.39**. What is largely responsible for the inversion barrier in AH_3 is the conversion of a nonbonding electron pair in a hybrid orbital $(2a_1)$ into a pair in a pure p orbital. In contrast, the inversion of AH_4 requires the conversion of a bonding electron pair $(1t_2)$ into a nonbonding electron pair.[1]

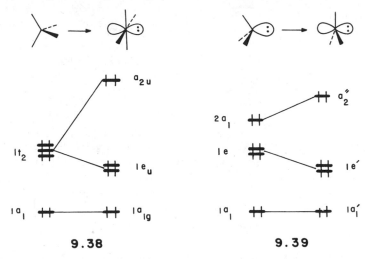

9.38   **9.39**

A square planar carbon configuration can be achieved only if the HOMO a_{2u} is substantially stabilized by good π-acceptor ligands. For example, trilithiomethane is calculated to adopt a planar structure **9.40a** instead of a tetrahedral structure **9.40b**.[18] As indicated in **9.40c**, the low-lying empty p orbitals of Li stabilize the two electrons in the a_{2u} orbital. Similarly, the eight-electron Li_2O is not bent like H_2O but linear in structure. However Cs_2O is bent although it is isoelectronic with Li_2O,[19] perhaps because the magnitude of π overlap associated with the p orbital of a heavier element is weaker. Recall from Section 9.3 that the nitrogen center of $H_2NCH{=}O$ is planar while the phosphorus center of $H_2PCH{=}O$ is pyramidal.

As the electronegativity of A in an AH_4 molecule is decreased, the p orbital of A is raised in energy and the A–H distance tends to increase. Furthermore, the $1t_2$ set becomes more concentrated on the hydrogens, and $2t_2$ more concentrated on A. Consequently, in a square planar AH_4 molecule with an electropositive atom, b_{1g} may become lower in energy than a_{2u} as shown in **9.41**. At an intermediate geometry the symmetry of the molecule is D_{2d} and both orbitals are of b_2 symmetry, so that an avoided crossing occurs. In fact, the HOMO of a square planar BH_4^-, SiH_4, or PH_4^+ molecule is calculated to be the b_{1g} level.[17] One important consequence of

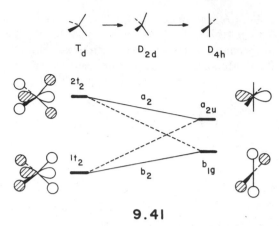

$$T_d \quad \longrightarrow \quad D_{2d} \quad \longrightarrow \quad D_{4h}$$

9.41

such a level ordering is that AL_4 systems with electronegative ligands such as halogens or alkoxides should have a lower barrier for configuration inversion at A than CH_4. This comes about because the HOMO b_{1g} carries electron density only on the ligand atoms, and will be stabilized by electronegative atoms. In addition, electronegative atoms carry nonbonding electron pairs. A two-orbital–two-electron stabilization results via interaction with the central atom p orbital (a_{2u}) of the planar structure as depicted in **9.42**.

It should be noticed that there is a geometric way to stabilize a square planar structure when the a_{2u} level is the HOMO. We saw in Section 9.2B that pyramidalization stabilizes an eight-electron AH_3 system: The HOMO, a_2'', is lowered in energy upon pyramidalization. The same thing occurs for square-planar CH_4. Distortion to a pyramidal, C_{4v} structure stabilizes a_{2u}. The $2a_{1g}$ level (see Figure 9.9) mixes into a_{2u}, yielding **9.43**. Computationally, the C_{4v} structure with *cis*- HCH angle of 72.5°

9.42   **9.43**

is about 23 kcal/mol more stable than the D_{4h}, square planar one.[17] One must be a little careful here. Obviously, good π-acceptor substituents around A will cause the structure to remain planar, just as in AL_3. For example, square planar oxygen is known in the solid-state structure of NbO and TiO.[20] Here the square planar geometry is stabilized by the presence of π-acceptor Nb (or Ti) d orbitals. Likewise, molecules in which a_{2u} is empty are expected to be more stable at the square planar rather than tetrahedral or C_{4v} pyramidal geometry. Examples include BH_4^+, CH_4^{2+}, or the doubly excited state of CH_4 in which a_{2u} is empty and b_{1g} is filled.[21] We will return to this problem of tetrahedral–square planar–C_{4v} pyramidal structure interconversion in Chapter 14, where AL_4 molecules with more than eight electrons are considered.

REFERENCES

1. B. M. Gimarc, *Molecular Structure and Bonding*, Academic Press, New York (1979).
2. A. D. Walsh, *J. Chem. Soc.*, 2260, 2266, 2288, 2296, 2301, 2306, 2318, 2321, 2325, 2330 (1953).
3. C. C. Levin, *J. Am. Chem. Soc.*, **97**, 5649 (1975).
4. B. M. Gimarc, *J. Am. Chem. Soc.*, **95**, 1417 (1973).
5. A. Rauk, L. C. Allen, and E. Clementi, *J. Chem. Phys.*, **52**, 4133 (1970).
6. R. Ahlrichs, F. Keil, H. Lischka, W. Kutzelnigg, and V. Staemmler, *J. Chem. Phys.*, **63**, 455 (1970).
7. D. A. Dixon and D. S. Marynick, *J. Am. Chem. Soc.*, **99**, 6101 (1977).
8. T. A. Claxton and N. A. Smith, *J. Chem. Phys.*, **52**, 4317 (1970).
9. L. J. Aarons, M. F. Guest, M. B. Hall, and I. H. Hillier, *J. Chem. Soc. Dalton II*, **69**, 643 (1973); L. J. Aarons, I. H. Hillier, and M. F. Guest, *J. Chem. Soc. Faraday II*, **70**, 167 (1974).
10. D. S. Marynick, *J. Chem. Phys.*, **73**, 3939 (1980).
11. K. Mislow, *Trans. N.Y. Acad. Sci.*, **35**, 227 (1973).
12. D. E. Dougherty, K. Mislow, and M.-H. Whangbo, *Tetrahedron Lett.* 2321 (1979); J. R. Damewood, Jr. and K. Mislow, *Monatshefte für Chemie*, **111**, 213 (1980).
13. V. J. Klimkowski, H. L. Sellers, and L. Schäfer, *J. Mol. Struct.*, **54**, 299 (1979).
14. S. Wolfe, H. B. Schlegel, I. G. Csizmadia, and F. Bernardi, *Can. J. Chem.*, **53**, 3365 (1975); P. G. Mezey, A. Kresge, and I. G. Csizmadia, ibid., **54**, 2526 (1976).
15. R. A. Moss, W. Guo, D. Z. Denny, K. N. Houk, and N. G. Rondan, *J. Am. Chem. Soc.*, **103**, 6164 (1981); P. H. Mueller, N. G. Rondan, K. N. Houk, J. F. Harrison, D. Hooper, B. H. Willen, and J. F. Liebman, ibid; **103**, 5049 (1981); N. G. Rondan, K. N. Houk, and R. A. Moss, ibid., **102**, 1770 (1980).
16. G. V. Gibbs, E. P. Meagher, M. B. Newton, and D. K. Swanson, *Structure and Bonding in Crystals*, Vol. 1, M. O'Keeffe and A. Navrotsky, editors, Academic Press, New York (1981), p. 195.
17. M.-B. Krogh-Jespersen, J. Chandrasekhar, E.-U. Würthwein, J. B. Collins, and P.v.R. Schleyer, *J. Am. Chem. Soc.*, **102**, 2263 (1980).
18. T. Clark and P.v.R. Schleyer, *J. Am. Chem. Soc.*, **101**, 7747 (1979).
19. A. F. Wells, *Structural Inoranic Chemistry*, 4th edition, Clarendon, Oxford (1975).
20. J. K. Burdett and T. Hughbanks, *J. Am. Chem. Soc.* **106**, 3101 (1984).
21. J. B. Collins, P.v.R. Schleyer, J. S. Binkley, J. A. Pople, and L. Radom, *J. Am. Chem. Soc.*, **98**, 3436 (1976).

CHAPTER TEN

Molecules with Two Heavy Atoms

10.1. INTRODUCTION

In the previous chapters, the MOs of AH_n systems were analyzed in some detail. These orbitals are convenient building blocks in constructing the MOs of large molecules such as A_2H_6, A_2H_4, and so on, that may be envisaged as being made up of AH_n fragments. Extension to the series A_2L_{2n} is also straightforward and covers a good bit of organic and main group chemistry. For instance, the MOs of C_2H_6 can be easily constructed from the MOs of two CH_3 fragments. It is important to realize that the use of such molecular fragments implies nothing concerning their existence as stable chemical species. They are just conceptual building blocks convenient for analyzing the properties of large molecules.[1] The MOs of molecular fragments are often referred to as fragment orbitals. In this chapter, the structural properties of several molecules are examined by constructing their MOs in terms of the appropriate fragment orbitals.

10.2. A_2H_6 SYSTEMS[2]

Molecules of formula A_2H_6 include ethane C_2H_6 and diborane B_2H_6. Important structures of ethane are the staggered and eclipsed conformations, **10.1** and **10.2**, respectively. Diborane adopts the bridged structure **10.3**, as noted in Section 9.2A.

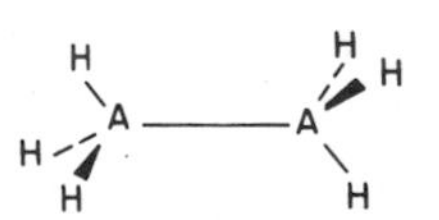

10.1	10.2	10.3

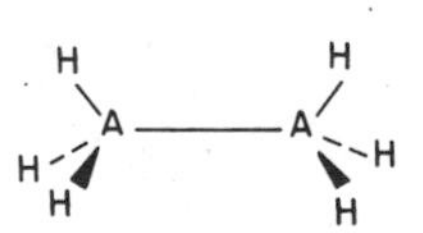

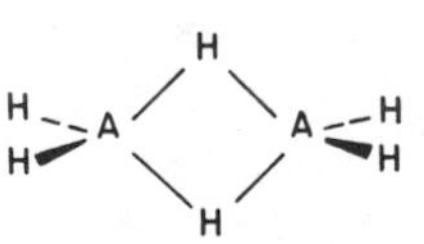

The interaction between the orbitals of two pyramidal AH$_3$ fragments, shown in Figure 10.1, leads to the MOs of staggered A$_2$H$_6$ depicted in **10.4**. For simplicity, the antibonding orbitals of AH$_3$ are not included in Figure 10.1. The $1a_{1g}$ and $1a_{2u}$ MOs are primarily the in-phase and out-of-phase combinations of the two σ_s fragment orbitals, respectively, while the $2a_{1g}$ and $2a_{2u}$ MOs are primarily the in-phase and out-of-phase combinations of the two n_σ fragment orbitals, respectively. The π_x^+ and π_x^- MOs are largely given by the in-phase and out-of-phase combinations of the two π_x fragment orbitals, respectively. Likewise, the π_y^+ and π_y^- MOs are largely given by the in-phase and out-of-phase combinations of the two π_y fragment orbitals, respectively. Note that the splitting for the n_σ fragment orbitals is larger than that for σ_s and π. This is due to a larger overlap between the n_σ orbitals. They are hybridized toward each other.

A.   ETHANE

With 14 electrons, the **HOMO** of staggered ethane, e_g, is doubly degenerate and completely filled. The energy of eclipsed ethane differs from that of the staggered conformer, since the magnitudes of the orbital interactions between two CH$_3$ groups

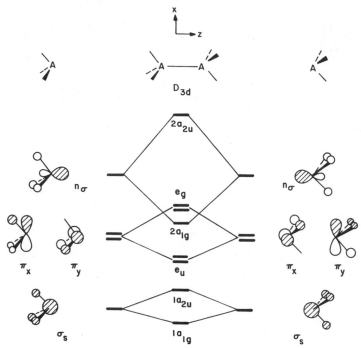

FIGURE 10.1.   Construction of the MOs of staggered A_2H_6 in terms of the orbitals of two pyramidal AH_3 fragments.

are different in the two conformations. Figure 10.2 shows the π_y and π_y^* (the anti-bonding analog of π_y shown in **9.7**) orbitals of two CH_3 groups in the staggered and eclipsed arrangements.[3] Each conformation gives rise to one destabilizing inter-action $(\pi_y - \pi_y)$, which is a two-orbital-four-electron situation since π_y is filled. There are also two stabilizing interactions $(\pi_y - \pi_y^*)$ and $(\pi_y^* - \pi_y)$.

The diagram in **10.5** shows that 1,4-overlap in the $(\pi_y - \pi_y)$ interaction, denoted by a double-headed arrow, is less positive in the staggered conformation. Thus the

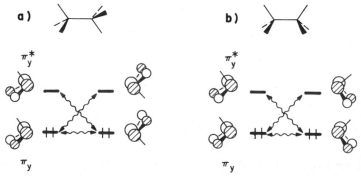

FIGURE 10.2.   Frontier orbital interactions between the two AH_3 fragments in staggered and eclipsed A_2H_6: (a) staggered and (b) eclipsed.

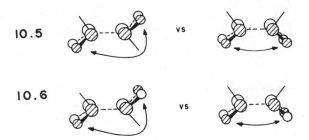

10.5 ... vs ...

10.6 ... vs ...

total overlap between the two π_y orbitals is smaller, and the magnitude of the destabilizing $(\pi_y - \pi_y)$ interaction is smaller in the staggered conformation. 1,4-overlap in the $(\pi_y - \pi_y^*)$ interaction is less negative in the staggered conformation as shown in **10.6**. Since the total overlap between the π_y and π_y^* is greater, with a correspondingly larger $(\pi_y - \pi_y^*)$ stabilization in the staggered conformation, this geometry is predicted to be more stable than the eclipsed conformer. The same conclusion results by considering the π-type interactions associated with the π_x and π_x^* orbitals of two CH$_3$ groups, because π_x and π_x^* are degenerate with π_y and π_y^*, respectively. Consequently the HOMO of ethane is lower in energy (less destabilized) in staggered ethane, as shown in **10.7**, compared to the eclipsed conformation. The rotational barrier in ethane has been determined to be 2.93 kcal/mol.[4]

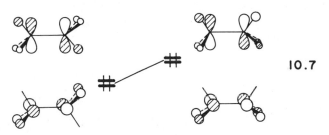

10.7

B. DIBORANE

With respect to ethane, diborane (B$_2$H$_6$) has two electrons less so that the HOMO of a staggered ethane-like B$_2$H$_6$ structure would be half-filled as shown in the left side of Figure 10.3. Upon the staggered to bridged distortion[5] (i.e., **10.1** → **10.3**), the π_y^- level is lowered in energy since the antibonding 1,3- and 1,4-interactions in the π_y^- (see **10.8**) are reduced. Furthermore, the nonbridging H–B–H angle increases

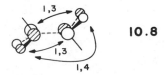

1,3

1,3

1,4

10.8

from ~110 to ~120° which increases the overlap between the hydrogens and the p_y atomic orbital on each boron atom (Figure 1.4, Section 7.3). On the other hand, the π_x^- level is raised in energy since bonding between A and H is lost upon the dis-

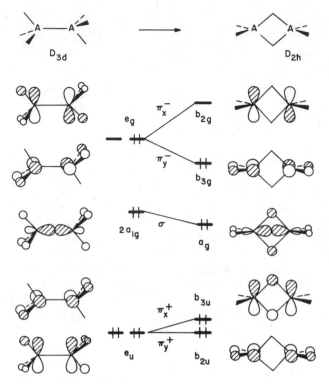

FIGURE 10.3. Correlation of the MO levels of staggered and bridged A_2H_6.

tortion. The degeneracy of e_g in staggered diborane is therefore lifted upon distortion to a bridged structure. In short, the **10.1** → **10.3** distortion relieves the Jahn–Teller instability associated with the staggered D_{3d} structure. This conclusion is also valid when there are three electrons to fill the HOMO level of the D_{3d} structure as in $C_2H_6^+$. In this case the π_x^- level is singly occupied (Figure 10.3) during the **10.1** → **10.3** distortion, and the driving force for the distortion is reduced compared with the case of B_2H_6. Therefore, $C_2H_6^+$ is predicted to have a C_{2h} structure **(10.9)**, intermediate between **10.1** and **10.3**.[6]

Diborane is two electrons short of a saturated compound like ethane. In the bond orbital approach, two three-center–two-electron bonds **(10.10)** are considered to be formed between the bridging hydrogens and the borons. In terms of MOs, the filled orbitals are the a_g and b_{3u} levels in Figure 10.3. One can see in them the essence of the a_1' level in the cyclic H_3^+ system (Section 7.4C). The b_{2g} orbital corresponds clearly to one component of the empty e' set in H_3^+.

10.9 C_{2h} **10.10**

C. DIMERIZATION OF AH$_3$

Let us consider A$_2$H$_6$ as a dimerization product of two AH$_3$ units. **10.11** shows the least-motion approach of two ·CH$_3$ radicals to form ethane C$_2$H$_6$. The MO correlation–interaction diagram in **10.12** shows that, given conservation of D_{3d} symmetry, the occupied MOs of two ·CH$_3$ radicals lead only to those of the ground state of ethane. Hence **10.11** is a symmetry-allowed reaction. If dimerization of

10.11

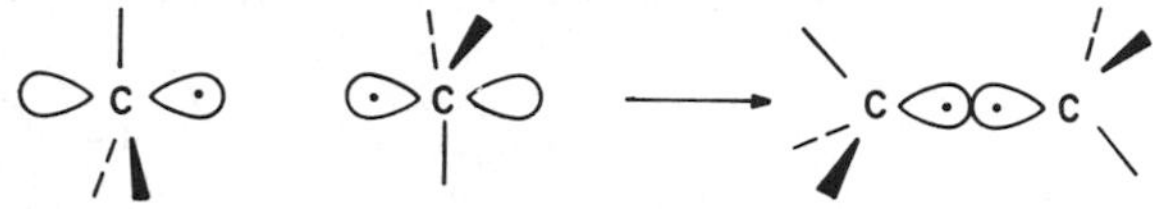

10.12

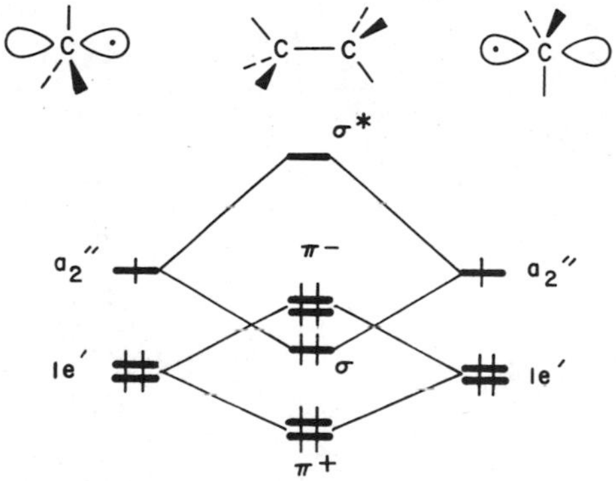

two BH$_3$ units were to follow a similar path as shown in **10.13**, not all the occupied MOs of two BH$_3$ units can be correlated with those of the ground state diborane in a staggered structure as shown in **10.14**. That is, **10.13** is a symmetry-forbidden process (see Section 4.7). As already discussed in Section 9.2A, dimerization of BH$_3$ proceeds via a reaction path that maximizes the HOMO–LUMO stabilizing

10.13

10.14

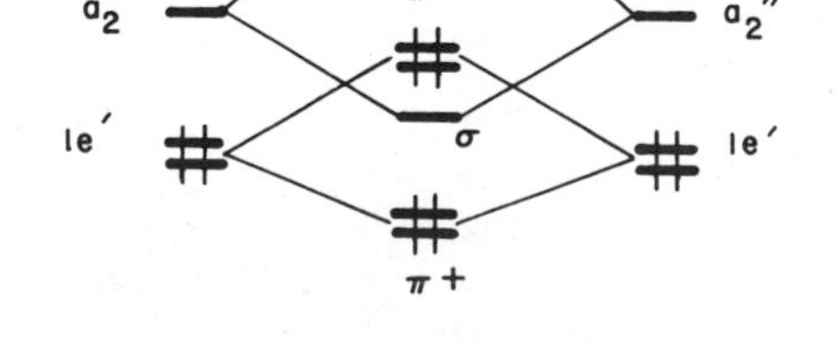

interactions that lead to a bridged structure. One can show that this reaction path is indeed symmetry allowed.[5]

10.3.   TWELVE-ELECTRON A_2H_4 SYSTEMS

Some structures of interest for A_2H_4 molecules are shown in **10.15**. Pyramidalization at each atomic center A leads the planar structure **10.15a** to the anti structure **10.15c**, and the perpendicular structure **10.15b** to the gauche structure **10.15d**. The

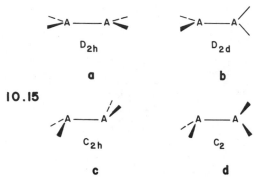

10.15

orbital interaction between two AH_2 units in Figure 10.4 gives rise to the MOs of planar A_2H_4 shown in **10.16**. With 12 electrons as in C_2H_4, the π and π^* levels

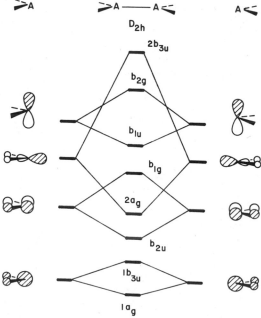

FIGURE 10.4.   Construction of the MOs of planar A_2H_4 in terms of the orbitals of two AH_2 fragments.

D_{2h} A ——— A

$2b_{3u}$ σ^*

b_{2g} π^*

b_{1u} π

10.16 b_{1g} π_y^-

$2a_g$ σ

b_{2u} π_y^+

$1b_{3u}$ σ_s^-

$1a_g$ σ_s^+

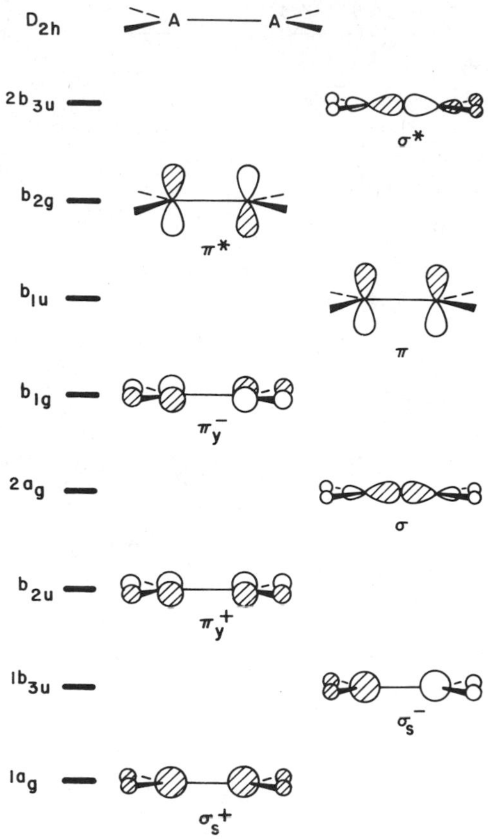

become the HOMO and LUMO of planar A$_2$H$_4$, respectively. At a given A—A distance, π bonding in the lower of these two levels is maximized when A$_2$H$_4$ is planar. It is the occupation of this level that leads to planar 12-electron A$_2$H$_4$ systems when A is from the first or second row of the periodic table, but the anti structure for heavier A atoms, as we will see later.

Occupation of the π orbital is also responsible for the large barrier to rotation around the C=C double bond (approximately 65 kcal/mol in C$_2$H$_4$) in alkenes,[7,8] since π bonding is completely lost upon rotation as shown in 10.17. Some electronic states of importance for a 12-electron A$_2$H$_4$ system are given in 10.18. Inspection of the HOMOs in 10.18 shows the diradical state 10.18c to be more stable than the ($\pi \rightarrow \pi^*$) excited state. Thus the ($\pi \rightarrow \pi^*$) excitation in alkenes provides a driving force for twisting around the C=C bond to a perpendicular structure. This is why cis–trans isomerization occurs in alkenes upon ($\pi \rightarrow \pi^*$) excitation.[9]

A.   SUDDEN POLARIZATION[10]

A perpendicular (D_{2d}) A$_2$H$_4$ molecule with 12 electrons is a typical diradical system (see Sections 8.8 and 8.10) 10.18c, with the triplet state lying very close in

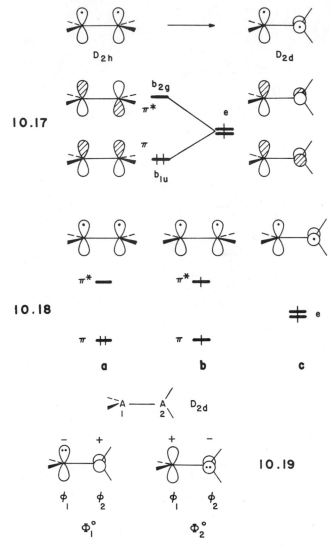

energy to the singlet state. An alternative to the diradical state is obtained by asymmetrically occupying the p orbitals ϕ_1 and ϕ_2 with two electrons as shown in **10.19**. The resulting electron configurations Φ_1^0 and Φ_2^0 are strongly dipolar because of the formal positive and negative charges created on adjacent atoms. In general, strongly dipolar electron configurations such as Φ_1^0 and Φ_2^0, obtained as an alternative to a nonpolar diradical configuration, are referred to as zwitterionic configurations. Φ_1^0 and Φ_2^0 are equivalent but differ in the way the two p orbitals are doubly occupied. Because of electron–electron repulsion arising from the orbital double occupancy (see Section 8.8), Φ_1^0 and Φ_2^0 are less stable than the diradical state. As the atomic centers A_1 and A_2 are equivalent, the state functions appropriate for perpendicular A_2H_4 are given by linear combinations of Φ_1^0 and Φ_2^0, namely,

$$\Psi_1^0 = \frac{\Phi_1^0 - \Phi_2^0}{\sqrt{2}}$$

$$\Psi_2^0 = \frac{\Phi_1^0 + \Phi_2^0}{\sqrt{2}} \tag{10.1}$$

Since these states have equal weights on Φ_1^0 and Φ_2^0, they have equal electron densities on the two carbon atoms and hence no charge polarization. As noted in Section 8.10, Ψ_1^0 is more stable than Ψ_2^0 by $-2K_{12}$, where

$$K_{12} = \langle\Phi_1^0|H|\Phi_2^0\rangle = (\phi_1\phi_2|\phi_1\phi_2) \tag{10.2}$$

The exchange integral K_{12} originates from the overlap density distribution $\phi_1\phi_2$, and is extremely small in magnitude (e.g., $1 \sim 2$ kcal/mol in perpendicular C$_2$H$_4$),[10] because ϕ_1 lies in the nodal plane of ϕ_2 and vice versa. Consequently, the energy difference between the two states Ψ_1^0 and Ψ_2^0 is small. The relative stability of these two states is reversed in a large-scale CI calculation, although the energy difference between the two still remains very small.[11] Similarly, a large-scale CI calculation shows that the singlet diradical state is only slightly more stable than the triplet diradical state against the prediction of Hund's rule. In the following we neglect the effect of a large-scale CI calculation on the relative stability of Ψ_1^0 and Ψ_2^0, because it is only the small energy difference between the two states that matters in our discussion.

Let us introduce a slight geometry perturbation to make the A$_1$ and A$_2$ sites nonequivalent. As an example, one may consider a slight pyramidalization at A$_1$ as shown in **10.20**. Three and four electron pairs around a given atom tends to make

10.20

that center planar and pyramidal, respectively. Thus Φ_1 is more stable than Φ_2, that is,

$$E_1 < E_2 \tag{10.3}$$

where

$$E_1 = \langle\Phi_1|\hat{H}|\Phi_1\rangle$$

$$E_2 = \langle\Phi_2|\hat{H}|\Phi_2\rangle \tag{10.4}$$

If the extent of the pyramidalization at A$_1$ is small, it is valid to use the following approximation

$$\langle\Phi_1|\hat{H}|\Phi_2\rangle \simeq \langle\Phi_1^0|\hat{H}|\Phi_2^0\rangle = K_{12} \tag{10.5}$$

Therefore, with the condition that

$$K_{12} \ll E_2 - E_1 \tag{10.6}$$

the state functions appropriate for **10.20** are given by

$$\Psi_1 \simeq \Phi_1 + \frac{K_{12}}{E_1 - E_2}\, \Phi_2$$

$$\tag{10.7}$$

$$\Psi_2 \simeq \Phi_2 + \frac{K_{12}}{E_2 - E_1}\, \Phi_1$$

Since K_{12} is very small in a perpendicular structure, equation 10.7 is valid when the energy difference $E_2 - E_1$ is small, that is, when the extent of pyramidalization at A_1 is small. Now the state Ψ_1 is strongly zwitterionic since Φ_2 is only a small component. Namely, the electron density is concentrated on the pyramidal center A_1 and diminished on the planar center A_2. Similarly, the state Ψ_2 shows zwitterionic character in which the electron density is concentrated on the planar center A_2.

The above discussion shows that a small geometry perturbation can induce a strong charge polarization. This kind of phenomenon is generally referred to as a sudden polarization.[10] It has been postulated to occur in photochemical cyclization of conjugated dienes and trienes.[12] The stereospecificity observed in these reactions can be explained by a zwitterionic intermediate containing an allyl anion moiety which in turn undergoes a conrotatory ring closure to a substituted cyclopropyl carbanion. Finally, collapse of the zwitterion leads to a bicyclic product. As examples, **10.21** shows the photochemical conversion of a 1,3-butadiene to a bicyclo[1.1.0]butane, and **10.22** that of a *cis, trans*-1,3,5-hexatriene to a bicyclo[3.1.0]hexene. In the above diene and triene, twisting of the $C_3 = C_4$ bonds

10.21

10.22

generates the appropriate initial zwitterions. A similar charge polarization is thought to occur in retinal, **10.23a**. The primary excitation of the retinal skeleton involves rotation around the $C_{11} = C_{12}$ bond leading to the all-*trans* form. At the halfway point of this rotation, the retinal skeleton is transferred into two pentadienylic moieties (i.e., the 7–11 and 12–16 fragments). Since these fragments are nonequivalent, charge polarization occurs in the excited state. Due to the positive charge on

10.23a   **10.23b**

the protonated imino group, a negative charge moves toward the 12–16 fragment (which therefore becomes neutral), and a positive charge toward the 7–11 fragment. This is shown in **10.23b**. The net result of excitation is the transformation of a photon into an electrical signal, as the positive charge migrates from the 12–16 to the 7–11 fragment. This sudden polarization is considered to be crucial for the mechanism of vision.[10,13]

B.   SUBSTITUENT EFFECTS[14]

Consider now an alkene with a π donor D or a π acceptor A on the C=C bond as shown in **10.24**. As usual, the C=C bond is described by the π and π^* orbitals. For simplicity, we describe a π donor by a filled orbital ϕ_d and a π acceptor by an empty orbital ϕ_a. **10.24a** represents a case in which ϕ_a lies below π. In **10.24b**, ϕ_d lies above π but much closer to π than π^*. **10.24c** shows a case in which ϕ_a lies below π^* but much closer to π^* than π. Finally, **10.24d** represents a case in which ϕ_a lies above π^*. By employing the rules of orbital mixing in Chapter 3, the $\pi_1, \pi_2,$

10.24

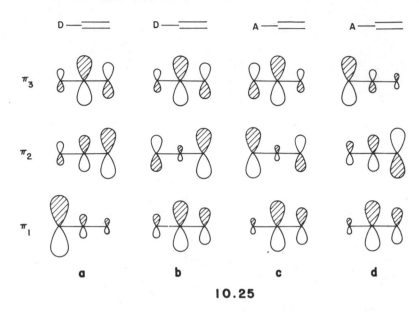

10.25

and π_3 MOs of **10.24a–10.24d** can be easily derived. Summarized in **10.25a–10.25d** are the relative weights of these MOs on the vinyl carbon atoms and donor or acceptor substituent (represented by a single p atomic orbital for simplicity) of these MOs for the case of **10.24a–10.24d**, respectively. It is noted from **10.25a** and **10.25b** that, for a π-donor substituted alkene, the HOMO has less weight but the LUMO has more weight on the carbon bearing the substituent. In contrast, **10.25c** and **10.25d** show that, for a π-acceptor substituted alkene, the HOMO has more weight but the LUMO has less weight on the carbon bearing the substituent. In all cases the resulting form of the orbitals resembles that of the allyl system. For the donor-substituted cases the π_2, "nonbonding," level is occupied and, like the allyl anion, electron density is concentrated on the donor and β-vinyl carbon atoms. The opposite occurs with an acceptor-substituted alkene which is analogous to an allyl cation. Consequently, the charge densities of the C=C bonds in π-donor and π-acceptor substituted alkenes are expected to polarize as shown in **10.26**, a result consistent with ^{13}C NMR chemical shifts in substituted alkenes.[15] Notice that the increase or decrease of electron density at the β-vinyl carbon is not solely due to the donation or acceptance of electron density by an electron donor or acceptor, respectively. By second-order orbital mixing, π^* mixes into π to polarize the charge distribution in π_1 (and π_2). The charge polarization effect of a π donor is opposite to that of a π acceptor. Thus when π donors and π acceptors are substituted as in **10.27**, the charge distribution of the C=C bond is strongly polarized (**10.28**). Thus

10.26   10.27

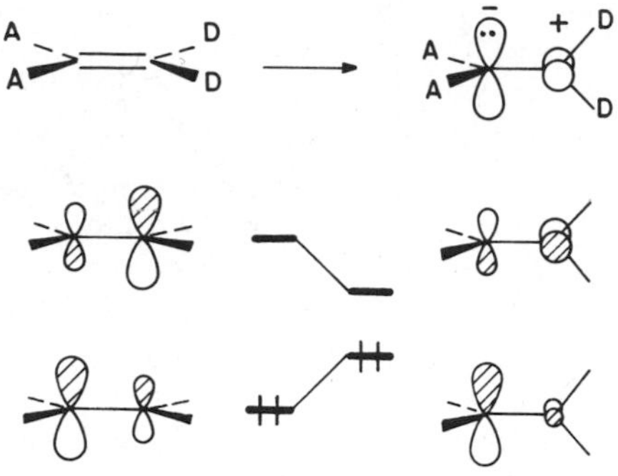

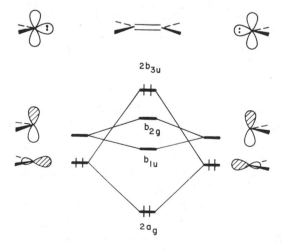

10.28   10.29

π bonding is weak in the planar structure of **10.27**, while the zwitterionic state of the perpendicular structure **(10.29)** is stabilized significantly by the π acceptors at the anion site and by the π donors at the cation site. For the perpendicular structure of **10.27**, the zwitterionic state becomes more stable than its alternative, diradical state. Consequently, the rotational barrier around the C=C bond is substantially reduced. For example, the C=C rotational barrier of **10.27** is less than 8 kcal/mol for A $= -COCH_3$ and D $= -N(CH_3)_2$.[16]

C.   DIMERIZATION OF AH_2[7]

Ethylene, $CH_2=CH_2$, may be obtained as a dimerization product of singlet carbene :CH_2. The least-motion approach, which maintains D_{2h} symmetry in this reaction, is symmetry forbidden as shown by the MO correlation–interaction diagram in **10.30**. This approach is energetically unfavorable since it maximizes the HOMO–

$2b_{3u}$

b_{2g}

b_{1u}

$2a_g$

10.30

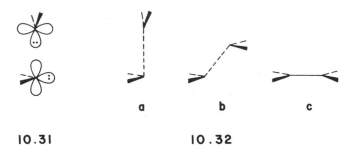

10.31   10.32

HOMO interaction (a two-orbital–four-electron destabilization) and minimizes the HOMO–LUMO interactions (a two-orbital–two-electron stabilization). Two :CH$_2$ units can approach in a more favorable way such that the HOMO of one :CH$_2$ unit is directed toward the LUMO of the other as shown in **10.31**. As the dimerization progresses, two :CH$_2$ units gradually tilt away from this perpendicular arrangement to become planar CH$_2$=CH$_2$. This is illustrated in **10.32**.

D. PYRAMIDALIZATION OF A$_2$H$_4$

It was pointed out that, as the atomic number of A increases in 12-electron A$_2$H$_4$ systems, the stability of the anti structure increases relative to that of the planar structure. Thus C$_2$H$_4$ and Si$_2$H$_4$ are predicted to be planar while Ge$_2$H$_4$ and Sn$_2$H$_4$ have the anti structure.[18,19] Although Si$_2$H$_4$ may be planar, the potential energy surface for the planar to anti distortion is calculated to be very soft. Experimentally, Sn$_2$R$_4$ [R = —CH(SiMe$_3$)$_2$] is found to have the anti structure shown in **10.33**.[18] In addition, it is also found for isoelectronic species such as Ge$_2$P$_4^{8-}$ and Ge$_2$As$_4^{8-}$ present in a crystalline environment with Ba^{2+} counterions.[20]

$$R = CH(SiMe_3)_2$$
$$Sn-Sn = 2.764 \text{ Å}$$
$$Sn-C = 2.28 \text{ Å}$$
$$C-Sn-C = 112°$$
$$\theta = 41°$$

10.33

We will use two different approaches to view this interesting problem. The first uses an argument similar to that employed in Section 9.2B to view the pyramidalization of eight-electron AH$_3$ systems. The MOs of A$_2$H$_4$ that result from the in-phase and out-of-phase combinations of σ_s^* on each AH$_2$ (i.e., σ_{s+}^* and σ_{s-}^*), omitted in Figure 10.4 for simplicity, are shown in **10.34** together with the π and π^* levels. Let us consider how π and π^* of planar A$_2$H$_4$ mix with the σ_{s+}^* and σ_{s-}^* levels during the planar to anti distortion. In the anti structure, the overlap between π and σ_{s-}^* (both are of b_u symmetry) and that between π^* and σ_{s+}^* (both are of a_g symmetry) are nonzero, see **10.35**. Consequently, orbital mixing occurs between π and σ_{s-}^* and between π^* and σ_{s+}^* on bending. An orbital correlation diagram is shown in **10.36**, where the n_+ and n_- orbitals of anti A$_2$H$_4$ are derived as in **10.37**.

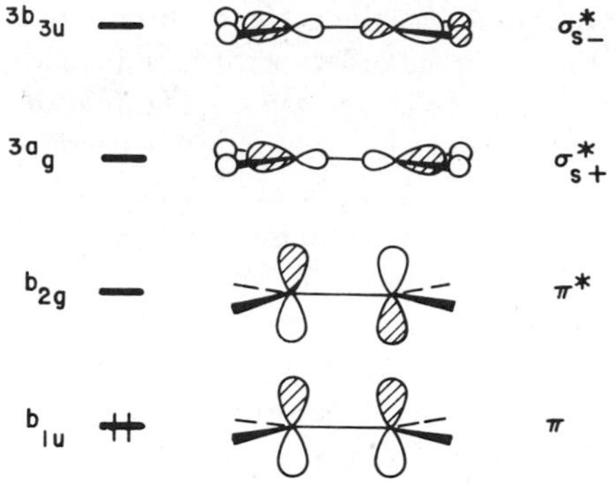

$3b_{3u}$ — σ_{s-}^*

$3a_g$ — σ_{s+}^*

10.34

b_{2g} — π^*

b_{1u} ⧺ π

$$\left\langle \;\middle|\; \right\rangle = \left\langle \;\middle|\; \right\rangle \;>\; 0$$

10.35

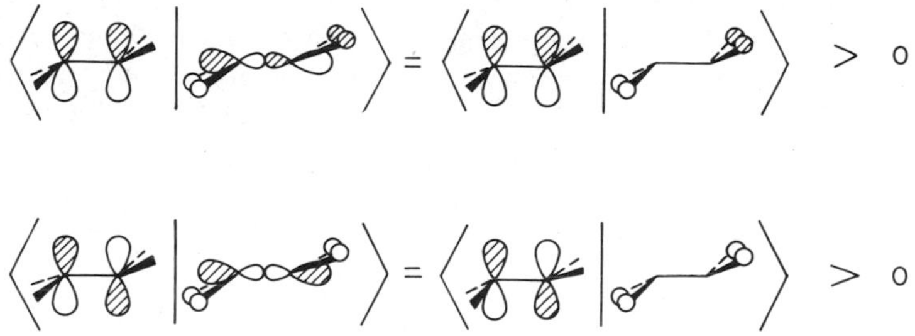

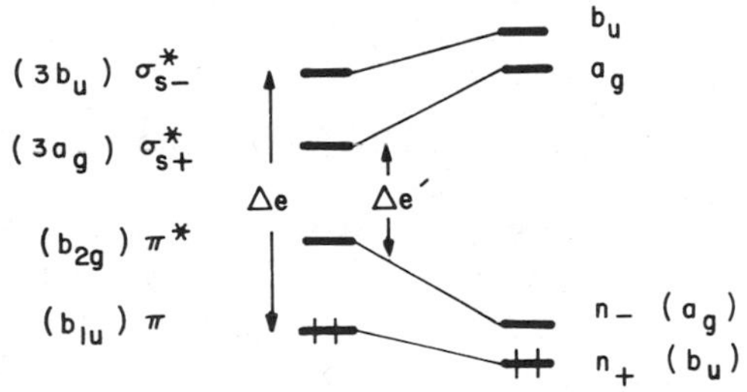

$$\left\langle \;\middle|\; \right\rangle = \left\langle \;\middle|\; \right\rangle \;>\; 0$$

$D_{2h} \longrightarrow C_{2h}$

$(3b_u)\; \sigma_{s-}^*$ b_u

$(3a_g)\; \sigma_{s+}^*$ a_g

Δe $\Delta e'$ 10.36

$(b_{2g})\; \pi^*$

$(b_{1u})\; \pi$ $n_-\;(a_g)$

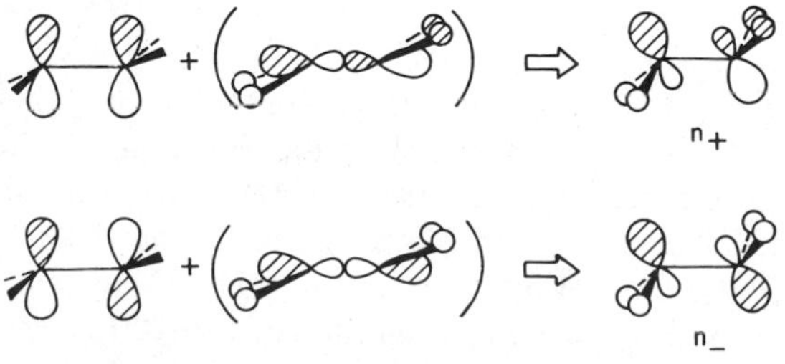

$n_+\;(b_u)$

$$+ \left(\right) \Rightarrow$$

n_+ 10.37

$$+ \left(\right) \Rightarrow$$

n_-

167

Due to this orbital mixing, the π and π^* levels are lowered upon the planar to anti distortion. From second-order perturbation theory the stabilization of π is inversely proportional to the energy gap Δe, and that of π^* to $\Delta e'$. With an increase in the atomic number of A, the orbitals of A become more diffuse. Thus σ-anti-bonding in A—H is reduced in σ_{s+}^* and σ_{s-}^*, so these levels are lowered in energy. In addition, the π and π^* levels are raised in energy because of the electronegativity of A decreases. Most importantly, the energy gap between π and π^* becomes quite small, since π-interaction in A—A is reduced by increasing diffuseness of the p atomic orbitals on A. The planar to anti distortion in a 12-electron A_2H_4 system is therefore a second-order Jahn–Teller distortion, which becomes increasingly stronger upon going down a column in the periodic table.

The electronic structure of anti A_2H_4 may also be described in terms of the dimerization of two AH_2 units as shown in **10.32** and **10.38**. The structure of **10.33** may then be regarded as a stannylene caught in the act of dimerizing. When AH_2 units cannot achieve strong π bonding because of a long A—A bond, they adopt an anti structure so as to maximize their mutual HOMO–LUMO interactions. It is noted from **10.36** that the HOMO–LUMO gap becomes smaller upon $D_{2h} \rightarrow C_{2h}$ distortion, and thus the corresponding second-order Jahn–Teller instability should actually increase. In the C_{2h} point group, $\Gamma_{a_g} \times \Gamma_{b_u} = \Gamma_{b_u}$ so that a distortion mode of b_u such as **10.39** would bring **10.32b** toward **10.32a**. There must be a whole range of geometries from **10.32b** to **10.32a** for 12-electron A_2L_4 systems of third and fourth row atoms A.

10.38   10.39

10.4.   FOURTEEN-ELECTRON AH_2BH_2 SYSTEMS

With 14 valence electrons, the HOMO of planar AH_2AH_2 is the π^* level of **10.16**. On rotation around the A—A bond π^* is lowered in energy as shown in **10.40**. Furthermore, this stabilization is greater than the amount of energy that the π level is destabilized. The total stabilization decreases as the A—A distance increases. The HOMO of perpendicular AH_2AH_2 is further stabilized by pyramidalization at each center A. This reduces the D_{2d} symmetry of the perpendicular geometry to C_2 in the resulting gauche structure. The Walsh diagram in Figure 10.5 reveals that σ_{s+}^* mixes into p_+ to give n_+, and σ_{s-}^* mixes into p_- to give n_-.

The relative stability of the gauche and anti conformations in AH_2AH_2 may be examined in terms of the interactions leading to the generation of their HOMOs. As far as each center A is concerned, a 14-electron AH_2—AH_2 can be regarded as an 8-electron AH_2—L (or L—AH_2) as depicted in **10.41**. The HOMO of AH_2L is the nonbonding orbital n_σ, and that of anti or gauche AH_2AH_2 is mainly composed of n_σ from each center A. The n_σ and $\pi_{AH_2}^*$ orbitals of each AH_2 unit are arranged

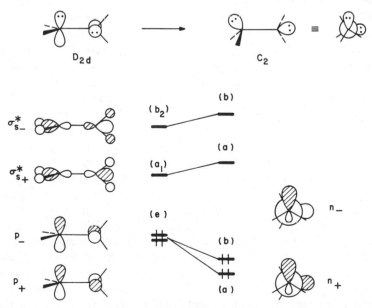

in anti and gauche conformations in **10.42** and **10.43**, respectively. The overlap between the two n_σ levels is substantial in the anti conformation, but essentially nonexistent in the gauche conformation. Therefore, the destabilizing interaction between the two n_σ orbitals is negligible in the gauche geometry. Further, the gauche conformation provides a nonzero overlap between the n_σ and $\pi^*_{\mathrm{AH_2}}$ orbitals, thereby leading to the $(n_\sigma - \pi^*_{\mathrm{AH_2}})$ stabilizing interaction, **10.44**. Both effects lead to the preference of the gauche conformation over the anti conformation in a 14-

FIGURE 10.5. Correlation of the MO levels of bisected and gauche A_2H_4.

10.42

10.43

10.44

electron AH_2AH_2 system such as hydrazine NH_2NH_2. If the A—A distance of AH_2AH_2 is large, as in diphosphine PH_2PH_2, then the interaction between the two n_σ orbitals is weak even in the anti conformation so that the energetic variation of the HOMO is not an important factor governing the conformational preference. In such a case, the anti conformation becomes comparable in energy to, or may become more stable than, the gauche conformation.[21,22] Note that we have conveniently singled out the $(n_\sigma - \pi^*_{AH_2})$ interaction as stabilizing the gauche conformation. The A—H bonding counterpart, π_{AH_2}, will also interact with n_σ. This is a two-orbital–four-electron destabilizing interaction which is maximized at the gauche geometry and minimized in the anti structure. Therefore, it does not cost much energy to rotate from one gauche structure to another by way of the anti conformation. On the other hand, rotation through a *syn* geometry will require a far greater activation energy.[21] In the *syn* geometry the four A—H bonds eclipse each other, and the two filled n_σ orbitals maximize their overlap at this geometry.

A sulfonium ylide $\bar{C}H_2\overset{+}{S}H_2$, **10.45**, is an example of a 14-electron AH_2BH_2 system. Since the valence orbitals of sulfur are more diffuse than those of carbon, the $\pi^*_{SH_2}$ level lies lower in energy than in $\pi^*_{CH_2}$ level. Based solely upon the ionization potentials of sulfur and carbon, one might expect the nonbonding orbital of sulfur (n_S) to lie higher in energy than that of carbon (n_C). However, orbital levels are lowered and raised upon introducing formal positive and negative charges respectively (Section 8.9). This effect raises the n_C level above n_S in sulfonium ylides[23] and consequently makes the carbanion center more nucleophilic than the sulfonium

10.45   10.46

ion center. Therefore, the relative orderings of the n_S, $\pi^*_{SH_2}$, n_C, and $\pi^*_{CH_2}$ may be approximated as in **10.46**. The energy gap between n_C and $\pi^*_{SH_2}$ is small compared with that between n_S and $\pi^*_{CH_2}$. This gives rise to a strong $(n_C - \pi^*_{SH_2})$ interaction in the gauche conformation. The interaction is further enhanced if the carbanion center becomes planar, because the p orbital of a planar carbanion center is closer in energy to $\pi^*_{SH_2}$ and overlaps better with $\pi^*_{SH_2}$ as shown in **10.47**. Thus a sulfonium ylide is expected to have a planar carbanion center, thereby leading to a bisected structure **10.48**.[23] This structure is also found for aminophosphine (PH$_2$NH$_2$, iso-electronic with $\overset{+}{S}H_2\overset{-}{C}H_2$) derivatives which have a planar nitrogen center and exist in this bisected geometry.[21, 24]

Notice that we have focused upon the $\pi^*_{SH_2}$ fragment orbital as a π acceptor. An empty d orbital on sulfur also plays the same role, as shown in **10.49**. Which is a

10.47   10.48   10.49

more accurate picture, that in **10.47** or **10.49**? The symmetry of the $(n_C - \pi^*_{SH_2})$ interaction is the same as that of the $(n_C - d)$ interaction, so there is no way to make a choice short of a calculation. Unfortunately, this does not provide a clear-cut answer, either. Inclusion of d-type functions is essential for a proper quantitative description of the structure and energetics of a molecule such as $\overset{+}{S}H_2\overset{-}{C}H_2$. In such a calculation, however, electron density transferred to the d-type functions is not significant. In other words, d-type functions act as polarization functions to accurately tailor the wavefunctions but are not important in the customary sense of p_π-d_π bonding as depicted in **10.49**.

10.5.   AH$_3$BH$_2$ SYSTEMS

A 12-electron system C$_2$H$_5^+$ may adopt a classical structure **10.50** or a nonclassical one **10.51**. To probe the transition between the two possibilities, we will first consider the orbital interaction between the CH$_3$ and CH$_2^+$ units in **10.50**. The $1a_1$ and $1b_2$ orbitals of the CH$_2^+$ fragment will interact with $1a_1$ and one component of $1e$ on CH$_3$, respectively, to form in-phase and out-of-phase combinations which are

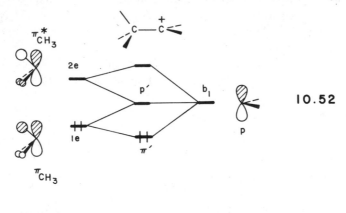

10.50   **10.51**

filled. The $2a_1$ level of CH_2^+ and $2a_1$ on CH_3 will interact strongly to form a filled σ and empty σ^* MO. This pattern is like that found for C_2H_6 (see Figure 10.1) and C_2H_4 (see Figure 10.4). A difference comes from the interaction between the b_1 fragment orbital (p_{CH_2}) and the other component of $1e$ and $2e$ on CH_3 (π_{CH_3} and $\pi_{CH_3}^*$, respectively) which overlap in a π fashion **(10.52)**. According to the interaction diagram **10.52**, the π_{CH_3} level is perturbed by $\pi_{CH_3}^*$ and p_{CH_2} as shown in **10.53**. Electron density from filled π_{CH_3} is transferred to empty p_{CH_2} via π overlap.

Since the π_{CH_3} orbital is C—H bonding, this interaction weakens the C—H bonds. To a lesser extent the in-phase mixing of $\pi_{CH_3}^*$ with p_{CH_2} also increases C—C bonding and decreases C—H bonding. In particular, the C—H bond periplanar to the carbonium ion p orbital is preferentially weakened since the coefficients in π_{CH_3} and $\pi_{CH_3}^*$ are largest for this hydrogen. This may also be seen quite clearly from the bond orbital descriptions of π_{CH_3} and $\pi_{CH_3}^*$ **(10.54)**. Thus the MO picture of **10.53**

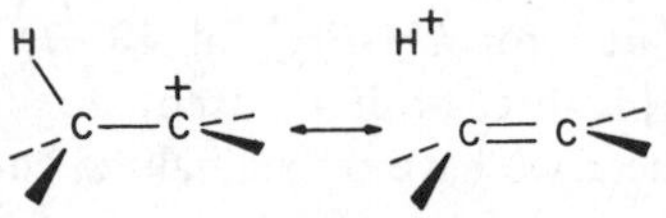

10.55

is analogous in nature to hyperconjugation, **10.55**. An electron-deficient group tends to migrate into an electron-rich region; we have seen this for H$_3^+$ in Section 7.4C. So the orbital correction in **10.53** facilitates the structural change which takes **10.50** to **10.51**. Along this distortion, the π' level stays relatively constant in energy as shown in **10.56**. Alternatively, one may view the bridged, nonclassical

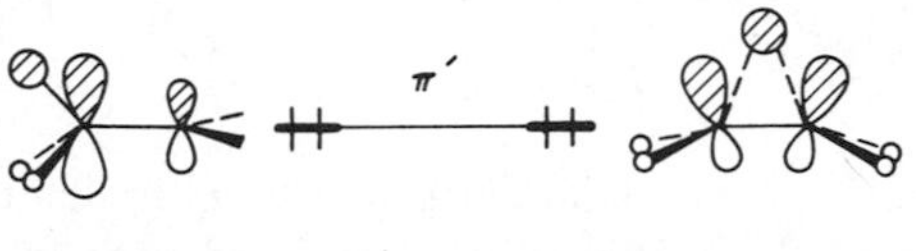

10.56a   10.56b

structure as a protonated ethylene. As a proton attacks ethylene, the empty s orbital of the proton will interact with the π orbital of ethylene. Maximum overlap occurs when the proton is centered over the carbon–carbon bond. This interaction directly leads to **10.56b**. The stability difference in the C$_2$H$_5^+$ isomers, **10.50** and **10.51**, is calculated to be very small even at a computational level beyond *ab initio* SCF MO calculations.[25] The orbital shape of **10.56b** is topologically equivalent to filled a_1' in the cyclic form of H$_3^+$. The π^* level of this protonated ethylene "complex" and the C—H antibonding analog of **10.56b** are similar to the empty e' set in H$_3^+$. Exactly the same pattern evolves from the transition state for 1,2-alkyl shifts in carbonium ions. The migrating alkyl group possesses a hybrid orbital that overlaps with the π level in the same way as the hydrogen s orbital does in **10.56b**.

Two structures of interest for a 14-electron system in CH$_3$CH$_2^-$ are shown in **10.57**. The interaction diagram **10.58** for the staggered structure **10.57a** is not

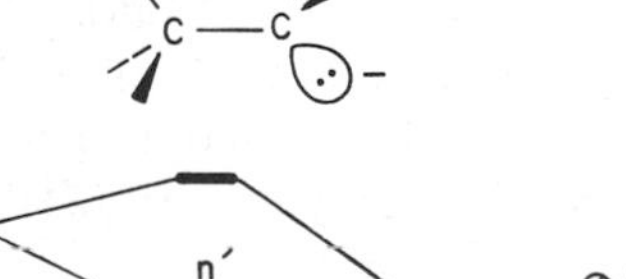

10.57a   10.57b

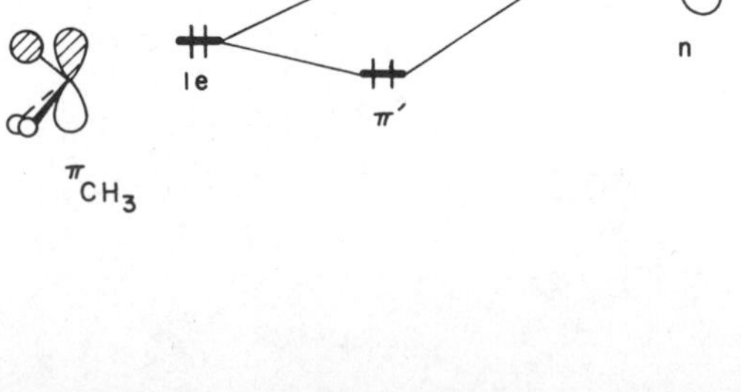

10.58

much different from that presented for $CH_3CH_2^+$ in **10.52**. We now focus our interest on the middle level of this three-orbital pattern. A simplified interaction diagram is presented in **10.58** where we have pyramidalized the carbanion center. There is still significant π-type overlap of n_σ with π_{CH_3} and $\pi^*_{CH_3}$. The form of π' is basically the same as that given in **10.53**. In the HOMO, n', the n_σ orbital combines out-of-phase with π_{CH_3} but in-phase with $\pi^*_{CH_3}$ so that methyl hydrogen character is enhanced at the expense of methyl carbon character (see **10.59**). As a result, the

10.59

hydrogen atom in the C—H bond antiperiplanar to n_σ has more weight and hence a greater charge accumulation. Charge transfer from n to $\pi^*_{CH_3}$, arising from the $(n_\sigma - \pi^*_{CH_3})$ interaction, **10.60**, weakens primarily the C—H bond antiperiplanar to n_σ. The same phenomenon is also observed in methylamine, **10.61**. The presence of a C—H bond antiperiplanar to a nitrogen nonbonding orbital n_σ is signaled by a characteristic infrared band in the C—H stretching region, known as the Bohlmann band.[26]

10.60   10.61

As in the case of staggered $CH_3CH_2^-$, the HOMO of the eclipsed structure **10.57b** can be easily derived. The HOMOs of the staggered and eclipsed structures are compared in **10.62**. The HOMO of $CH_3CH_2^-$ lies lower in the staggered structure since it

10.62a   10.62b

avoids antibonding between the two large orbital lobes. Thus the staggered geometry is energetically favored over the eclipsed one. In connection with nucleophilic addition reactions to multiple bonds, it is of interest to consider $CH_3CH_2^-$ as derived from a nucleophilic addition of H^- to ethylene **(10.63)**. The orbital interaction dia-

10.63

gram in **10.64** leads to the HOMO as shown in **10.65**. Note that p orbital character

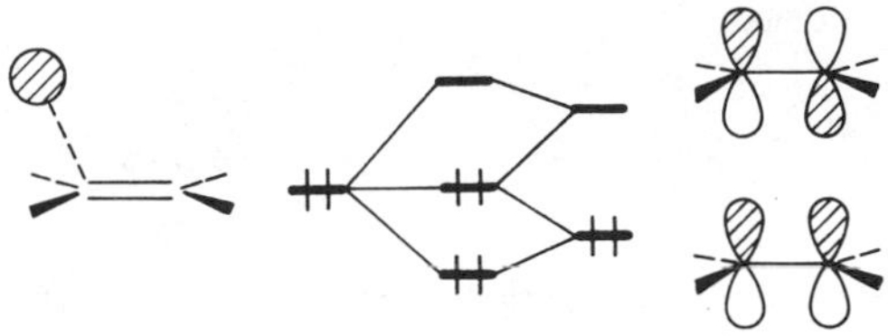

10.64

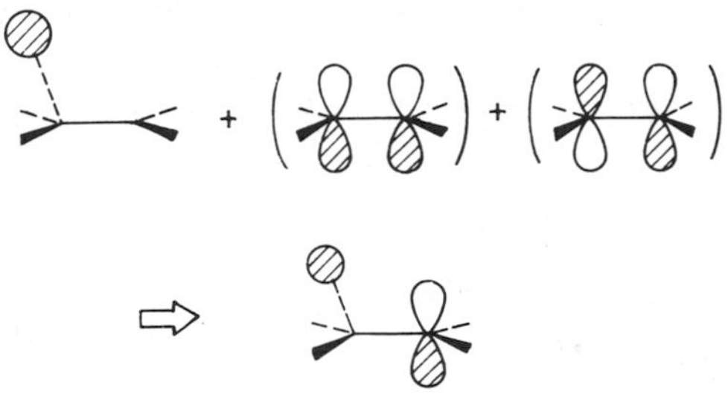

10.65

is reduced on C1 (not shown) but enhanced on C2, that is, carbanion character develops on C2 at the expense of weakening π bonding between C1 and C2. The HOMO of **10.65** is further stabilized upon pyramidalization at the C1 and C2 centers. As expected from **10.62** and **10.66**, this distortion should favor an antiperi-

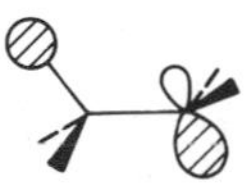

10.66a    10.66b

planar arrangement of the developing lone pair on C2 to the newly forming bond at C1. The hydride in **10.63** directly attacked ethylene C1. It will make little sense to have the hydride attack between the C1—C2 bond as was predicted for the approach of an electrophile (e.g., H$^+$) toward ethylene. Such a path maximizes hydrogen s-ethylene π overlap which in the present case (see **10.64**) constitutes a two-orbital–four-electron destabilization. The stabilizing term, hydrogen s mixing with ethylene π^*, is zero by symmetry when the hydride bisects C1—C2. There are still a large range of paths that can be considered allowing for the fact that the hydride should directly attack C1. Three geometries are illustrated in **10.67**. Again the

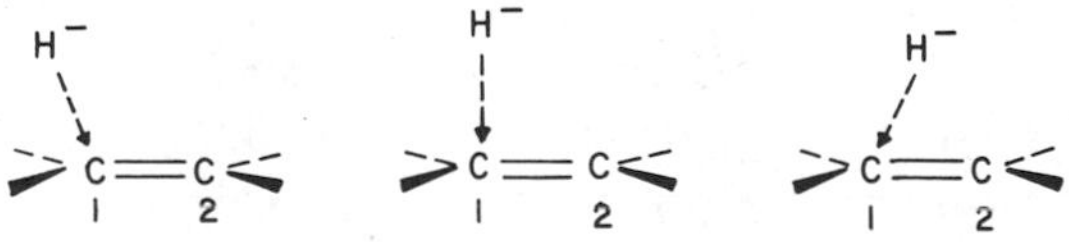

a   b   c

favored path minimizes overlap between hydrogen s and ethylene π and maximizes overlap between hydrogen s and ethylene π^*. For the former case the three possibilities are shown in **10.68**. The smallest overlap occurs in geometry **10.67a** where

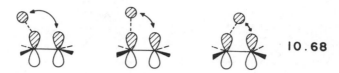

10.68

the overlap between hydrogen s and ethylene π (as indicated by the double-headed arrow) is minimized. Thus, the two-orbital–four-electron interaction is least desta-bilizing in **10.67a**. As shown in **10.68** the overlap between hydrogen s and ethylene π^* is maximized at **10.67a** since now the 1,3-overlap between hydrogen s and the p atomic orbital at C2 is negative. Hence the two-orbital–two-electron interaction is most stabilizing in **10.69a**. Consequently the hydride is expected to approach

10.69

ethylene C1 at an oblique angle when the overlap between the reactants becomes appreciable. It can be easily seen that this is nothing more than a restatement of the linear vs. bent H_3^- problem. For the same reasons that mandate a linear H_3^- geometry over a bent one, the H^- + ethylene system prefers to adopt an oblique H—C1—C2 angle along the reaction path. Furthermore, the hydrogen s-ethylene π^* interaction results in charge transfer from the filled hydride s orbital to the empty ethylene π^* orbital. As shown in Section 9.2 this may be made more sta-bilizing by pyramidalization at C1 and C2, a process that occurs as the reaction proceeds.

Now the reaction path and rationale for it can readily be extended to any reac-tion where a nucleophile attacks a π bond. The lone pair HOMO of a nucleophile (n) plays the same role and is topologically analogous to the hydride s orbital. However, a nucleophile does not normally react with ethylene itself. The desta-bilization between n and ethylene π is not compensated by the stabilization be-tween n and π^*. Perturbations that bring about the following two changes will render the nucleophilic attack more favorable: One is to provide low-lying π HOMO and π LUMO so as to decrease the energy gap between n and π LUMO and increase that between n and π HOMO. The other is to increase the orbital coefficient of the carbon under nucleophilic attack (i.e., C1 in **10.63**) in the π LUMO but decrease it in the π HOMO. Such an orbital polarization decreases the overlap between n and π HOMO. All of these can be achieved either by having a π-acceptor substituent at C2 (see **10.25c**) or by substituting a more electronegative atom for C2 in **10.63** (see Section 6.4). The electronegativity perturbation encountered upon going from $CH_2=CH_2$ to $CH_2=O$ not only lowers π and π^* but also polarizes these orbitals so that π is concentrated at the more electronegative oxygen atom and π^* is concen-trated on carbon. This is indicated on the right side of the interaction diagram in **10.70**. Consequently, the stabilizing ($n - \pi^*$) and destabilizing ($n - \pi$) interactions are increased and decreased, respectively, compared to ethylene. Notice that the

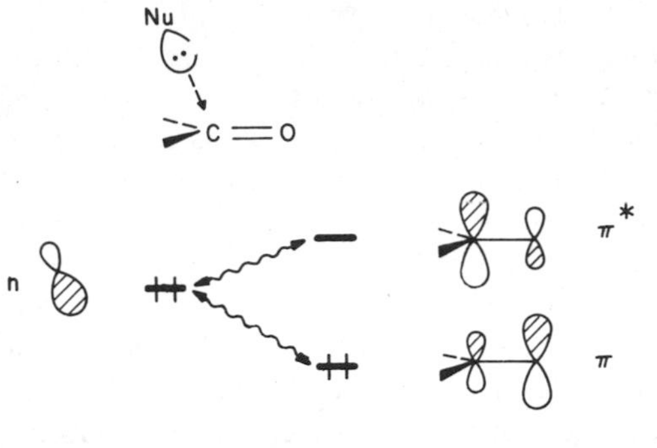

10.70

nucleophile still prefers an oblique approach and the carbon center will pyramidalize for the same reasons as discussed for the ethylene case.[27] A crystallographic mapping of the reaction path for nucleophilic attack on carbonyl containing compounds has been provided.[28, 29] There exist a large number of X-ray structures where a nucleophilic center is forced to be in close proximity to a carbonyl group. This may occur as a result of intra- and/or intermolecular contacts. The geometry around the nucleophile and carbonyl group is then adjusted to correspond to the energetically most favorable situation. Consequently, these X-ray structures effectively chart the path of least energy for the reaction. The features of the reaction that we have discussed are experimentally found; namely, as the nucleophile begins to appreciably interact with the carbonyl carbon it does so with an oblique Nu—C—O angle as shown in **10.70**. Furthermore, the carbonyl carbon center becomes progressively more pyramidal as the Nu—C distance decreases.[28, 29]

Let us now return to the CH$_3$CH$_2^-$ system. Of the three methyl hydrogens in CH$_3$CH$_2^-$, the one in the C—H bond antiperiplanar to n_σ carries the highest charge density as already pointed out. Therefore, this hydrogen is the preferred site for substitution by an electronegative ligand X, and an anti structure **10.71** is expected for XCH$_2$CH$_2^-$. Furthermore, due to the electronegativity perturbation provided by X, the σ^*_{CX} orbital is found lower in energy than σ^*_{CH}. Therefore, σ^*_{CX} is closer in energy to n_σ as indicated in **10.72**. The σ^*_{CX} orbital is also more concentrated at

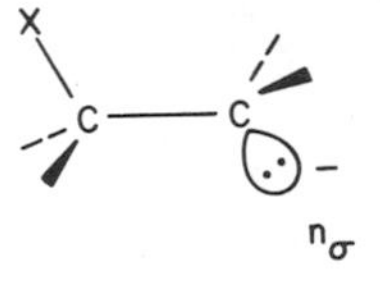

10.71

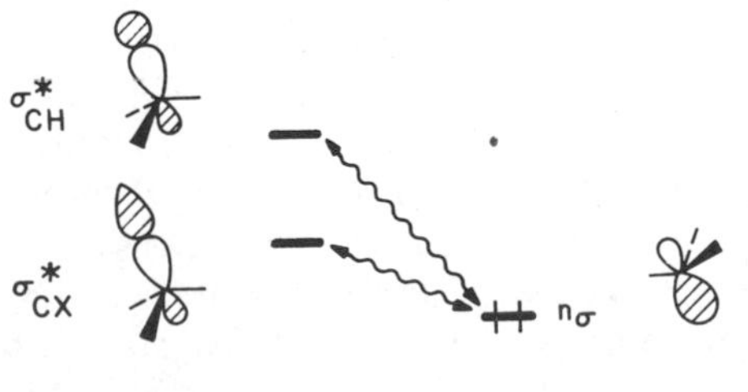

10.72

carbon than is σ^*_{CH}. This is again due to electronegativity differences. Consequently, the overlap of σ^*_{CX} with n_σ is expected to be larger than that between σ^*_{CH} and n_σ. Thus, both the energy gap and overlap factors favor the $(n_\sigma - \sigma^*_{CX})$ interaction over $(n_\sigma - \sigma^*_{CH})$. The picture we have developed for the optimum conformation in $XCH_2CH_2^-$ contains all of the elements that determine the stereochemistry of $E2$ (and $E1cb$) elimination reactions.[30,31] In an alkyl halide the C—H bond antiperiplanar to the C—X bond is attacked by a base. At the transition state for an $E2$ process, substantial negative charge is developed at the carbon atom that is being deprotonated. In orbital terms, the weakened C—H bond becomes localized on carbon and is similar in shape to n_σ in **10.71**, which is the intermediate for an $E1cb$ process. The favorable interaction with σ^*_{CX} is, of course, maximized at an antiperiplanar conformation. The interaction also transfers electron density from the carbon atom being deprotonated to the σ^*_{CX} orbital. Therefore, the C—X bond is weakened. Ultimately the C—X and C—H bonds are totally broken and the π orbital of ethylene is formed.

The $(n_\sigma - \sigma^*_{AH_3})$ interaction in AH_3BH_2 can be enhanced by lowering the energy of the $\pi^*_{AH_3}$ level. As illustrated in **10.73**, the $\pi^*_{AH_3}$ level is lowered by substituting

10.73

electronegative ligands for hydrogens, by moving down a column in the periodic table for A (the overlap between p atomic orbital on A and hydrogen s becomes smaller), and by introducing a formal positive charge on AH_3. In the phosphonium ylide $\bar{C}H_2\overset{+}{P}H_3$, **10.74**, the $\pi^*_{PH_3}$ level is low in energy and the $(n_\sigma - \pi^*_{PH_3})$ interaction **(10.75)** is maximized by having a planar carbanion center.[23,32] Similarly, the nitro-

10.74   **10.75**

gen center of $N(SiMe_3)_3$ is found to be planar[33] which maximizes overlap between the lone pair on nitrogen with π^*_{Si-C}. The traditional explanation for the stability of phosphonium ylides and planarity of the carbanion center has been ascribed to the intervention of d orbitals on phosphorus [or d orbitals on silicon for $N(SiMe_3)_3$]. The interaction shown in **10.75** is topologically equivalent to one utilizing d orbitals. When electronegative groups (e.g., RO, F, etc.) are substituted at phosphorus, the $\pi^*_{PH_3}$ level is lowered in energy and becomes a better π acceptor since it is more localized on phosphorus. It should be noted from **9.7** that there are two degenerate $\pi^*_{PH_3}$ orbitals. We have focused upon the $2e_x$ component, as shown in **9.7**. Rotation of the methylene group in **10.74** by $90°$ causes n_σ to interact with the $2e_y$ component. Since the two acceptor orbitals on PH_3 are degenerate, the overlap of n_σ to them is equivalent. In reality this only needs to be a $30°$ rotation. At any intermediate geometry, n_σ will interact with a linear combination of $2e_x$ and $2e_y$. Therefore, even if there is strong π bonding in **10.74**, the barrier to rotation about the P—C bond is quite small.[23, 34]

10.6.   AH₃BH SYSTEMS

An example of a 14-electron AH₃BH system is methanol, CH_3OH. Its stable conformation **10.76** has the three electron pairs of CH_3 staggered with respect to those of OH in the bond orbital description. With a modified methyl group XCH_2, in which X refers to an electronegative atom or group (e.g., X = halogen, OH, or NH_2), XCH_2—OH gives rise to two staggered conformational possibilities **10.77** and **10.78**. In the anti conformation **10.77**, each hybrid lone pair of oxygen is antiperiplanar to an adjacent σ^*_{CH} orbital. In the gauche conformation **10.78**, one lone pair of oxygen is antiperiplanar to a σ^*_{CH} orbital while another is antiperiplanar to the σ^*_{CX} orbital. The σ^*_{CX} level is closer in energy to the oxygen lone pair, n_O, than is σ^*_{CH} so that the $(n_O - \sigma^*_{CX})$ interaction is favored over the $(n_O - \sigma^*_{CH})$ interaction. This makes the gauche conformation more stable in XCH_2—OH. This bond orbital description is a straightforward application of the $XCH_2CH_2^-$ problem.

10.76

10.77

10.78

In the MO description of the problem, oxygen lone pairs of XCH_2OH are represented by the n_π and n_σ orbitals **(10.79)**.[35, 36] The n_π level is higher lying in energy than n_σ. Of the possible interactions between these two lone pairs and the σ^*_{CX} and σ^*_{CH} orbitals of XCH_2, the $(n_\pi - \sigma^*_{CX})$ interaction is the most favorable one in terms of the energy gap. This interaction is maximized when the X—C—O—H dihedral angle θ is $90°$ as indicated in **10.80**. The presence of the $(n_\pi - \sigma^*_{CH})$, $(n_\sigma - \sigma^*_{CX})$, and $(n_\sigma - \sigma^*_{CH})$ interactions, which attain their maximum stabilization at θ values

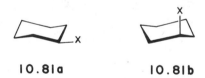

10.79

10.80

other than 90°, makes the actual θ value smaller than 90°. With the bond orbital description, the gauche conformation of XCH_2OH is predicted by simply requiring one hybrid lone pair of oxygen to be antiperiplanar to σ^*_{CX}. In specifying various conformations of low-symmetry molecules containing OH groups, we will find it convenient to adopt a bond orbital description.

A monosubstituted cyclohexane prefers an equatorial conformation **10.81a** since its alternative, an axial conformation **10.81b**, leads to unfavorable 1,3-diaxial inter-

10.81a   10.81b

actions. The conformational preference for the equatorial over the axial structure is diminished significantly in a tetrahydropyran ring which has an electronegative ligand X attached at a carbon adjacent to the oxygen (**10.82**). This kind of prefer-

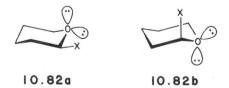

10.82a   10.82b

ential stabilization of the axial over the equatorial conformation, known as the anomeric effect,[36-39] can be readily explained. The equatorial and axial conformations of 2-X-tetrahydropyran **10.82** are simply equivalent to the anti and gauche conformations, respectively, of XCH_2OH, as depicted in **10.83**.

Besides this conformational preference, an $(n_O - \sigma^*_{CX})$ interaction brings about an important bond length change. The $(n_O - \sigma^*_{CX})$ interaction, **10.84**, is bonding between carbon and oxygen so that the C—O bond is strengthened and hence

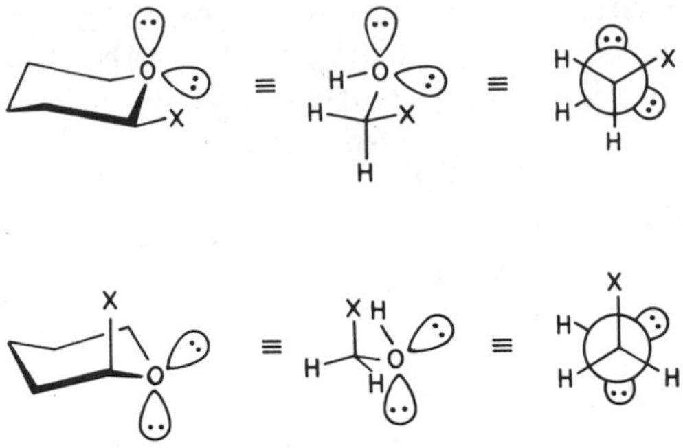

10.83

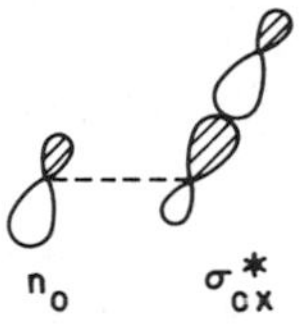

10.84

n_O σ^*_{CX}

shortened. It also leads to charge transfer into σ^*_{CX}, which weakens and hence lengthens the C—X bond. Consider, for example, the C—O and C—Cl bond lengths in cis-2,3-dichloro-1,4-dioxane, **10.85**.[38] The C2—Cl bond is axial and is antiperiplanar to a lone pair of O1. The C3—Cl bond is equatorial and is not antiperiplanar to either lone pair of O4. Consequently, the C2—O1 bond is shorter than O4—C3, while the C2—Cl bond is longer than C3—Cl.

Two conformationally locked 1,3-dioxanes are shown in **10.86**, where **10.86a**

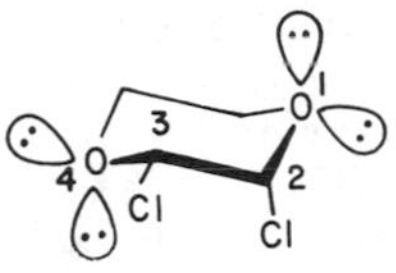

C2 – O1 $= 1.394 \overset{\circ}{A}$
C3 – O4 $= 1.428$ 10.85
C2 – Cl $= 1.819$
C3 – Cl $= 1.781$

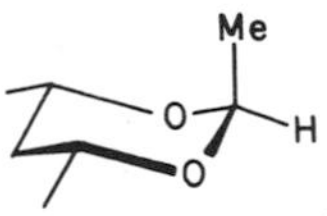

10.86a 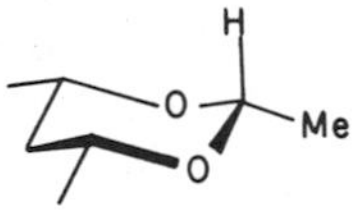   10.86b

has each oxygen lone pair antiperiplanar to the C2—H bond while **10.86b** has no oxygen lone pair antiperiplanar to the C2—H bond. Thus, the C2—H bond is weaker in **10.86a** than in **10.86b**. For the hydrogen abstraction of equation 10.8, the C2—H bond of **10.86a** is found to have a rate constant larger than that of

$$t\text{-BuO}\cdot + \text{R}—\text{H} \rightarrow t\text{-BuOH} + \text{R}\cdot \qquad (10.8)$$

10.86b by about an order of magnitude.[40] A similar observation has been noted for the cleavage of the tetrahedral intermediate in the hydrolysis of esters and amides.[41,42] Specific cleavage of a carbon–oxygen and a carbon–nitrogen bond in the tetrahedral intermediate is allowed only if the other two heteroatoms (oxygen or nitrogen) each provide a lone pair oriented antiperiplanar to the leaving O-alkyl or N-alkyl group. In general, determination of molecular reactivities by the relative orientation of the bond being broken or made and lone pairs on heteroatoms attached to the reaction center is known as stereoelectronic control.[37,41,42]

REFERENCES

1.   M.-H. Whangbo, H. B. Schlegel, and S. Wolfe, *J. Am. Chem. Soc.*, **99**, 1296 (1977).
2.   B. M. Gimarc, *Molecular Structure and Bonding*, Academic Press, New York (1979).
3.   J. P. Lowe, *J. Am. Chem. Soc.*, **92**, 3799 (1970); **96**, 3759 (1974).
4.   S. Weiss and G. Leroi, *J. Chem. Phys.*, **48**, 962 (1968).
5.   B. M. Gimarc, *J. Am. Chem. Soc.*, **95**, 1417 (1973).
6.   M. Iwasaki, K. Toriyama, and K. Nunome, *J. Am. Chem. Soc.*, **103**, 3591 (1981).
7.   L. Salem, *Electrons in Chemical Reactions*, Wiley, New York (1982).
8.   A. Lifschitz, S. H. Bauer, and E. L. Resler, *J. Chem. Phys.*, **38**, 2056 (1963).
9.   N. J. Turro, *Modern Molecular Photochemistry*, Benjamin/Cummings, Menlo Park (1978).
10. L. Salem, *Acc. Chem. Res.*, **12**, 87 (1979).
11. R. J. Buenker and S. D. Peyerimhoff, *Chem. Phys.*, **9**, 75 (1976); B. Brooks and H. F. Schaefer, *J. Am. Chem. Soc.*, **101**, 307 (1979).
12. W. G. Dauben, M. S. Kellog, J. I. Seeman, N. D. Wietmeyer, and P. H. Wendschuh, *Pure Appl. Chem.*, **33**, 197 (1973).
13. L. Salem and P. Bruckmann, *Nature*, **258**, 526 (1975).
14. L. Libit and R. Hoffmann, *J. Am. Chem. Soc.*, **96**, 1370 (1974).
15. G. C. Levy, R. L. Lichter, and G. L. Nelson, *Carbon-13 Nuclear Magnetic Resonance Spectroscopy*, 2nd edition, Wiley, New York (1980), pp. 81–86.
16. I. Wennerbeck and J. Sandstrom, *Org. Magn. Resonance*, **4**, 783 (1972), and see also L. M. Jackman, in *Dynamic Nuclear Magnetic Resonance Spectroscopy*, L. M. Jackman and F. A. Cotton, editors, Academic Press, New York (1975), p. 203.
17. R. Hoffmann, R. Gleiter, and F. B. Mallory, *J. Am. Chem. Soc.*, **92**, 1460 (1970).
18. T. Fjeldberg, A. Haaland, M. F. Lappert, B. E. R. Schilling, R. Seip, and A. J. Thorne, *J. Chem. Soc., Chem. Commun.*, 1407 (1982); D. E. Goldberg, D. H. Harris, M. F. Lappert, and K. M. Thomas, *J. Chem. Soc., Chem. Commun.*, 261 (1976); P. J. Davidson, D. H. Harris, and M. F. Lappert, *J. Chem. Soc., Dalton Trans.*, 2269 (1976).
19. J. Satge, *Adv. Organomet. Chem.*, **21**, 241 (1982).
20. B. Eisenmann and H. Schaefer, *Z. Anorg. Allg. Chem.*, **484**, 142 (1982).
21. A. H. Cowley, D. J. Mitchell, M.-H. Whangbo, and S. Wolfe, *J. Am. Chem. Soc.*, **101**, 5224 (1979).
22. A. H. Cowley, M. J. S. Dewar, D. W. Goodman, and M. C. Padolina, *J. Am. Chem. Soc.*, **96**, 2648 (1974).
23. D. J. Mitchell, S. Wolfe, and H. B. Schlegel, *Can. J. Chem.*, **59**, 3280 (1981).
24. P. Forti, D. Damiami, and P. G. Favero, *J. Am. Chem. Soc.*, **95**, 756 (1973).
25. K. Raghavachari, R. A. Whiteside, J. A. Pople, and P. v. R. Schleyer, *J. Am. Chem. Soc.*, **103**, 5649 (1981).
26. J. Skolik, P. J. Krueger, and M. Wieworowski, *Tetrahedron*, **24**, 5439 (1968); F. Bohlmann, *Chem. Ber.*, **91**, 2157 (1958); D. Kost, H. B. Schlegel, D. J. Mitchell, and S. Wolfe, *Can. J. Chem.*, **57**, 729 (1979).
27. A. J. Stone and R. W. Erskine, *J. Am. Chem. Soc.*, **102**, 7185 (1980).

28. F. H. Allen, O. Kennard, and R. Taylor, *Acc. Chem. Res.*, **16**, 146 (1983).
29. H.-B. Bürgi and J. D. Dunitz, *Acc. Chem. Res.*, **16**, 153 (1983); H.-B. Bürgi, J. D. Dunitz, and E. Sheffer, *Acta Crystallogr., Sect. B*, **30**, 1517 (1974).
30. J. P. Lowe, *J. Am. Chem. Soc.*, **94**, 3718 (1972); N. T. Anh and O. Eisenstein, *Nouv. J. Chim.*, **1**, 61 (1977).
31. T. II. Lowry and K. S. Richardson, *Mechanism and Theory in Organic Chemistry*, 2nd edition, Harper & Row, New York (1981).
32. A. W. Johnson, *Ylid Chemistry*, Academic Press, New York (1966).
33. K. Hedberg, *J. Am. Chem. Soc.*, **77**, 6491 (1955).
34. H. Lischka, *J. Am. Chem. Soc.*, **99**, 353 (1977) and references therein.
35. O. Eisenstein, N. T. Anh, Y. Jean, A. Devaquet, J. Cantacuzene, and L. Salem, *Tetrahedron*, **30**, 1717 (1974).
36. S. Wolfe, M.-H. Whangbo, and D. J. Mitchell, *Carbohyd. Res.*, **69**, 1 (1979); G. A. Jeffrey, J. A. Pople, and L. Radom, ibid., **25**, 117 (1972).
37. A. J. Kirby, *The Anomeric Effect and Related Stereoelectronic Effects at Oxygen*, Springer-Verlag, New York (1983).
38. C. Romers, C. Altona, H. R. Buys, and E. Havinga, *Top. Stereochem.*, **4**, 39 (1969).
39. N. S. Zefirov and N. M. Shektman, *Russ. Chem. Rev.*, **40**, 315 (1971).
40. A. L. J. Beckwith and C. J. Easton, *J. Am. Chem. Soc.*, **103**, 615 (1981).
41. P. Deslongchamps, *Tetrahedron*, **31**, 2463 (1975).
42. P. Deslongchamps, *Stereoelectronic Effects in Organic Chemistry*, Pergamon, Oxford (1983).

CHAPTER ELEVEN

Orbital Interactions Through Space and Through Bonds

11.1. INTRODUCTION

In the previous few chapters, we showed how to construct the orbitals of a molecule in terms of orbital interaction diagrams. Many structural and reactivity problems can be rationalized by using arguments based upon the form and energy of the frontier orbitals. Thus a primary purpose of an orbital interaction diagram is to identify and characterize these orbitals and to understand the nature of those orbital interactions that control the form of the frontier orbitals. The magnitude of an orbital interaction depends not only upon through-space, direct interaction but also upon through-bond, indirect interaction.[1-4] In the following some structural and reactivity problems of organic molecules that utilize these concepts will be examined.

11.2. IN-PLANE σ ORBITALS OF SMALL RINGS[5-7]

A. CYCLOPROPANE

In Section 5.3 we derived the orbitals of triangular H_3 and in Section 5.7 the tangential p orbitals of cyclopropenium. As shown in **11.1**, cyclopropane may be considered as made up of three methylene units. Each CH_2 unit uses its $1a_1$ and b_2 orbitals for C—H bonding. The remaining $2a_1$ and b_1 orbitals, n_σ and n_π respectively, are used to form the C—C bonds.[8,9] In terms of the n_σ and n_π orbitals of each methylene unit, the in-plane σ MOs of cyclopropane can be constructed as discussed in these earlier chapters. The composite result is shown in **11.2**, where the

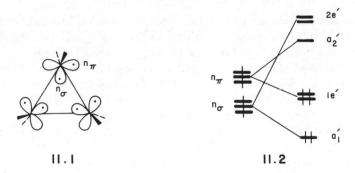

11.1   **11.2**

a_1' and $2e'$ orbitals result from the three n_σ orbitals as shown in **11.3**, and $1e'$ and a_2' result from the three n_π orbitals as shown in **11.4**. Since the $2e'$ orbitals of **11.3** and the $1e'$ set of **11.4** have the same symmetry, the two may mix together as shown in **11.5** and **11.6**.

11.3   **11.4**

11.5

11.6

The in-plane σ orbitals of **11.2** (occupied by three electron pairs) describe the three C—C bonds of a cyclopropane ring, so that the a_1' and $1e'$ orbitals are each doubly occupied. The $1e'$ set lies at a high energy compared to most C—C σ orbitals. They are composed almost totally of carbon p character. Furthermore, the p orbitals are not directed toward each other on the internuclear axis as is the case in other cycloalkanes with larger dimensions. Therefore, the $1e'$ set behaves as a good elec-tron donor for cyclopropyl substituents. For the same reasons the a_2' orbital lies lower in energy compared to most C—C σ^* orbitals and can serve as an electron ac-ceptor orbital. The three C—C bonds of cyclopropane are equivalent since both of the $1e'$ pair of orbitals are occupied. The two components, however, have different capabilities of interaction with a ring substituent. For example, dimethylcyclopropyl carbonium ion is more stable in the bisected conformation **11.7a** than in **11.7b**.[10] In fact, the rotational barrier between the two, as determined by NMR methods, is very large ($\sim$14 kcal/mol).[11] As shown in **11.8**, this arises simply because the car-

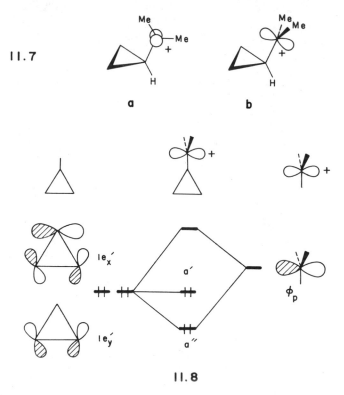

I1.7

11.8

bonium ion p orbital, ϕ_p, interacts effectively only with the $1e_x'$ orbital.[12] The resulting (ϕ_p - $1e_x'$) interaction is of the charge-transfer type, and leads to electron removal from the $1e_x'$ orbital as shown in **11.9**. A direct result is a reduction in the antibonding interaction between C2 and C3 and in the bonding interactions between C1 and C2 and between C1 and C3. As a result, a good electron acceptor such as —$\overset{+}{C}R_2$, —C≡N or —CO_2R strengthens the C2—C3 bond while weakening the

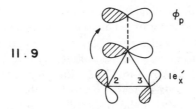

11.9

C1—C2 and C1—C3 bonds. The most direct demonstration of this phenomenon comes from crystal structures.[13] A typical example is shown in **11.10**. A related effect allows us to understand why the norcaradiene–cycloheptatriene equilibrium, shown in **11.11**, is shifted toward the side of norcaradiene when the substituent R

11.10   11.11

is a good electron acceptor. The cyclopropyl portion of the norcaradiene molecule is stabilized by electron withdrawing groups. When R = H the compound exists solely as the cycloheptatriene tautomer, and when R = CN the equilibrium lies totally on the norcaradiene side.[5]

Protonation of cyclopropane seems to occur via edge attack as shown in **11.12**.[14] This may be understood by considering the interaction of one component of the HOMO (1e′) with the empty 1s orbital of the proton (**11.13**). Notice that the form

11.12   11.13

of **11.13** is little different from the bridged isomer of $C_2H_5^+$ (see **10.51**) where the bridging hydrogen s orbital interacts with the π orbital of ethylene. **11.12** is another example of closed three-center–two-electron bonding. Bicyclo[3.1.0]hexyl tosylates undergo solvolysis at a much faster rate than cyclohexyl tosylates.[15] Labeling studies in the first set of compounds have shown that the resulting cation in **11.14** has C_{3v}

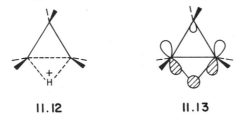

11.14

symmetry. In other words, each methine carbon possesses a formal $\frac{1}{3}$ positive charge and there is a formal bond order of $\frac{1}{3}$ between each nonadjacent methine carbon. A bonding orbital is formed, which resembles the y component of $1e'$ and allows a stabilization to be associated with the delocalization of the positive charge in **11.14**. The form of the orbital is shown in **11.15**. Note that it is the equatorial tosylate isomer in **11.14** that exhibits increased reactivity. The C—OTs σ^* orbital is then ideally situated for overlap with the filled $1e'_y$ level of the cyclopropane portion which facilitates loss of the tosylate anion in this solvolysis reaction.

11.15

Let us examine the attack of a singlet carbene on ethylene which proceeds to give cyclopropane.[16] Shown in **11.16** are two possible approaches that a carbene might undertake. The important interactions **(11.17)** are the destabilizing interaction $(n_\sigma - \pi_{CC})$ and also the stabilizing interactions $(n_\pi - \pi_{CC})$ and $(n_\sigma - \pi^*_{CC})$. In the least-motion attack, **11.16a**, the orbitals associated with the $(n_\sigma - \pi_{CC}), (n_\pi - \pi_{CC})$, and $(n_\sigma - \pi^*_{CC})$ interactions are arranged spatially as in **11.18a, 11.18b**, and **11.18c**,

11.16

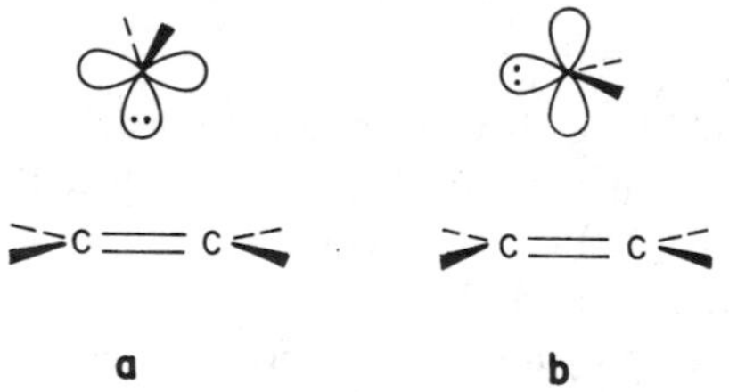

11.17

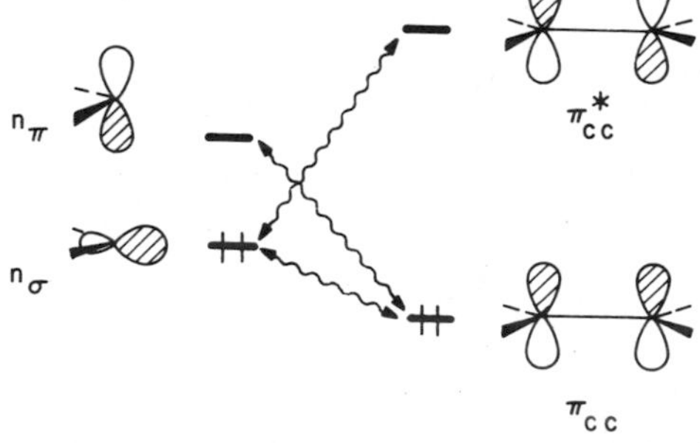

11.18

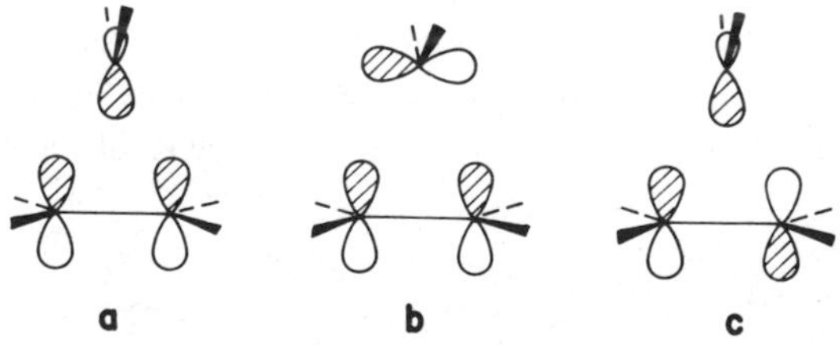

respectively. It is clear that in this geometry both of the stabilizing interactions vanish by symmetry, while the destabilizing interaction remains. In the non-least-motion attack **11.16b**, the orbitals associated with the $(n_\sigma - \pi_{CC})$, $(n_\pi - \pi_{CC})$, and $(n_\sigma - \pi_{CC}^*)$ interactions are arranged as in **11.19a**, **11.19b**, and **11.19c**, respectively. Here

11.19

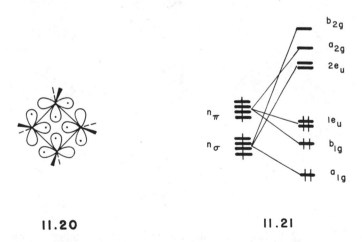

neither of the stabilizing interactions vanishes, while the destabilizing interaction is diminished in magnitude. Consequently, the initial approach of a carbene to ethylene is predicted to undergo a non-least-motion approach such as **11.16b**.

B. CYCLOBUTANE

By analogy with cyclopropane, cyclobutane may be constructed from four methylene units **(11.20)**.[17] For convenience, the molecule will be considered to have a square planar rather than puckered structure. In terms of the n_σ and n_π orbitals of each methylene, the σ MOs of cyclobutane are obtained by extending the results of Sections 5.4 and 5.7. This is summarized in **11.21**. The a_{1g}, $2e_u$, and b_{2g} orbitals

11.20

11.21

b_{2g}

a_{2g}

$2e_u$

n_π

$1e_u$

n_σ

b_{1g}

a_{1g}

arise largely from the four n_σ orbitals as shown in **11.22**, while the b_{1g}, $1e_u$, and a_{2g} orbitals arise from the four n_π orbitals as shown in **11.23**. The $2e_u$ orbital of **11.22** and the $1e_u$ orbital of **11.23** have the same symmetry, and so mix together as shown in **11.24** and **11.25**. Because of the flexibility allowed in the description of degenerate sets of orbitals, the $1e_u$ set of **11.24** may be combined to give the alternative set of **11.26**. Similarly the $2e_u$ pair of **11.25** may be manipulated to give the

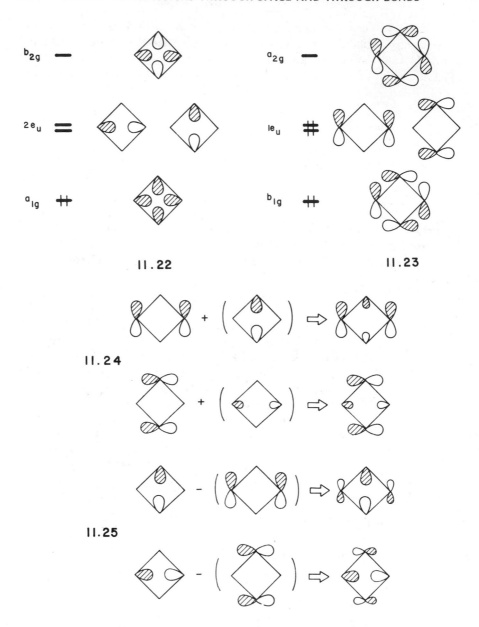

orbitals of **11.27**. With four electron pairs, the four σ-bonding orbitals a_{1g}, b_{1g}, and $1e_u$ are occupied and the four C—C σ bonds of cyclobutane result.

Consider now the concerted dimerization of ethylene in **11.28**. The frontier orbital interactions of this reaction are the $(\pi_{CC} - \pi_{CC})$, $(\pi_{CC} - \pi_{CC}^*)$, and $(\pi_{CC}^* - \pi_{CC})$ interactions shown in **11.29**. The one destabilizing interaction $(\pi_{CC} - \pi_{CC})$ is nonzero as can be seen from **11.30a**, while the stabilizing interactions $(\pi_{CC} - \pi_{CC}^*)$ and $(\pi_{CC}^* - \pi_{CC})$ vanish as shown in **11.30b** and **11.30c**, respectively. As two ground

11.26

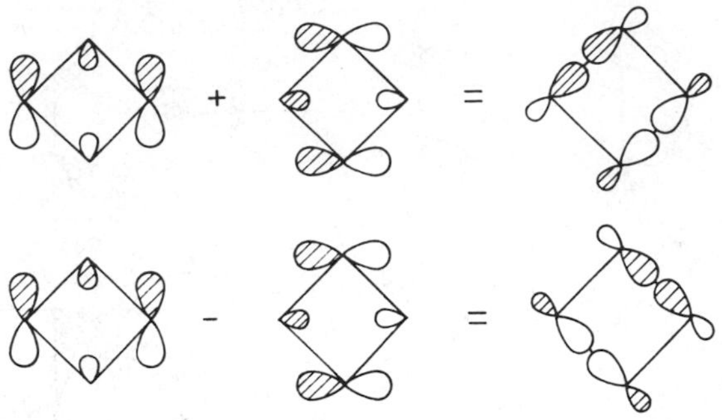

11.27

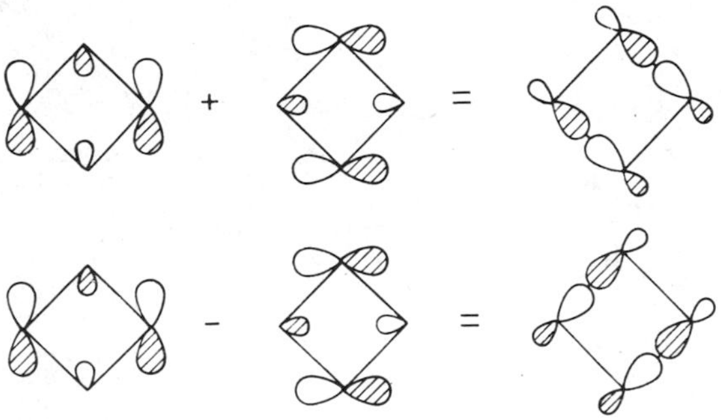

11.28

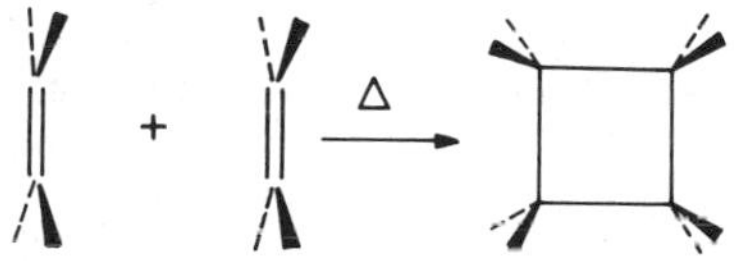

11.29

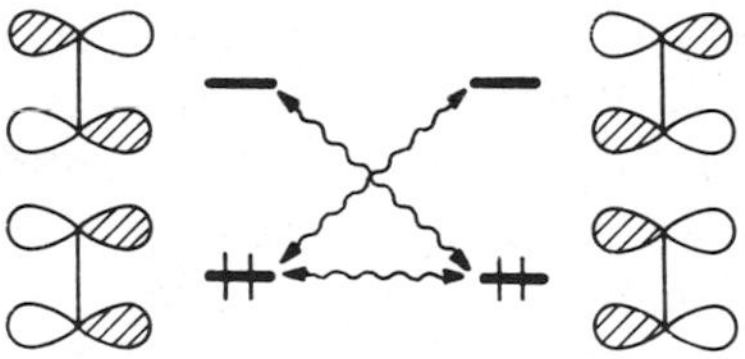

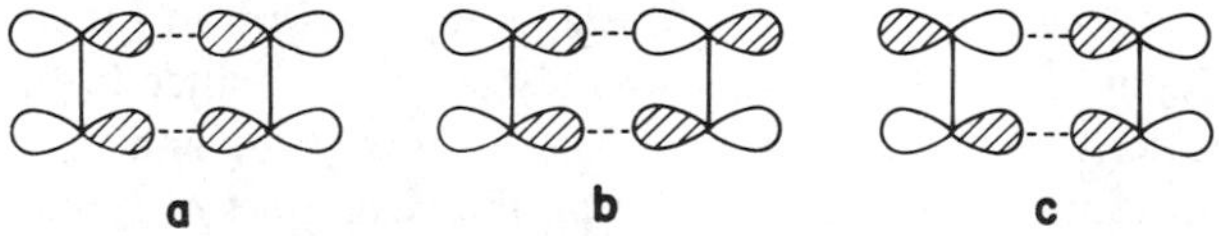

11.30

state ethylene molecules approach each other in **11.28**, a strongly repulsive barrier is encountered. The orbital correlation diagram for this reaction is shown in Figure 11.1 On the reactant side, the π_{CC} orbitals of two ethylenes lead to the π_+ and π_-

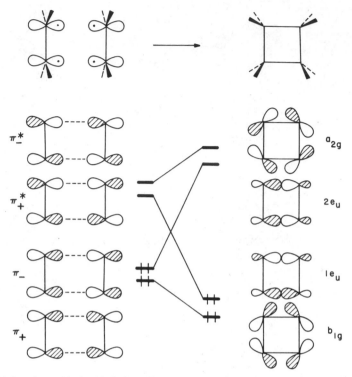

FIGURE 11.1. An orbital correlation diagram for the concerted dimerization reaction of ethylene.

MOs, while the two π^*_{CC} orbitals lead to the π^*_+ and π^*_- MOs. In the initial stage of the dimerization, the interaction between two ethylenes is weak so that π_+ and π_- lie far below the π^*_+ and π^*_- levels, so that only π_+ and π_- are occupied. Of the σ orbitals of cyclobutane described earlier, only those related to the π_+, π_-, π^*_+, and π^*_- levels by symmetry are shown in Figure 11.1. Not all the occupied MOs of the reactant lead to occupied orbitals in the product. In particular, π_- correlates with one component of the empty $2e_u$ set in cyclobutane. The π^*_+ combination ultimately becomes one component of the filled $1e_u$ set in cyclobutane. So the reaction is symmetry forbidden. The reader should carefully compare the correlation diagram for ethylene dimerization here with the $H_2 + D_2$ reaction in Figure 5.8. The two correlation diagrams are very similar, as they should be, since in this instance the spatial distributions of π and π^* are similar to those of σ_g and σ_u, respectively, in H_2. These two reactions are probably the premier examples of symmetry-forbidden reactions. A related symmetry-allowed example is the concerted cycloaddition of ethylene and butadiene, the Diels–Alder reaction. We shall not cover the orbital symmetry rules for organic, pericyclic reactions. There are several excellent reviews that the reader should consult.[18-23] But it should be pointed out that the orbital symmetry rules have stereochemical implications in terms of the reaction path and products formed. The development of these rules by Woodward and Hoffmann revolutionized the way organic chemists think about reactions. We shall return to some reactions related to ethylene dimerizations in later chapters where a transition metal ML_n unit is inserted between two methylene units of the olefins.

Tricyclooctadiene **11.31** readily rearranges to semibullvalene **11.32b** at room temperature. The experimental evidence for this reaction is consistent with the formation of the diradical **11.32a**. However, a similar molecule **11.33** is found to be quite stable. Note that **11.31** contains a cyclobutane ring 1,3-bridged by two ethylene units, and **11.33** contains the same four membered ring but 1,3-bridged by two butadiene units. The above-mentioned difference between **11.31** and **11.33** can be studied in terms of the simplified model systems **11.34** and **11.35**, respectively.[24]

11.31

11.32a

11.32b

11.33   11.34   11.35

The frontier orbital interactions between the ethylene and cyclobutane units in **11.34** are shown in Figure 11.2. It is easy to see that the HOMO–LUMO gap of **11.34** is smaller than that for ethylene itself because of interaction with the cyclobutane ring. The π orbital overlaps with one component of the filled $1e_u$ set on the cyclobutane portion and is destabilized, whereas π^* is stabilized by the empty a_{2g} orbital. The opposite effect occurs with the butadiene unit of **11.35**. The frontier orbital interactions between the butadiene and cyclobutane units in **11.35** are shown in Figure 11.3. The HOMO–LUMO gap of **11.35** is enhanced with respect to that of butadiene. Now the HOMO of butadiene, π_2, is of correct symmetry to be stabilized by a_{2g} on the cyclobutane fragment and the LUMO, π_3, is destabilized by one component of the $1e_u$ set. Notice that these different results for the two molecules emerge as a direct consequence of the nodal properties of the π orbitals of the ethylene and butadiene fragments. A molecule with a small energy gap between the HOMO and LUMO is susceptible to a distortion that allows orbital mixing between them (i.e., a second-order Jahn–Teller distortion). According to Figure 11.2, the LUMO of **11.34** has some contribution of a_{2g} which is antibonding between the carbon atoms of the cyclobutane ring. Thus any asymmetrical distortion of **11.34**

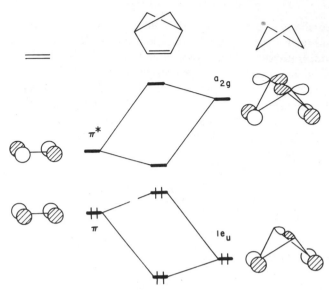

FIGURE 11.2. A simplified orbital interaction diagram for **11.34**.

that allows the mixing of the LUMO into the HOMO would effectively weaken the bonding between the carbon atoms of the cyclobutane ring. This is consistent with the cleavage of a C—C bond in **11.31** to produce the diradical **11.32a**. Since the HOMO–LUMO gap in **11.35** is predicted to be quite large, the driving force for an analogous second-order distortion is lost. Although **11.33** should have approximately the same strain energy as **11.31**, nonetheless it is thermally more stable.[24]

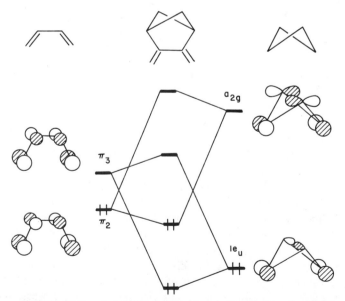

FIGURE 11.3. A simplified orbital interaction diagram for **11.35**.

11.3. THROUGH-BOND INTERACTION[1-4]

A. THE NATURE OF THROUGH-BOND COUPLING

Conceptual fragmentation of a molecule allows us to readily construct its MOs in terms of linear combinations of fragment orbitals. How a given molecule should be fragmented depends upon the simplicity of the resulting orbital interaction picture. Given a pair of orbitals located on two molecular fragments, direct through-space interaction between them leads to two energy levels described in the usual in-phase and out-of-phase combinations. Typically, the in-phase combination is lower in energy than the out-of-phase one. However, this level ordering is not always found to be correct if the fragment orbitals involved further interact with the orbitals of a third fragment. We encountered a similar situation in Chapter 7.6.

As an example, let us consider the two lone pair levels in diazabicyclooctane **11.36**. The direct, through-space interaction between the hybrid lone pairs of nitrogen lead to the symmetry-adapted levels n_+ and n_- shown in **11.37**. Because of the

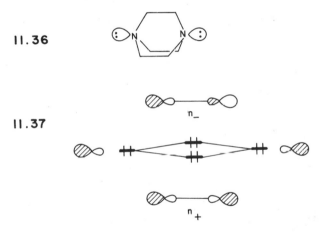

11.36

11.37

large distance between the two hybrid lone pairs, the energy difference between n_+ and n_- is expected to be small. However, both calculation and experiment show a large splitting between the n_+ and n_- levels. The first two bands in the photoelectron spectrum of **11.36** (maxima at 7.52 and 9.65 eV), to which n_+ and n_- is the major contributor, show vibrational fine structure.[25,26] Comparison of this with the normal vibrational coordinates[27] of **11.36** shows that the vibrational frequencies of the bands centered at 7.52 and 9.65 eV are primarily associated with C—C stretching and CNC bending, respectively.[26] In addition, analysis of the overlap populations of the C—C and C—N bonds in **11.36** and its cation suggests that electron removal from n_+ and n_- should mainly induce C—C stretching and CNC bending deformations of the molecular frame of **11.36**, respectively.[26] Consequently, the n_+ and n_- levels are responsible for the bands centered at 7.52 and 9.65 eV, respectively,[26] namely, n_+ lies higher in energy than n_-. This counterintuitive ordering is due to the fact that each hybrid lone-pair orbital of nitrogen cannot be considered in isolation. They do interact with the σ and σ^* orbitals associated with the intervening C—C bonds. This kind of indirect effect is called through-bond interaction. The essence of through-bond interaction for the diazabicyclooctane system can be analyzed by a simplified system **11.38**. Shown in **11.39** are the interactions of the n_+ and n_- orbitals with the C—C σ and σ^* orbitals. Compared with the through-space 1,4-interaction between the nitrogen centers, the 1,2-interactions $(n_+ - \sigma)$ and $(n_- - \sigma^*)$ are strong. As a result, the energy of the n_+ and n_- levels are raised and lowered, respectively, by the σ and σ^* levels. This leads to a large splitting between the levels which we can still describe approximately as n_+ and n_-. A closely related example is the pyrazine molecule, **11.40**. The n_+ combination of the nitrogen lone pairs has been experimentally shown to lie 1.72 eV higher in energy than the n_- combination.[28]

11.38

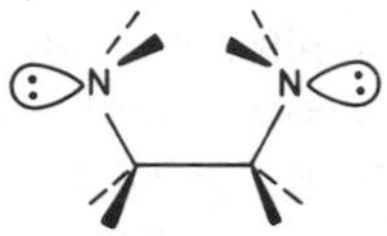

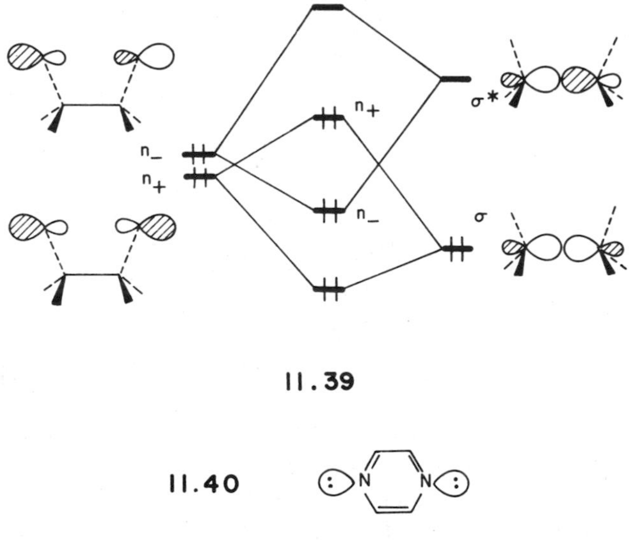

11.39

11.40

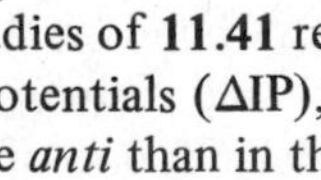

Competition between through-space and through-bond interactions often gives rise to interesting results. For example, let us consider tricyclo-3,7-octadiene which can adopt either the *anti* or the *syn* structure shown in **11.41a** and **11.41b**, respectively. In the *anti* structure the through-space interaction between the double bonds is negligible. However, photoelectron studies of **11.41** reveal that the difference between the first and second ionization potentials (ΔIP), which is a measure of the extent the π levels are split, is larger in the *anti* than in the *syn* structure (ΔIP = 0.97 and 0.36 eV for **11.41a** and **11.41b**, respectively).[29] As shown in **11.42**, the through-space interaction in **11.41a** is small so that π_+ is only slightly lower than π_-. Following our discussion of Section 11.2B, the in-plane σ orbitals of the cyclobutane ring

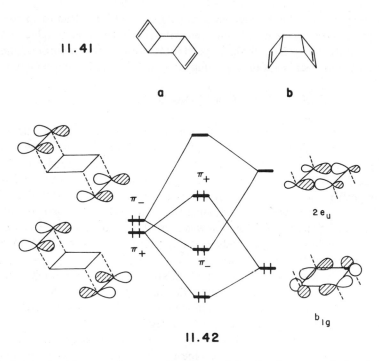

11.41

a   b

11.42

that can interact with π_+ and π_- are the b_{1g} and $2e_u$ levels, respectively, shown in **11.42**. Due to the through-bond interactions $(\pi_+ - b_{1g})$ and $(\pi_- - 2e_u)$, the π_+ and π_- levels are raised and lowered, respectively. The through-space interaction of **11.41b** is large as shown in **11.43**, which gives rise to a large splitting between π_+

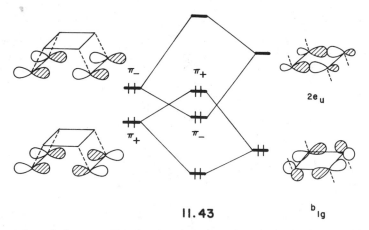

11.43

and π_-, with π_+ below π_-. This level ordering is altered by the through-bond interactions $(\pi_+ - b_{1g})$ and $(\pi_- - 2e_u)$. What determines the magnitudes of these interactions in **11.41a** or **11.41b** are the 1,2-interactions of the double bond carbon atoms with the cyclobutane ring. Thus the through-bond interactions are nearly the same in magnitude in **11.41a** and **11.41b**. Consequently, the raising of π_+ and the lowering of π_- by the through-bond interactions in **11.41b** lead to an effectively smaller gap between π_- and π_+, compared with the corresponding value of **11.41a**. In other words, the greater through-space interaction in the *syn* isomer induces a smaller ΔIP value because of the opposing effect of through-bond interaction.

B. OTHER THROUGH-BOND COUPLING UNITS

As discussed above, the concept of through-bond interaction arises when a molecule is regarded as being composed of three fragments. Two of these fragments typically carry lone pair orbitals, radical p orbitals or double bond π orbitals, the in-phase and out-of-phase combination of which lead to the frontier orbitals of the whole molecule. The energy ordering of these combinations is affected by the 1,2-interactions associated with the remaining fragment, a through-bond coupling unit. Our discussion of the previous section was limited to through-bond interactions occurring via three intervening σ bonds. It is important to examine how through-bond interactions are affected by the length of the coupling unit.

Schematically shown in **11.44** and **11.45** are two p orbitals coupled via the σ-bond framework of four and five single bonds, respectively. The 1,2-interactions associated with both ends of each coupling unit are indicated by dashed lines. Thus **11.44a** is a simplified representation of 2,7-dehydronaphthalene, **11.46a**; **11.45a** is analogous to the extended conformation of 1,3-diaminopropane **11.46b**; and the coupling of the π orbitals in the basketane precursor **11.46c** can be represented by

11.44c. In the examples of **11.46** the σ-bond framework which is involved in the through-bond coupling (i.e., the coupling unit) is highlighted by a thickened line. For simplicity, we may represent each σ bond of a coupling unit by σ and σ^* bond orbitals. Then the HOMO and LUMO of the coupling unit are approximated by the most antibonding combination of the σ orbitals and by the most bonding combination of the σ^* orbitals, respectively, as summarized in **11.47**.[30] By considering only

11.47

a   b

the 1.2-interactions with these frontier orbitals, through-bond interactions in **11.44–11.45** can be easily estimated. For example, in **11.44c**, the through-space and through-bond interactions reinforce each other as shown in **11.48**. In **11.44b**, the through-space interaction is negligible but the through-bond interaction is as strong as that in **11.44c** to a first approximation (see **11.49**). Consequently, the

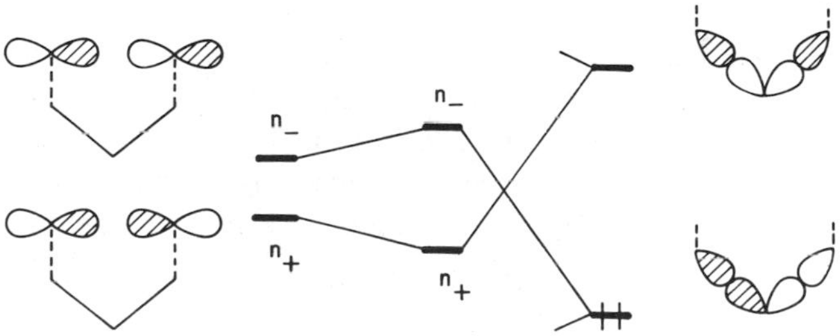

11.48

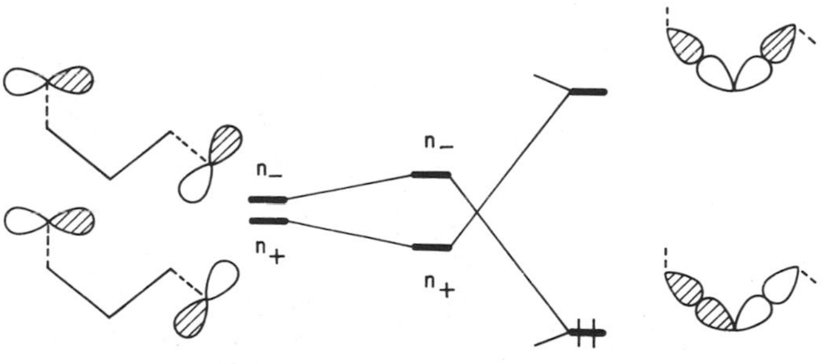

11.49

energy gap between n_+ and n_- is larger in **11.44c** than in **11.44b**. This conclusion remains valid when the p orbitals of **11.44–11.45** are replaced by hybrid lone pair orbitals or by double bond π orbitals. Photoelectron studies show that the ΔIP values of **11.50a** and **11.50b** are 1.26 and 0.44 eV, respectively,[13] consistent with the analysis given above.

11.50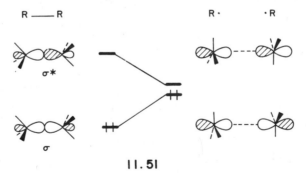

a

b

The magnitude of the through-space interaction decreases sharply with the distance between the interacting groups. However, the magnitude of a through-bond interaction attenuates slowly with increasing the length of its coupling unit. This is due to the fact that a through-bond interaction is governed primarily by the 1,2-interactions associated with both ends of a coupling unit. In the HOMO and LUMO of a coupling unit, the weights on both ends diminish slowly as the length of a coupling unit increases.

11.4. BREAKING A C—C BOND

In previous chapters we discussed the ways reagents attack organic substrates (e.g., nucleophilic substitution, addition reactions and elimination reactions). Let us examine some interesting ways that C—C bonds can be broken in a homolytic manner which, in turn, lead to unusual predictions concerning reaction paths or intermediate structures. We start this discussion with a simple example of homolytic C—C bond cleavage in an alkane. A correlation diagram for this process is displayed in **11.51**.

R —— R   R ·   · R

$\sigma*$

σ

11.51

The σ bonding level between the two alkyl groups rises in energy as the C—C bond distance increases and $\sigma*$ is stabilized. When the distance between the two radicals is large, the overlap between the two p orbitals is negligible so we again have a typical diradical situation of two closely spaced orbitals with two electrons to put in them.[32] We will sidestep the issue of how to correctly describe the electronic states (see Section 8.10) that are created on the product side of **11.51**, but note that the σ level at no time crosses $\sigma*$ along the reaction path.

11.52a is an example of a class of molecules called propellanes. The strain energy of the molecule has been estimated to be about 90 kcal/mol[33] so one might think that the molecule is incapable of existence. The central C—C bond is expected to

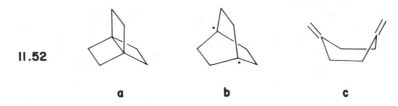

II.52

a   b   c

be extraordinarily weak and rupture of it to the diradical **11.52b** should require little, if any, activation energy. Yet an amide derivative of **11.52a** has been isolated at $-30°C$.[34] It decomposes, presumably via **11.52b**, with an activation energy of 22 kcal/mol. The reason behind the unexpectedly high activation energy associated with this homolytic cleavage is outlined in **11.53**.[35] In diradical **11.52b** the through-bond interaction provided by the three C—C bonds parallel to the two p orbitals ensures that the n_- level lies below n_+, with a large splitting between them. This is identical to the lone pair-splitting problem in diaza[2.2.2]bicyclooctane, **11.36**. By correlating orbitals that have the same symmetry in **11.53** (a mirror plane which bisects the C—C linkage), one can see that this reaction is symmetry forbidden. Therefore, **11.52a** is separated from singlet **11.52b** by a sizable barrier.

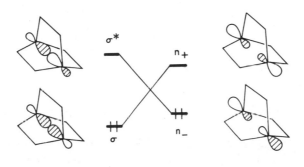

II.53

When two or more stable conformations of a molecule, related to each other by a simple bond stretching, differ in their electronic configuration, this is called bond-stretch isomerization. In the **11.52a** → **11.52b** bond-stretch isomerization the singlet diradical is actually not stable. It undergoes a symmetry-allowed fragmentation to 1,4-bismethylenecyclohexane, **11.52c**.[35] But there are other ways to stabilize a diradical. Solvolysis of **11.54** or protonation of **11.55** initially generates the carbonium ion **11.56**.[36] The central C—C bond in the bicyclobutane portion of the molecule is again weakened by strain and so bond-stretch isomerization of **11.56** to **11.57** should be possible. Now the antisymmetric combination of the two p orbitals which form the diradical is markedly stabilized by a through-space interaction with the empty p orbital on the carbonium carbon. This is illustrated in **11.58**. The symmetric combination, **11.59**, cannot interact with the carbonium carbon p orbital. So there is a large splitting between filled **11.58** and empty **11.59**. A correlation diagram for the $11.56 \rightarrow 11.57$ interconversion in the parent compound $C_5 H_5^+$ is illustrated in Figure 11.4.[37] On the left side are the σ and σ^* orbitals of the C—C bond which will be broken along with the p orbital at C1. These orbitals are classified as being symmetric (S) or antisymmetric (A) with respect to a mirror plane which contains C1, C2, and C4 and bisects the C3—C5 bond. The right side of this figure displays the relevant three orbitals of diradical 11.57. A bond-stretch process that conserves this mirror plane is symmetry forbidden. Calculations at all levels[38,39] indicate that the "closed" bond-stretch isomer (analogous to 11.56) is less stable than the "open" form (11.57). All experimental evidence[38,39] is also consistent with this, so the through-space interaction provided by 11.58 does indeed greatly stabilize this diradical.

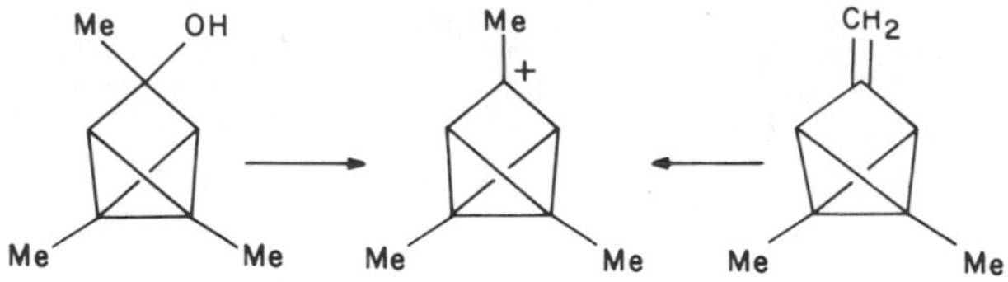

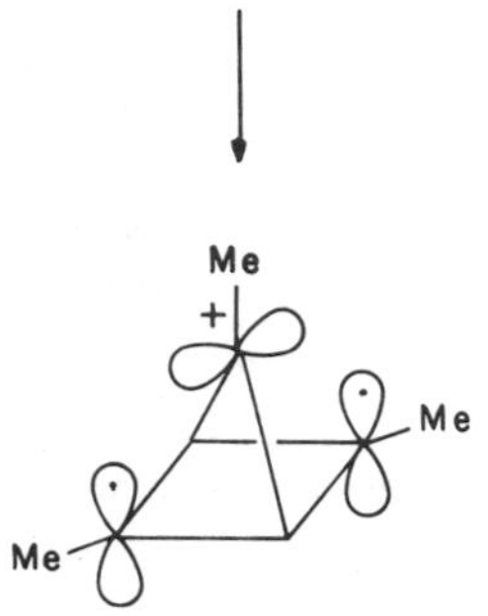

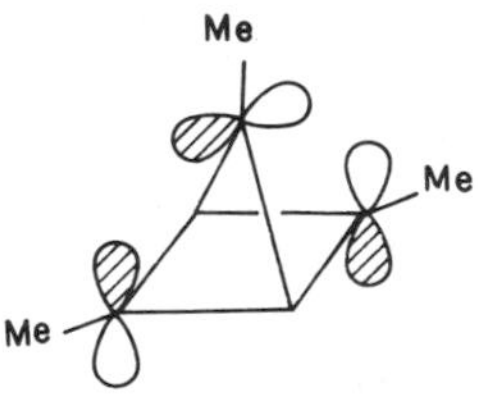

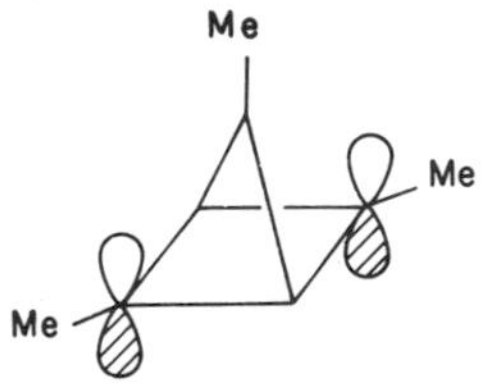

Let us now consider in more detail how the $11.56 \rightarrow 11.57$ rearrangement is likely to proceed. A measure of how much the C3—C5 bond is stretched can be given by the angles α and β, as defined in 11.60. Implicit in the correlation diagram of Figure 11.4 was that $\alpha = \beta = {\sim}140°$ for the "closed" isomer and ${\sim}90°$ for the "open" one. Conservation of the mirror plane requires that $\alpha = \beta$ for all points along the reaction path. When $\alpha = \beta = 120°$ the geometry of the molecule is D_{3h}. The three MOs used in Figure 11.4 then transform as a_2' and e' symmetry, 11.61. This is the geometry where the HOMO—LUMO crossing occurs. The e' set is half-filled and

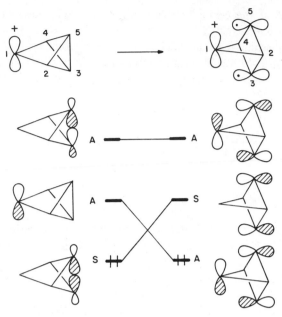

FIGURE 11.4. A simplified correlation diagram for bond stretch isomerization in $C_5H_5^+$.

there must be a non-least-motion way to avoid it. This requires that α and β in **11.60** will vary at different rates along the true minimum energy reaction path. The potential energy surface has the shape shown in Figure 7.5. The coordinates along two edges of the triangle in this figure are α and β. Structures A, B, and C are three possible (equivalent) "open" isomers in $C_5H_5^+$. The transition state(s) which interconnects them, D, is actually the "closed" bond-stretch isomer. Finally the high

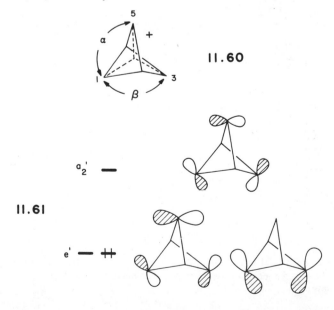

energy point E at the center of this surface represents the Jahn–Teller unstable structure where $\alpha = \beta = 120°$.

There is something more unusual about the electronic structure of the "open" isomer, redrawn in **11.62**. A bond-switching process converts **11.62** into **11.63** by way of **11.64**. This is a symmetry-allowed reaction.[37] Notice that the symmetry of

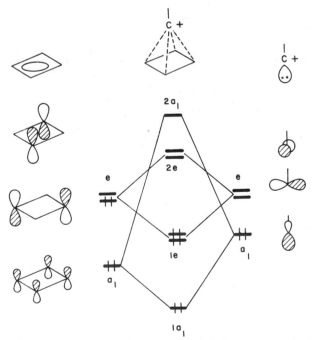

II.62   **II.64**   **II.63**

11.62 and **11.63** is C_{2v} while that in **11.64** is C_{4v}. All experimental evidence on derivatives of $C_5H_5^+$ indicate that the symmetry of the parent cation is C_{4v}.[38,39] In other words, **11.64** is the ground state of this stabilized diradical, not **11.62**. The bonding in **11.64** is best described in terms of the interaction diagram in Figure 11.5. On the left side are the π orbitals of cyclobutadiene (see Sections 5.4 and 5.7) and on the right are the fragment orbitals for CH^+ (see Section 9.1A). The fragment orbitals are given symmetry labels consistent with the C_{4v} symmetry of the molecule. Both fragment orbitals of a_1 symmetry combine to produce a bonding ($1a_1$) and

FIGURE 11.5. An orbital interaction diagram for $C_5H_5^+$. The fragment orbitals are labeled according to the C_{4v} symmetry of the molecule. The highest energy C_4H_4 orbital has been left off for simplicity.

antibonding ($2a_1$) MO. Likewise the e set on both fragments overlap substantially to produce a bonding ($1e$) and antibonding ($2e$) set. The highest π level of cyclobutadiene remains nonbonding. There are a total of six electrons in these valence orbitals which fill the $1a_1$ and $1e$ levels. These filled bonding MOs are shown in **11.65**.

II.65

$1a_1$ $1e$

It can be seen that $1a_1$ and one member of the $1e$ set correspond to symmetry-adapted combinations of the two C—C σ bonds between the apical C—H unit and the four-membered ring as were explicitly drawn out in the classical structures of **11.62** or **11.63**. The other member of the $1e$ set is then identical to the lowest (filled) orbital of Figure 11.4. So this delocalized picture in Figure 11.5 suggests that there is actually little difference between **11.62** (and **11.63**) and **11.64**. In fact, **11.62** and **11.63** can be regarded as resonance structures which contribute to the electronic structure of the $C_5H_5^+$ isomer with C_{4v} symmetry. Such delocalized pictures of cage molecules will form the basis of Chapter 22. The C_{4v} structure **11.64** of $C_5H_5^+$ is not unusual when viewed in the context of cage and cluster molecules in general. There are clearly four equivalent C—C distances between the apical carbon and the remaining four basal carbon atoms and a total of six electrons in the bonding MOs (**11.65**). Therefore, one can view each apical–basal interaction as containing 1.5 electrons. In other words, this is another example of electron deficient bonding. Two-center–two-electron bonding between the apical and basal carbons is not going to be an energetically favorable situation. There would then be eight electrons in the interaction diagram of Figure 11.5. The extra two electrons will need to be placed in the antibonding $2e$ set.

A number of other carbocations similar to $C_5H_5^+$ have been studied, the most notable being $C_6Me_6^{2+}$. It has been demonstrated[40,41] that the structure of this cation is C_{5v}, **11.66**, rather than a rapidly equilibrating series of classical structures, **11.67**, where a bond-switching process permutes three two-center–two-electron

II.66   **II.67**

bonds around the five-membered ring. An orbital interaction diagram for **11.66** where all methyl groups have been replaced by hydrogen atoms is shown in Figure 11.6. The π orbitals of a cyclopentadienyl cation have been taken from Sections

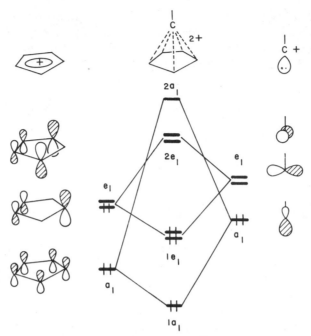

FIGURE 11.6. An orbital interaction diagram for $C_6H_6^{2+}$. The fragment orbitals are labeled according to the C_{5v} symmetry of the molecule.

5.6 and 5.7. The bonding in $C_6H_6^{2+}$ has several features in common with $C_5H_5^+$. Again the three fragment orbitals of the CH^+ "capping" unit find good matches in overlap and energy with the three lowest π orbitals in the cyclopentadienyl cation fragment. Three bonding MOs ($1a_1 + 1e_1$) are produced. Six electrons nicely fill these orbitals. A compound with two more electrons, C_6H_6, cannot have this C_{5v} structure because the antibonding $2e_1$ is half-filled. The intracluster bonding between the apical CH unit and the five basal carbons is again electron deficient.

The pattern we have presented here for $C_5H_5^+$ and $C_6H_6^{2+}$ can be extended to any type or number of capping units and any size of carbocyclic ring.[42] A stable electronic configuration is achieved when there are six intracluster electrons partitioned between the capping unit and carbocyclic ring. The C_{4v} structure of $C_5H_5^+$ and even the existence of $C_6H_6^{2+}$ may be unsettling to organic chemists. We have noted above that such "unusual" arrangements fit in to quite a general pattern when considered in the wider context of cage and cluster molecules. In fact, $C_5H_5^+$ and $C_6H_6^{2+}$ are a small subset of fully categorized compounds that have identical shapes and bonding features. We shall return to $C_5H_5^+$ and $C_6H_6^{2+}$ to explore these relationships in the last two chapters of this book.

REFERENCES

1.   R. Hoffmann, *Acc. Chem. Res.*, **4**, 1 (1971).
2.   R. Gleiter, *Angew Chem. Intern Ed. Engl.*, **13**, 696 (1974).
3.   M. N. Paddon-Row, *Acc. Chem. Res.*, **15**, 245 (1982).

4. J. W. Verhoeven, *Rec. J. Roy. Neth. Chem. Soc.*, **99**, 369 (1980).
5. R. Hoffmann, *XXIIIrd International Congress of Pure and Applied Chemistry*, Vol. 2, Butterworths, London (1971), p. 233.
6. W. L. Jorgensen and L. Salem, *The Organic Chemist's Book of Orbitals*, Academic Press, New York (1973).
7. R. Gleiter, *Topics Current Chem.*, **86**, 199 (1979).
8. A. D. Walsh, *Trans. Faraday Soc.*, **45**, 179 (1949).
9. E. Honegger, E. Heilbronner, A. Schmelzer, and W. Jian-Qi, *Isr. J. Chem.*, **22**, 3 (1982).
10. C. U. Pittman, Jr. and G. A. Olah, *J. Am. Chem. Soc.*, **87**, 2998 (1965).
11. D. S. Kabakoff and E. Namanworth, *J. Am. Chem. Soc.*, **92**, 3234 (1970).
12. W. J. Hehre, *Acc. Chem. Res.*, **8**, 369 (1975).
13. M. A. M. Meester, H. Schenk, and C. H. MacGillavry, *Acta Cryst.*, **B27**, 630 (1971).
14. C. J. Collins, *Chem. Rev.*, **69**, 543 (1969).
15. S. Winstein and J. Sonnenberg, *J. Am. Chem. Soc.*, **83**, 3235, 3244 (1961).
16. R. Hoffmann, *J. Am. Chem. Soc.*, **90**, 1475 (1968); N. G. Rondan, K. N. Houk, and R. A. Moss, ibid, **102**, 1770 (1980); P. H. Mueller, N. G. Rondan, K. N. Houk, J. F. Harrison, D. Hooper, B. W. Willen, and J. F. Leibman, ibid., **103**, 5049 (1981); R. A. Moss, W. Guo, D. Z. Denney, K. N. Houk, and N. G. Rondan, ibid., **103**, 6164 (1981).
17. R. Hoffmann and R. B. Davidson, *J. Am. Chem. Soc.*, **93**, 5699 (1971).
18. R. B. Woodward and R. Hoffmann, *The Conservation of Orbital Symmetry*, Academic Press, New York (1970).
19. T. L. Gilchrist and R. G. Storr, *Organic Reactions and Symmetry*, Cambridge Univ. Press, London (1972).
20. A. P. Marchand and R. E. Lehr, editors, *Pericyclic Reactions*, Vols.1 and 2, Academic Press, New York (1977).
21. I. Fleming, *Frontier Orbitals and Organic Chemical Reactions*, Wiley, London (1976).
22. K. N. Houk, *Acc. Chem. Res.*, **8**, 361 (1975).
23. Nguyen Trong Anh, *Die Woodward-Hoffmann-Regeln und ihre Anwendung*, Verlag Chemie, Weinheim (1972).
24. W. L. Jorgensen and W. T. Borden, *J. Am. Chem. Soc.*, **95**, 6649 (1973).
25. P. Bischof, J. A. Hashmall, E. Heilbronner, and V. Hornung, *Tetrahedron Lett.*, 4025 (1969).
26. E. Heilbronner and K. A. Muszkat, *J. Am. Chem. Soc.*, **92**, 3818 (1970).
27. P. Bruesch, *Spectrochim. Acta*, **22**, 861, 867 (1966); P. Bruesch and Hs. H. Gunthard, ibid., **22**, 877 (1966).
28. K. A. Muszkat and J. Schaublin, *Chem. Phys. Lett.*, **13**, 301 (1972).
29. R. Gleiter, E. Heilbronner, M. Heckman, and H.-D. Martin, *Chem. Ber.*, **106**, 28 (1973).
30. J. W. Verhoeven and P. Pasman, *Tetrahedron*, **37**, 943 (1981).
31. H.-D. Martin and R. Schwesinger, *Chem. Ber.*, **107**, 3143 (1974).
32. W. T. Borden, editor, *Diradicals*, Wiley, New York (1982).
33. M. D. Newton and J. M. Schulman, *J. Am. Chem. Soc.*, **94**, 4391 (1972).
34. P. E. Eaton and G. H. Temme, *J. Am. Chem. Soc.*, **95**, 7508 (1973).
35. W.-D. Stohrer and R. Hoffmann, *J. Am. Chem. Soc.*, **94**, 779 (1972).
36. V. I. Minkin, N. S. Zefirov, M. S. Korobov, N. V. Averina, A. M. Boganov, and L. E. Nivorozhkin, *Zh. Org. Khim.*, **17**, 2616 (1981).
37. W.-D. Stohrer and R. Hoffmann, *J. Am. Chem. Soc.*, **94**, 1661 (1972).
38. H. Schwartz, *Angew. Chem.*, **93**, 1046 (1981); *Angew. Chem. Intern. Ed. Engl.*, **20**, 991 (1981).
39. V. I. Minkin and R. M. Minyaev, *Usp. Khim.*, **51**, 586 (1982); *Engl. Trans.*, 332 (1982).
40. H. Hogeveen and P. W. Kwant, *Acc. Chem. Res.*, **8**, 413 (1975).
41. H. Hogeveen and E. M. G. A. van Kruchten, *J. Org. Chem.*, **46**, 1350 (1981).
42. E. D. Jemmis and P. v. R. Schleyer, *J. Am. Chem. Soc.*, **104**, 4781 (1982).

CHAPTER TWELVE

Polyenes and Conjugated Systems

12.1. ACYCLIC POLYENES

Here we will build up in general the π orbitals of a conjugated chain of N carbon atoms. We can start from the simple case of ethylene shown in Figure 12.1 and add on an extra orbital to get to allyl (Figure 12.2). We will not spend any time describing how the form of these orbitals actually come about since this three orbital problem is identical to the H_3 problem in Section 4.8. The result is a low energy orbital, bonding between each pair of adjacent atoms, a higher energy nonbonding orbital with a node at the central atom, and a higher lying orbital which is antibonding between both pairs of adjacent atoms. Figure 12.3 shows the orbitals of the lowest few polyenes which may be built up in a similar way. Although we have drawn the carbon backbone in a straight line for simplicity, these orbitals are applicable to real systems with perhaps very different geometries. For example, the pentadienyl orbitals apply to all the species in 12.1. There are some general rules which guide us in their derivation.

12.1

1. As their energy increases, the orbitals alternate in parity with respect to a mirror plane which bisects the π system. The lowest energy orbital is always symmetric with respect to this plane.

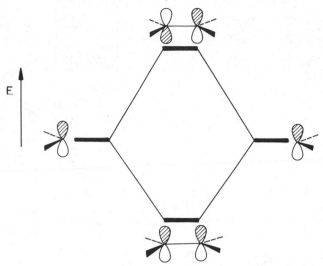

FIGURE 12.1. The π orbitals of ethylene, assembled from two CH_2 units.

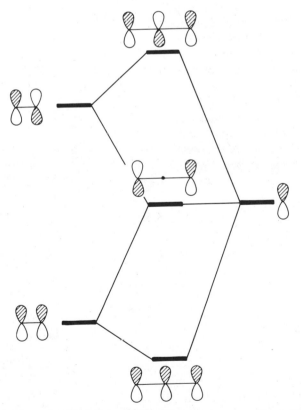

FIGURE 12.2. The π orbitals of allyl, assembled from the corresponding orbitals of a two carbon moiety (ethylene minus a hydrogen atom) and a CH_2 group.

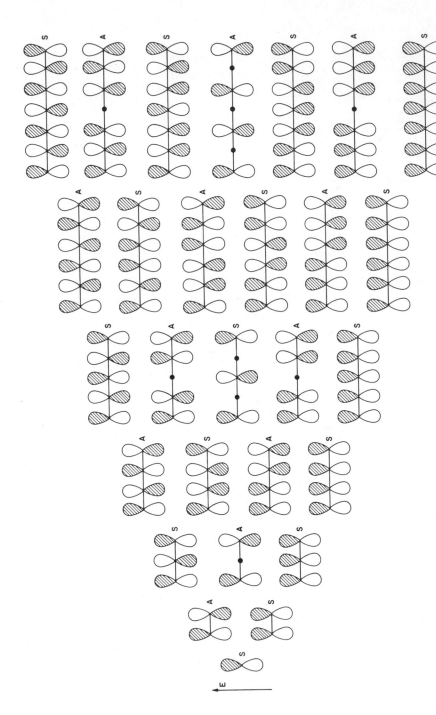

FIGURE 12.3. The $p\pi$ orbitals of the first few linear chain polyenes. No attempt has been made to represent the different orbital coefficients. The orbitals are labeled either symmetric or antisymmetric with respect to the mirror symmetry which bisects the length of the chain.

2. The number of nodes increases by one on going from one orbital to the one next highest in energy. The lowest energy orbital always has zero nodes (bonding between each pair of adjacent atoms) and the highest energy orbital always has nodes between every adjacent pair (i.e., it is antibonding between all such pairs).

3. Nodes must always be symmetrically located with respect to the central mirror plane.

4. In systems with an odd number of atoms, the antisymmetric levels always have a node at the central carbon atom.

In the next section we use simple Hückel theory to quantify these results.

12.2. HÜCKEL THEORY[1-3]

Conjugated π systems in carbon compounds represent one of the few areas where simple algebraic expressions may be easily derived for the orbital energies and wavefunctions. The basis of Hückel's approach is very simple indeed. Since all the $p\pi$ orbitals are antisymmetric with respect to reflection in the plane of the molecule, and the σ type orbitals by definition are symmetric with respect to this symmetry operation, then there is no overlap between the two sets. So one can treat the π orbitals alone when using a one-electron theory as implicitly assumed in Section 12.1. In the approach interactions are ignored between $p\pi$ orbitals located on atoms which are not directly linked via the σ framework, and the overlap integrals between all pairs of $p\pi$ orbitals, whether directly linked or not, are set equal to zero. The energy of each carbon p orbital before interaction (Coulomb integral) is put equal to α $(= e_i^0)$ and the interaction energy between two adjacent $p\pi$ orbitals equal to β $(= H_{ij})$, the resonance integral. The secular equations (Section 1.3) describing the interaction of the two $p\pi$ orbitals in ethylene are then

$$(\alpha - e)c_1 + \beta c_2 = 0$$
$$\beta c_1 + (\alpha - e)c_2 = 0 \tag{12.1}$$

where e is the energy of the resulting molecular orbital and c_1 and c_2 the coefficients of the atomic orbitals on atoms 1, 2, that is, $\psi = c_1 \chi_1 + c_2 \chi_2$. The secular determinant is then

$$\begin{vmatrix} \alpha - e & \beta \\ \beta & \alpha - e \end{vmatrix} = 0 \tag{12.2}$$

with roots

$$e = \alpha \pm \beta \tag{12.3}$$

for the energies of the two orbitals. α, β are both negative so the plus sign refers to the bonding level and the minus sign to the antibonding one. Substitution of $e = \alpha + \beta$ into either of the equations 12.1 gives the relationship $c_1 = c_2$ and substitution of $e = \alpha - \beta$ gives $c_1 = -c_2$. Since interatomic overlap has been neglected the normalization condition is very simple and leads directly to the numerical values for

$$1 = \int \psi^* \psi \, d\tau = c_1^* c_1 \int \chi_1^* \chi_1 \, d\tau + c_2^* c_2 \int \chi_2^* \chi_2 \, d\tau$$

$$= c_1^2 + c_2^2 \tag{12.4}$$

the orbital coefficients (equation 12.5).

$$\psi_{\text{bonding}} = \frac{1}{\sqrt{2}} \, (\chi_1 + \chi_2)$$

$$\tag{12.5}$$

$$\psi_{\text{antibonding}} = \frac{1}{\sqrt{2}} \, (\chi_1 - \chi_2)$$

This then is the very simplest type of one-electron model we could imagine with a very basic form of the wavefunctions and a two-parameter form for the energies. In principle it is then a straightforward matter to generate the energy levels and orbital coefficients for any conjugated system, whether acyclic, cyclic, polycyclic, or generally complex, by solution of the relevant determinant. Equation 12.6 shows the secular determinant for allyl **12.2**. Since atoms 1 and 3 are not directly con-

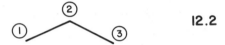

12.2

nected, a zero appears in the 1, 3 and 3, 1 positions. The roots, obtained by solution of the resulting cubic equations are shown in Figure 12.4a. The orbitals of bu-

$$\begin{vmatrix} \alpha - e & \beta & 0 \\ \beta & \alpha - e & \beta \\ 0 & \beta & \alpha - e \end{vmatrix} = 0 \tag{12.6}$$

tadiene are shown in Figure 12.4b and are obtained by solving the secular determinant (a quartic equation this time) for the four-orbital problem. The orbitals of this species may also be derived along exactly analogous lines to those used for the linear H_4 problem of Section 5.3. Similarly the pattern for linear H_3 looks just like that for allyl.

Simple functions describe both the energy levels and orbital coefficients for these acyclic systems. The energy of the jth MO for a system with $N\,p\pi$ orbitals is given by

$$e_j = \alpha + 2\beta \cos \frac{j\pi}{N+1} \tag{12.7}$$

As the number of orbitals in the chain increases, the energy of the highest energy molecular orbital tends to $\alpha - 2\beta$ and that of the lowest to $\alpha + 2\beta$. The orbital coefficient for the rth atomic orbital in molecular orbital ψ_j where j runs from 1 to

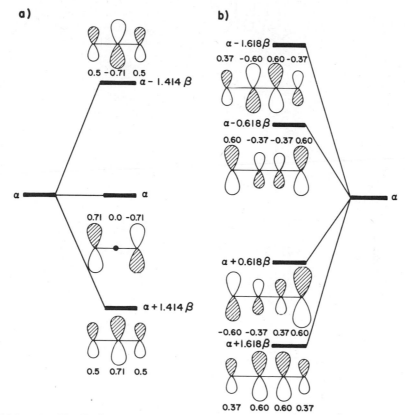

FIGURE 12.4. The Hückel π energy levels and coefficients of the (a) allyl and (b) butadiene systems.

N, is given by

$$c_{rj} = \left(\frac{2}{N+1}\right)^{1/2} \sin\left(\frac{rj\pi}{N+1}\right) \tag{12.8}$$

The corresponding functions for cyclic systems are described in the next section. The energy levels and orbital coefficients for more complex systems are to be found in the mammoth compilation of Streitweiser and Coulson.[4]

12.3. CYCLIC SYSTEMS

Just as the π orbitals of the linear three- and four-atom chains formed molecular orbital patterns identical to those of linear H_3 and H_4 in Chapter 5, so there is a one-to-one correspondence between the orbitals of cyclic polyenes and those derived for cyclic H_n units. Figure 12.5 shows the orbitals of cyclic C_3—C_6 systems for comparison. There are several patterns which emerge.

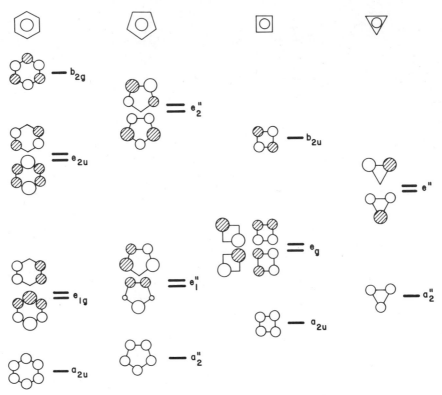

FIGURE 12.5.   The π orbitals of the first few cyclic polyenes.

1.   The number of nodes increases by one on going from one orbital to the one next highest in energy, as in the linear case. The lowest energy orbital has no nodes; each degenerate pair of orbitals has the same number of nodes.

2.   The lowest energy orbital is always nondegenerate. All other orbitals come as degenerate pairs except in even-membered rings where the highest energy orbital is also nondegenerate **(12.3)**. In group theoretical terms for a cyclic N atom ring

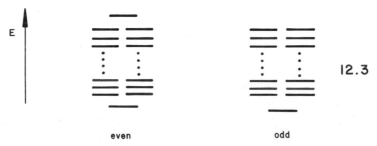

there are N orbitals, each corresponding to a different representation of the cyclic group of order N, as we will see below.

Algebraically the levels of the cyclic polyenes may be derived using simple Hückel theory. The general result is given in equation 12.9 for the energy of the jth level for a cyclic system containing N atoms

$$e_j = \alpha + 2\beta \cos \frac{2j\pi}{N} \tag{12.9}$$

j runs from $0, \pm1, \pm2 \ldots [\pm N/2$ for N even or $\pm(N-1)/2$ for N odd]. The very simple form of this equation leads to a useful mnemonic for remembering the energy levels of these molecules. Draw a circle of radius 2β and inscribe an N-vertex polygon such that one vertex lies at the six o'clock position. The points at which the two figures touch define the Hückel energy levels as in **12.4**. This construction is

called a Frost circle. The form of the coefficients of the pth atomic orbital in the wavefunction with an energy set by equation 12.9 is given by equation 12.10

$$\psi_j = \sum_{p=1}^{N} c_{pj}\chi_p = \frac{1}{\sqrt{N}} \sum_{p=1}^{N} \left[\exp\left(\frac{2\pi ij(p-1)}{N}\right)\right]\chi_p \tag{12.10}$$

Here i is the square root of -1. As in equation 12.9, j runs from $0, \pm1, \pm2 \ldots$. We shall see below, and, very importantly, in the next chapter, that this complex form of the wavefunction is very useful. It is interesting to see where this expression comes from. Group theory provides the answer. The molecular point group of, for example, benzene is D_{6h}. However, the group C_6 is the simplest one we can use which will generate the π orbitals of the molecule. Table 12.1 shows its character table. The reducible representation for the basis set of six π orbitals is

$$
\begin{array}{c|cccccc}
 & E & C_6 & C_3 & C_2 & C_3^2 & C_6^5 \\
\hline
D & 6 & 0 & 0 & 0 & 0 & 0 \\
\end{array}
\tag{12.11}
$$

which reduces to $a + b + e_1 + e_2$, that is, two nondegenerate orbitals and two degenerate pairs as shown in Figure 12.5. Now symmetry-adapted linear combinations

TABLE 12.1   Character Table for the C_6 Group

C_6	E	C_6	C_3	C_2	C_3^2	C_6^5		$\epsilon = \exp(2\pi i/6)$
A	$+1$	$+1$	$+1$	$+1$	$+1$	$+1$	z	$x^2 + y^2, z^2$
B	$+1$	-1	$+1$	-1	$+1$	-1		
E_1	$\begin{cases}+1 \\ +1\end{cases}$	$\begin{array}{c}\epsilon \\ \epsilon^*\end{array}$	$\begin{array}{c}-\epsilon^* \\ -\epsilon\end{array}$	$\begin{array}{c}-1 \\ -1\end{array}$	$\begin{array}{c}-\epsilon \\ -\epsilon^*\end{array}$	$\begin{array}{c}\epsilon^* \\ \epsilon\end{array}$	$\begin{array}{c}x+iy \\ x-iy\end{array}$	(xz, yz)
E_2	$\begin{cases}+1 \\ +1\end{cases}$	$\begin{array}{c}-\epsilon^* \\ -\epsilon\end{array}$	$\begin{array}{c}-\epsilon \\ -\epsilon^*\end{array}$	$\begin{array}{c}+1 \\ +1\end{array}$	$\begin{array}{c}-\epsilon^* \\ -\epsilon\end{array}$	$\begin{array}{c}-\epsilon \\ -\epsilon^*\end{array}\Big\}$		$(x^2 - y^2, xy)$

may be generated by using the characters of Table 12.1 and equation 4.8. They become [with $\epsilon = \exp(2\pi i/6)$]

$$\psi(a) \propto \chi_1 + \chi_2 + \chi_3 + \chi_4 + \chi_5 + \chi_6$$

$$\psi(b) \propto \chi_1 - \chi_2 + \chi_3 - \chi_4 + \chi_5 - \chi_6$$

$$\psi(e_1) \propto \chi_1 + \epsilon\chi_2 - \epsilon^*\chi_3 - \chi_4 - \epsilon\chi_5 + \epsilon^*\chi_6$$

$$\psi(e_1)' \propto \chi_1 + \epsilon^*\chi_2 - \epsilon\chi_3 - \chi_4 - \epsilon^*\chi_5 + \epsilon\chi_6 \qquad (12.12)$$

$$\psi(e_2) \propto \chi_1 - \epsilon^*\chi_2 - \epsilon\chi_3 + \chi_4 - \epsilon^*\chi_5 - \epsilon\chi_6$$

$$\psi(e_2)' \propto \chi_2 - \epsilon\chi_2 - \epsilon^*\chi_3 + \chi_4 - \epsilon\chi_5 - \epsilon^*\chi_6$$

which are identical to the form of the functions from equation 12.10, without the normalization constant, for this case. Thus the exponential in equation 12.10 represents the character of the jth irreducible representation of the cyclic group of order N. Note that the complex description of the orbitals only shows up in equation 12.12 for the degenerate molecular levels. These may be rewritten in a simpler way. A linear combination of the wavefunctions of a pair of degenerate orbitals (e.g., $j = +1$, -1 or $j = +2$, -2 etc.) produces two new orbitals which are equivalent in every respect. We can then recast the functions of equation 12.10, by making use of the trigonometric identity $\exp(ix) = \cos x + i \sin x$. The result is given by equation 12.13.

$$\psi_{j'} = \frac{1}{2}(\psi_j + \psi_{-j}) = \frac{1}{2\sqrt{N}} \sum_{p=1}^{N} \left[\cos \frac{2\pi j(p-1)}{N} \right] \chi_p$$

$$\psi_{j''} = \frac{1}{2i}(\psi_j - \psi_{-j}) = \frac{1}{2\sqrt{N}} \sum_{p=1}^{N} \left[\sin \frac{2\pi j(p-1)}{N} \right] \chi_p \qquad (12.13)$$

Let us return, however, to equation 12.10 and see how this wavefunction simply leads to the energies of equation 12.9. From **12.5** it is easy to see that the energy

12.5

of the orbital is given by

$$e_j = \langle \psi_j | H^{\text{eff}} | \psi_j \rangle$$

$$= \alpha + \beta \sum_p [c_{pj}^* c_{(p+1)j} + c_{pj}^* c_{(p-1)j}] \qquad (12.14)$$

which represents the sum of all the interactions of each atom with its two neighbors. Substitution of c_{pj} from equation 12.10 leads to

$$e_j = \alpha + \beta \sum_p \left(\frac{1}{\sqrt{N}} \exp \frac{-2\pi ij(p-1)}{N} \right.$$

$$\left. \cdot \frac{1}{\sqrt{N}} \exp \frac{2\pi ij(p-1)}{N} \right) \left[\exp \frac{2\pi ij}{N} + \exp \frac{-2\pi ij}{N} \right] \quad (12.15)$$

The term in parenthesis is equal to $1/N$, and the term in brackets expressible as a simple cosine function. So

$$e_j = \alpha + \beta \sum_{p=1}^{N} \frac{1}{N} \cdot 2 \cos \frac{2\pi j}{N}, \qquad j = 0, \pm 1, \pm 2 \ldots \quad (12.16)$$

and, since there are N atoms in the ring,

$$e_j = \alpha + 2\beta \cos \frac{2\pi j}{N}, \qquad j = 0, \pm 1, \pm 2 \ldots \quad (12.17)$$

The number of electrons present in these cyclic systems has an important bearing on their stability, structure, and properties. In particular we might expect that some sort of stability would exist for those systems where all bonding and nonbonding orbitals (where they exist) are completely filled with electrons. This leads to Hückel's $4n + 2$ rule, nothing more than recognition of the special stability of a closed shell of electrons. From Fig. 12.5, we can see that after the first, lowest level, the orbitals always occur in pairs. If n pairs of these levels are occupied by electrons, a total of $2n + 1$ orbitals will be filled for a total of $(4n + 2)$ electrons ($n = 0, 1, 2,$ etc.). The cases we shall come across most frequently are cyclobutadiene^{-2}, cyclopentadienyl^{-1}, benzene, and cycloheptatriene$^+$ which all have $n = 1$. Cyclooctatetraene^{-2} is an example with $n = 2$. Such species with $4n + 2$ π electrons are aromatic systems. In cyclic $4n$ π systems, a degenerate pair of π levels will be half-filled. Consequently the molecule with all electrons paired will become either nonplanar or at least distort to a nonsymmetrical structure as anticipated for a Jahn–Teller instability. Cyclooctatetraene C_8H_8 is one such example of the first type. The molecule **(12.6)** is tub shaped and the four double bonds are not conjugated to each

 12.6

other. Cyclobutadiene is an example of the second type. This orbital problem is just like that of H_4 and is shown in Figure 12.6a. The situation is of the Jahn–Teller type where the molecule distorts so as to remove the orbital degeneracy. The result is an opening up of a HOMO–LUMO gap and an overall stabilization of occupied orbitals on distortion to the rectangle. An exactly analogous result is shown in Figure 12.6b for the distortion to a diamond or rhomboid geometry. (Note that we have chosen different ways to write the wavefunctions of the degenerate nonbonding π set in the two cases which make the results easier to understand.) Here the anti-

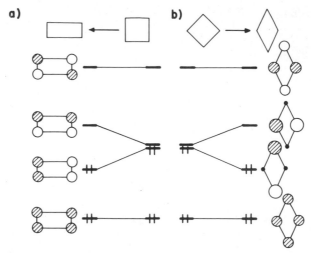

FIGURE 12.6.   Relief of the Jahn–Teller instability in singlet square cyclobutadiene by distortion to (a) a rectangle and (b) a diamond.

bonding 1,3-interaction is decreased for one component and increased for the other on distortion.

Planar $4n$ π systems are said to be antiaromatic. The stability of aromatic compounds and the lack of antiaromatic ones has been an important concept in organic chemistry.[5] There are a couple of interesting points to be made. Addition of two extra electrons to cyclobutadiene leads, overall, to no increase in orbital stabilization, since these electrons enter a nonbonding orbital. In fact, Coulomb repulsion may lead overall to a destabilizing effect. However, on distortion to the rectangle, for $C_4H_4^{2-}$ two electrons in this nonbonding orbital are destabilized. Quantitatively this is shown in **12.7** where the distortion has gone all the way to two double bonded units. Cyclobutadiene itself has the same π energy as two ethylene segments but the dianion loses 2β of π energy on distortion. Put another way, the zero HOMO–LUMO gap in the $4n$ π molecule signals a Jahn-Teller type of distortion and also increases reactivity by having a high lying HOMO and/or a low lying LUMO.

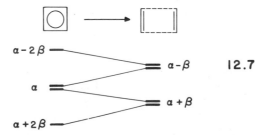

For cyclobutadiene there is another interesting possibility which we have explored before with the case of methylene in Section 8.8. Is it possible to produce a stable structure by allowing the two highest energy electrons of the $4n$ species to separately occupy the orthogonal pair of degenerate orbitals with their spins parallel? The result would be a triplet diradical species as in **12.8**. For such an electronic configuration there would be no obvious tendency from the orbital picture of Figure 12.6 to distort away from the square planar structure. So there exists the possibility of two structures **(12.9)** dependent on the spin state of the molecule.

12.8   **12.9**

Experimental evidence[6] indicates a distorted structure for the singlet state but no information is available for the structure of the diradical (triplet) state. The discussion in Chapter 8 leads us to expect that the triplet should be more stable than the singlet at the square planar geometry because of more favorable two-electron energy terms. Results of calculations which include configuration interaction indicate[7] a reversal of this energy ordering. Examples of square singlet species with $4n + 2 = 6\pi$ electrons, which are therefore stable at this geometry, include the chalcogenide ions A_4^{2+} (A = S, Se, Te) and the derivative S_2N_2 which we examine below.

12.4.   PERTURBATIONS OF CYCLIC SYSTEMS

The orbitals of the previous section may be used to understand the orbital structures of other systems. Just as the orbitals of cyclobutadiene were generated in Figure 5.4 by linking together the end atoms of butadiene, so the orbitals of naphthalene **(12.10)** and azulene **(12.11)** may be derived by linking together pairs of atoms in the cyclic 10-annulene **(12.12)** as in Figures 12.7 and 12.8. Whether an

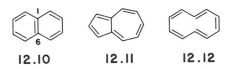

12.10   **12.11**   **12.12**

orbital goes up or down in energy during the process depends upon the relative phases of the coefficients on the linking pair of atoms in that orbital. Notice that some orbitals remain unchanged in energy in the naphthalene case. They are the orbitals with nodes running through the pair of atoms 1, 6 **(12.10)** and one partner of each degenerate pair is of this type. The new orbital energies may be derived numerically using first-order perturbation theory, as shown very nicely in Heilbronner and Bock's book.[3] With reference to equations 3.2, 3.4, and 3.5, the perturbation is simply one of increasing the value of $H_{\mu\nu}$ from zero to β for the interaction integral linking the orbitals μ and ν located on the atoms between which a bond is to be made. Recall that within the Hückel approximation, overlap inte-

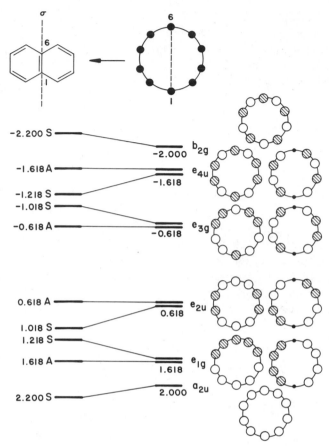

FIGURE 12.7. Generation of the π energy levels of napthalene by linking together a pair of atoms of 10-annulene. The energies are given in units of β and no attempt has been made to represent the actual AO coefficients.

grals between orbitals on different atoms are ignored. This leads to all $\delta S_{\mu\nu} = 0$, all $\delta H_{\mu\nu} = 0$ for both $\mu = \nu$, and $\mu \neq \nu$ *except* for the one case (let us call this $\delta H_{\kappa\lambda}$) which involves the bond formation itself. So in equations 3.4 and 3.5, $\widetilde{\Delta}_{ii} = c_{\kappa i}^0 \beta c_{\lambda i}^0$ and $\widetilde{S}_{ii} = 0$ leading to $e_i^{(1)} = c_{\kappa i}^0 c_{\lambda i}^0 \beta$. So the largest energy changes will be associated with the largest products of orbital coefficients $c_{\kappa i}^0 c_{\lambda i}^0$. This has guided our qualitative pictures of Figures 12.7 and 12.8.

Another perturbation of these orbitals occurs when the atoms of the carbon framework are replaced with others of different electronegativity. Figure 12.9 shows how the π orbitals of S_2N_2 are derived from those of cyclobutadiene^{2-}. The lower point symmetry removes the degeneracy of the middle pair of orbitals and these two new orbitals are either pure sulfur or pure nitrogen in character.

We will leave it as an exercise for the reader to use the same perturbation theoretic ideas as employed for H_3 in Figure 6.2 to generate these level shifts and the

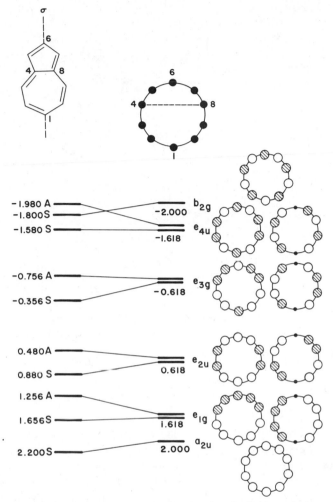

FIGURE 12.8.   Generation of the π energy levels of azulene by linking together a pair of atoms of 10-annulene. The energies are given in units of β and no attempt has been made to represent the actual AO coefficients.

form of the new wavefunctions. With a total of six π electrons the HOMO is a pure sulfur p orbital which lies above the mean value of N and S p atomic energies.

Figure 12.10 shows the orbitals of "inorganic benzene," the borazine molecule $B_3N_3H_6$. The degenerate benzene levels have not been split apart in energy but the three higher energy orbitals contain more boron character than nitrogen character while the opposite is true for the three lower energy orbitals. This is a result clearly in keeping with an electronegativity perturbation on benzene, as also shown pictorially in the figure. With a total of six π electrons, this collection of orbitals is filled through $1e$. $S_3N_3^-$ **(12.13)** has an analogous orbital pattern but has a total of

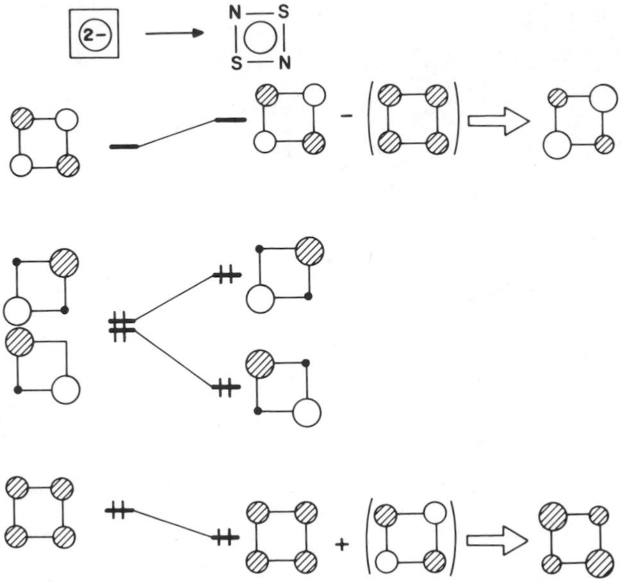

FIGURE 12.9.   Generation of the π orbitals of S_2N_2 from those of cyclobutadiene. For simplicity we assume here that the electronegativity of carbon lies midway between that of sulfur and nitrogen. As a result the old and new level patterns have a symmetry about the midpoint.

12.13

$10\,\pi$ electrons. For this species the levels are filled through $2e$. This is a feature of sulfur–nitrogen compounds in general–occupation of the lowest energy orbital (as in S_2N_2), or lowest energy pair of orbitals (as in $S_3N_3^-$) which lie above the midpoint of the π energy diagram. The molecule $S_3N_3^-$ is isoelectronic with the planar P_6^{4-} unit in Rb_4P_6.[8] In this species (isoelectronic with $C_6H_6^{4-}$) the P—P distances (2.15 Å) are longer than a typical P=P distance (2.0 Å) but slightly shorter than a typical P—P distance (2.2 Å).

There are two obvious ways (12.14, 12.15) to reduce the symmetry of square

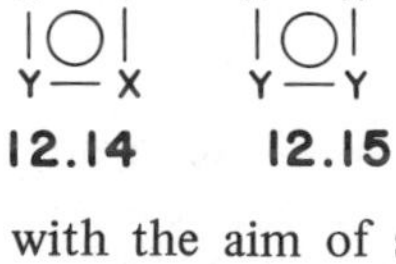

12.14   12.15

cyclobutadiene by substitution with the aim of stabilizing the singlet structure. 12.14 corresponds to the geometry of S_2N_2 which we know exists (but with two more electrons). What about[9] the alternative structure 12.15? The form of the wavefunctions in the substituted molecules give us good clues as to the HOMO–LUMO gaps for the two possibilities. Recall that for the degenerate pair of cyclobutadiene orbitals we have some flexibility in the choice of the wavefunctions. (See Figure

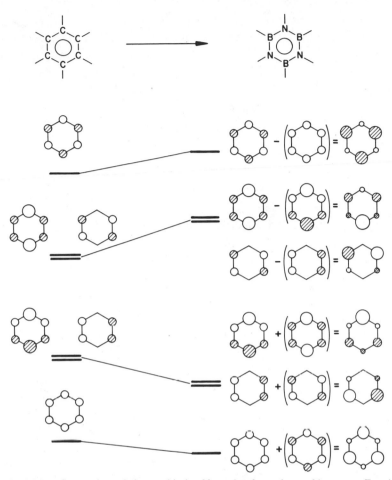

FIGURE 12.10. Generation of the π orbitals of borazine from those of benzene. For simplicity we have assumed that the electronegativity of carbon lies midway between those for boron and nitrogen.

12.5.) We will choose the degenerate pairs as in **12.16** and **12.17** which reflect the symmetry properties of **12.14** and **12.15**, respectively. In fact the HOMO and LUMO of the lower symmetry structures will look very much like these. The energies of the two functions **12.16** will differ by an amount which depends on the XY electronegativity difference alone, since these two orbitals are either completely X or completely Y located. The energies of **12.17** on the other hand are expected to be much like cyclobutadiene itself with a small gap between them. *Both* components

12.16   **12.17**

of **12.17** are stabilized when X is more electronegative than Y in **12.15**. In addition to $B_2N_2R_4$ (**12.18**) all cyclobutadienes containing π-donor and π-acceptor substituents which have been made (e.g., **12.19**) have a substitution pattern of the type **12.14**.

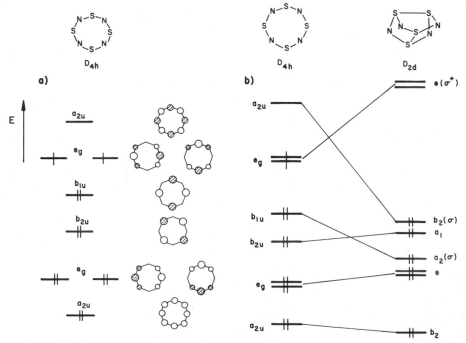

12.18   **12.19**

One interesting orbital derivation[10] is that of the unusual cradle-shaped molecule S_4N_4 from the perturbed 8-annulene **12.20**. Just as on moving to S_2N_2 from cyclo-

12.20

butadiene (Figure 12.9), the middle pair of orbitals of the 8-annulene split apart in energy on moving to S_4N_4, leading to the b_{2u} and b_{1u} levels of Figure 12.11. With

FIGURE 12.11.   (a) The π orbitals of planar S_4N_4 which may be obtained from those of the 8-annulene in a similar fashion to the generation of the levels of borazine in Figure 12.10. (b) Orbital correlation diagram for the formation of the S_4N_4 cage molecule.

a total of 12 π electrons, either a triplet planar molecule is expected with double occupancy of all the orbitals except the highest degenerate level, or a singlet species with some sort of distorted geometry. The latter will be necessary to remove the orbital degeneracy. One way this may be done (Figure 12.11) is to link two pairs of opposite atoms of the eight-membered ring as in **12.20**. This results in a dramatic stabilization of the highest energy π type orbital (labeled a_{2u}) and the generation of a substantial HOMO–LUMO gap. How the linking process occurs is an interesting question to answer. Just as in the derivation of naphthalene from 10-annulene, the largest orbital energy changes will be found when the coefficients on the linking atoms are largest. Since the a_{2u} orbital is an antibonding orbital, the largest coefficients will be associated with the least electronegative atom. The a_{2u} level is converted into a S—S bonding orbital **12.21** on forming the cradle. Likewise the b_{1u} level is converted into **12.22**. $S_4N_4^{2+}$ has two fewer electrons. It is clear from Fig-

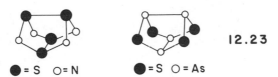

12.21   12.22

ure 12.11 that the planar molecule will have a large HOMO–LUMO gap and that the bicyclic molecule will have a small one. Inspection of the levels of Figure 12.11 shows that while the b_{1u} orbital is stabilized, the two members of the e_g orbital are destabilized on bending the planar molecule. $S_4N_4^{2+}$ is found as a planar species. In S_4N_4 there is the unusual result of a two-coordinate nitrogen atom and three-coordinate sulfur. In the isoelectronic As_4S_4 and As_4Se_4 the chalcogen is two coordinate **(12.23)** in accord with the relative electronegativities of arsenic and chalcogen.

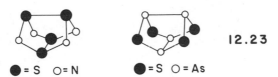

● = S   O = N   ● = S   O = As   12.23

   In Section 6.4 we saw how to predict the substitution pattern of molecules containing atoms of different electronegativity by making use of the charge distribution of the parent, unsubstituted molecule. The same approach may be used for polyenes containing inequivalent atoms. **12.24** shows the charge distribution of pentalene and **12.25** the "inorganic pentalene" made by replacing half of the carbon atoms with nitrogen and half with boron atoms. The more electronegative nitrogen atoms occupy the sites of highest charge density in the unsubstituted analog.

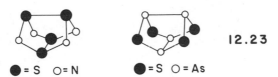

          –0.17

      –0.20   0.19

      12.24   12.25

12.5.   CONJUGATION IN THREE DIMENSIONS

Two acyclic polyene chains may be linked together by a single carbon atom to give a spiro-geometry as in **12.26**. The overall symmetry of two planar polyenes that

(CH)$_n$ (CH)$_n$ **12.26**

are connected by a single atom in this way is either D_{2h} or C_{2v}, depending upon whether the two ring sizes are identical or not. Take a case of C_{2v} symmetry. The π levels of the polyenes will then be of b_1, b_2, or a_2 symmetry. It can be shown by the insertion of the relevant phases in **12.26** that there will be nonzero overlap only when two orbitals of a_2 symmetry interact. Furthermore there is a simple rule to tell whether this linking process results in a stabilization (to give a spiro-aromatic molecule) or a destabilization (to give a spiro-antiaromatic molecule). First all $4n + 2$ systems are spiro-aromatic, as we can see by assembling the π picture for spiro-heptatriene in Figure 12.12. A stabilization of the HOMO of the butadiene fragment occurs by overlap with the LUMO of the ethylene fragment. The other

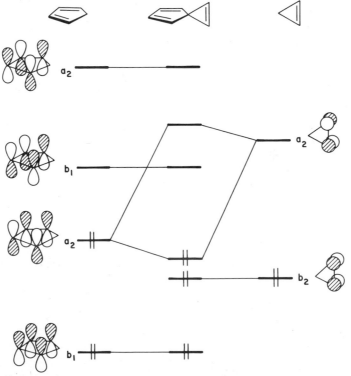

FIGURE 12.12.   Assembly of the π orbital diagram for spiro-heptatriene from those of a four-carbon and two-carbon fragment.

occupied orbitals of both molecules do not find a partner of the correct symmetry with which to interact. Although we have shown a large stabilization in Figure 12.12 the two sets of interacting orbitals are not close and the stabilization of the occupied a_2 level is certainly going to be small. The ionization potential determined[12] by photoelectron spectroscopy of an electron in the a_2 orbital is not that different from an analog where the three-membered ring portion of the molecule is saturated. Clearly a more favorable stabilization of the butadiene a_2 orbital will come about if the unoccupied a_2 orbital on its partner is lowered in energy. Placing π electron withdrawing groups on the cyclopropene portion and π donors on the cyclopentadiene side would partially accomplish this. An even more efficient way is to vary the ring size. The interaction diagram for the spiro-octatrienyl cation, **12.27** can readily be constructed along the lines of Figure 12.12. The important difference is that now the a_2 butadiene HOMO is stabilized by the LUMO of an allyl cation as shown in **12.28**. The nonbonding π level of allyl lies at a much lower energy than π^* of ethylene and consequently one expects that **12.27** should be stabilized more (in a relative sense) than spiro-heptatriene. Calculations have shown[13] this to be the case.

In contrast to the above, $4n$ systems are spiro-antiaromatic. This is easily shown by constructing the diagram (Figure 12.13) for spiro-nonatetraene. Compared to

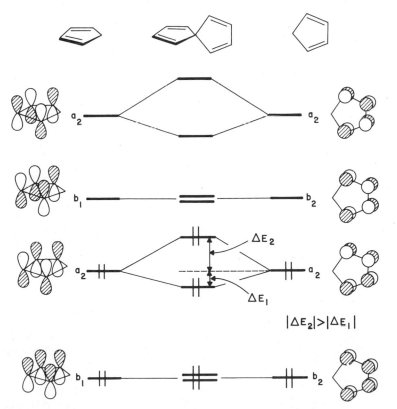

FIGURE 12.13.   Assembly of the π orbital diagram for spiro-nonatetraene from those of two four-carbon fragments.

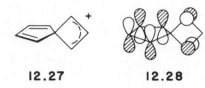

12.27   **12.28**

the previous case, the HOMOs of both systems are of the correct symmetry and energy to interact with one another. Just as in the case of the repulsion of two ground state helium atoms with closed shells of electrons (Section 2.2), this two-orbital–four-electron situation is, overall, a destabilizing one. Intrafragment mixing between the occupied and unoccupied sets of a_2 orbitals will be very small; in the isolated fragments they are orthogonal. As a result of this destabilization, spiro-nonatetraene is spiro-antiaromatic. It is also exceedingly reactive. The splitting between the bonding and antibonding combinations of the occupied a_2 set has been determined[14] to be 1.2 eV by photoelectron spectroscopy.

The potential for conjugation within polyene π "ribbons" has been examined for several other π topologies. **12.29** and **12.30** illustrate two bicyclic motifs. Interaction

12.30

12.29

diagrams can easily be constructed and generalized electron counting rules have been established.[15]

REFERENCES

1. A. Streitweiser, *Molecular Orbital Theory for Organic Chemists*, Wiley, New York (1961).
2. K. Yates, *Hückel Molecular Orbital Theory*, Academic Press, New York (1978).
3. E. Heilbronner and H. Bock, *The HMO-Model and its Application*, Wiley, New York (1976).
4. C. A. Coulson and A. Streitweiser, *Dictionary of π-electron Calculations*, Freeman, San Francisco (1965).
5. P. J. Garrett, *Aromaticity*, McGraw-Hill, New York (1971).
6. For example, D. W. Whitman and B. K. Carpenter, *J. Amer. Chem. Soc.*, **102**, 4272 (1980).
7. J. A. Jafri and M. D. Newton, *J. Amer. Chem. Soc.*, **100**, 5012 (1978).
8. W. Schmettow, A. Lipka, and H. G. von Schnering, *Angew. Chem.*, **86**, 379 (1974).
9. R. Hoffmann, *Chem. Comm.*, 240 (1969).
10. R. Gleiter, *Angew. Chem.* (Int. Ed.), **20**, 444 (1981).
11. R. J. Gillespie, J. P. Kent, J. R. Sawyer, D. R. Slim, and J. D. Tyrer, *Inorg. Chem.*, **20**, 3799 (1981).
12. P. Bischof, R. Gleiter, H. Dürr, B. Ruge, and P. H. Herbst, *Chem. Ber.*, **109**, 1412 (1976).
13. P. Bischof, R. Gleiter, and R. Haider, *J. Amer. Chem. Soc.*, **100**, 1036 (1978).
14. C. Batich, E. Heilbronner, and M. F. Semmelhack, *Helv. Chim. Acta*, **56**, 2110 (1973).
15. M. J. Goldstein and R. Hoffmann, *J. Amer. Chem. Soc.*, **93**, 6193 (1971).

CHAPTER THIRTEEN

Solids

13.1. ENERGY BANDS

In previous chapters we have examined the orbitals of molecules of finite extent. In this chapter we describe the case where there are, for all practical purposes, an infinite number of orbitals, namely those of a solid—a giant molecule. We will be exclusively concerned with crystalline materials, that is, those with a regularly repeating motif in all three dimensions. The results of earlier chapters, especially the previous one, will carry over quite naturally to this area. We start with a one-dimensional situation, that of an infinite chain of carbon $p\pi$ orbitals (**13.1**). From the re-

13.1

sults of Section 12.2 we know qualitatively what the orbitals of this infinite chain will look like. Simple Hückel theory provided an analytic expression for the orbitals of such linear polyenes and equation 13.1 gives the energy of the jth level for an N atom (orbital) chain:

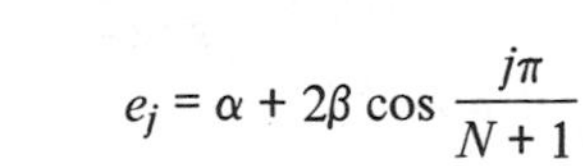

$$e_j = \alpha + 2\beta \cos \frac{j\pi}{N+1} \tag{13.1}$$

When N is very large, the lowest level ($j = 1$) will lie at $e \sim \alpha + 2\beta$ where there are bonding interactions between all adjacent atom pairs. The highest energy level ($j = N$) will be at $e \sim \alpha - 2\beta$ and contains antibonding interactions between all adjacent atom pairs. Between them lies a continuum of orbitals which we call an energy band with an energy spread of $(\alpha - 2\beta) - (\alpha + 2\beta) = -4\beta$ (**13.2, 13.3**). In the middle of this stack of levels at $e = \alpha$ there is a nonbonding situation (**13.3**) which may be written in several different ways. This is analogous to the choice we had for the de-

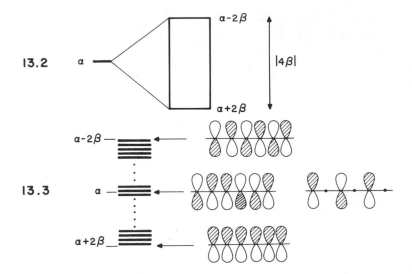

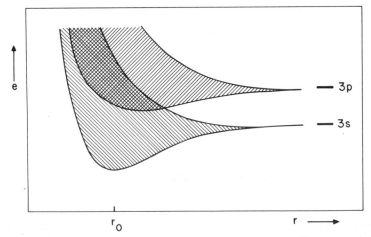

generate levels in the case of cyclic H_4 (Chapter 5) or cyclobutadiene in Figure 12.5. The number of nodes increases as the energy increases, just as for the finite case. The general result is the production of an energy band for each of the atomic orbitals located on the atoms which make up the chain.

In two and three dimensions a similar process occurs. The atomic energy levels of each of the atoms of, for example, elemental sodium are broadened into bands in the solid. The width of these bands depends upon the magnitude of the corresponding interaction integrals (the equivalent of the Hückel β for the $p\pi$ one-dimensional chain above) between the orbitals concerned. Figure 13.1 shows[1] how

FIGURE 13.1.   Dependence on internuclear separation of sodium atomic levels. Notice how as r decreases the collection of s and p orbitals broaden into bands. At the equilibrium internuclear distance (r_0) the s and p bands overlap.

the energy levels of a collection of sodium atoms varies with internuclear distan[ce]. The shaded areas represent the energy bands formed from 3s and 3p orbitals. Noti[ce] that the bottom of each band, at the equilibrium separation r_0, lies lower in energy than the corresponding atomic level at infinite separation (i.e., it is bonding) but the top of the band lies above this energy. Also notice that at large internuclear separation there are two separate "s" and "p" bands but as this distance decreases the two bands overlap. In general, the energetic relationship of the energy bands of a solid material and how many electrons are contained in each has an extremely important bearing on the properties of the system. If the highest occupied band (the valence band) is full then the solid is an insulator or semiconductor, depending on whether the energy gap, E_g, (the band gap) between the valence band and the lowest empty band (conduction band) is respectively large or small (13.4). If the valence band is only partially full, or full and empty bands overlap, then a typical metal results. In the notation used in 13.4 we imply that all the electrons in the occupied levels are paired. The case of 13.5 where the band is full of *unpaired* electrons gives rise to a magnetic insulator. The energetic considerations that control the stability of the alternatives 13.5 and 13.6 are very similar indeed to those used in Section 8.8 to view high and low spin arrangements in molecules.

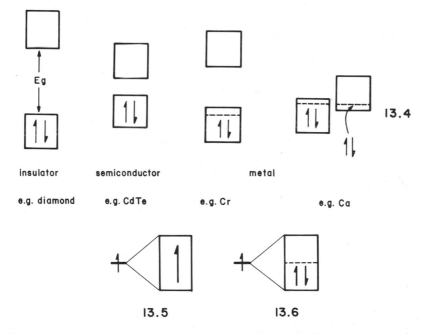

insulator semiconductor   metal

e.g. diamond e.g. CdTe e.g. Cr   e.g. Ca

13.4

13.5   13.6

How are we going to represent the complex situation of the giant molecule and handle this infinite collection of orbitals? We can make use of results from the previous chapter and assume that the atoms in the very long one-dimensional chain behave as if they were embedded in a very big ring. Alternatively, we can imagine imperceptibly bending the very long chain and tying the end atoms together (13.7) to make a cyclic system. Surely the overwhelming majority of the atoms of a real

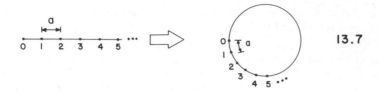

13.7

crystal are so far away from the edges so as not to know the difference. Obviously then our discussion will only be valid for macroscopic crystals, those where most of the atoms are "bulk" rather than surface atoms.

Now the values of the energy levels of the very long cyclic chain, (with N atoms) are given from equation 12.9 as

$$e_j = \alpha + 2\beta \cos \frac{2\pi j}{N} \qquad (13.2)$$

where j runs from 0 through ±1, ±2, . . . , . Since N is quite a large number, we can recast this equation to make it easier to handle. It was mentioned above that we will only study crystalline materials in this book. These are systems where a fundamental building block of atoms is regularly repeated in three dimensions. In **13.8** we

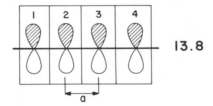

13.8

show a part of the infinite one-dimensional chain of carbon $2p$-orbitals with several unit cells outlined. The position of unit cell p is given by $R_p = (p - 1)a$. The unit cell (of length a in this case) contains the regularly repeating motif. We can define a new index $k = 2\pi j/Na$ which runs from 0 to $\pm\pi/a$ such that equation 13.2 becomes

$$e(k) = \alpha + 2\beta \cos ka \qquad (13.3)$$

Let us show this result pictorially.[2] Using equation 12.10 we may plot out the energy levels of, for example, C_5H_5 as shown in **13.9**. The allowed values of j are 0, ±1, . . . and in this case therefore to ±2. The reader can show that substitution of values of $|j|$ larger than 2 just leads to duplication of the values we have already derived, that is, use of $|j| > (N - 1)/2$ for N = odd (or $N/2$ for N = even) leads to redundant information. **13.10** shows an analogous plot for a ring containing 15 atoms. Here j runs from 0 through ±1, ±2, and so on to ±7. Finally **13.11** shows a diagram exactly analogous to those of **13.9** and **13.10** for the infinite system. Now k runs from 0 through $\pm\pi/a$ or j from 0 through $\pm(N - 1)/2$ where N is very large, just as in the finite case.

One important difference between the finite and infinite cases, of course, is that whereas j increases in discrete steps, k increases continuously. Also in a way closely

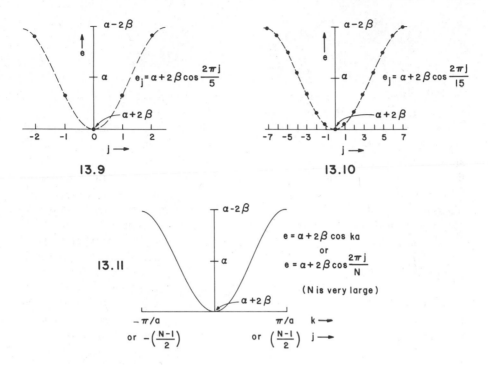

13.9

13.10

13.11

$e = \alpha + 2\beta \cos k a$

or

$e = \alpha + 2\beta \cos \dfrac{2\pi j}{N}$

(N is very large)

similar to the behavior of e_j in the finite case, when $|j| > (N-1)/2$, values of $|k| > \pi/a$ lead to redundant information in the solid state. In the crystal the region of k values between π/a and $-\pi/a$ is referred to as the first Brillouin zone, usually just called the Brillouin zone. The point $k = \pm \pi/a$ is called the zone edge and $k = 0$ the zone center.

Since the diagram **13.11** has mirror symmetry about $k = 0$ it will suffice just to use one-half of this diagram. We choose the right-hand half, that corresponding to positive k. The index k is called the wavevector. The variation in energy as a function of the wavevector k is called the dispersion of the band. In three-dimensional situations the vector nature of the wavevector becomes apparent and we will write it as **k**.

The wavefunctions describing the chain **13.7** may be generated by seeing how the wavefunctions of the finite ring change when N becomes large. As before we define $\chi(r - R_p)$ as the atomic orbital wavefunction located on the atom in the pth unit cell. From equation 12.10 the wavefunctions of the N atom chain are given by

$$\phi_j = \sum_{p=1}^{N} c_{pj}\chi(r - R_p) = \frac{1}{\sqrt{N}} \sum_{p=1}^{N} \left\{ \exp\left[\frac{2\pi ij(p-1)}{N}\right] \right\} \chi(r - R_p) \quad (13.4)$$

Substitution of $k = 2\pi j/Na$ leads to the expression

$$\phi(k) = \frac{1}{\sqrt{N}} \sum_{p=1}^{N} \{\exp[ik(p-1)a]\} \chi(r - R_p) \quad (13.5)$$

which may be rewritten as

$$\phi(k) = \frac{1}{\sqrt{N}} \sum_{p=1}^{N} [\exp (ikR_p)] \, \chi(r - R_p) \tag{13.6}$$

In three dimensions the exponential in this equation needs to be written as a vector dot product $\exp (ik \cdot R_p)$. Just as the vector R_p (with dimensions of length) maps out a direct space (x, y, z coordinates of points) with which we are familiar, so k [with dimensions of (length)$^{-1}$] maps out a reciprocal space. The functions $\phi(k)$ are called Bloch functions[3-5] and are nothing more than the symmetry-adapted linear combination of atomic orbitals, under the action of the infinite translation group, just as the orbitals of equation 12.10 are the symmetry-adapted linear combinations of orbitals under the action of the cyclic group of order N. In Section 12.3 we showed for the illustrative example of the π orbitals of benzene ($N = 6$) that the exponential in equations 12.10 and 13.4 was just the character of the jth irreducible representation of the cyclic group of order N. Similarly the exponential in equations 13.5 and 13.6 is related to the character of the kth irreducible representation of the cyclic group of infinite order, which, according to the picture of **13.7**, we may replace with an (infinite) linear translation group. Just as the wavefunctions of equation 12.10 with different j are orthogonal to each other so the wavefunctions of equation 13.6 are orthogonal for different k values. At $k = 0$ we can write, using equation 13.6,

$$\phi(k = 0) = \frac{1}{\sqrt{N}} [\cdots \chi(r) + \chi(r - a) + \chi(r - 2a) + \cdots] \tag{13.7}$$

where $\chi(r)$ is some arbitrary orbital located on some atom in the chain; $\chi(r - a)$ lies at a distance a along the chain (**13.8**), $\chi(r - 2a)$ at a distance $2a$, and so on, from $\chi(r)$. The coefficients from equation 13.7 are all equal. This wavefunction is shown in **13.12** and of course extends all the way through the crystal. The normalization

13.12   13.13

constant of $N^{-1/2}$ has been included as a result of the Hückel approximation of Section 12.2. We can easily calculate the energy associated with the wavefunction of equation 13.7 as

$$e(k = 0) = \langle \cdots \chi(a) + \chi(r - a) + \chi(r - 2a) \cdots | H^{\text{eff}} | \cdots \chi(a)$$
$$+ \chi(r - a) + \chi(r - 2a) \cdots \rangle \tag{13.8}$$

Using the same technique as in **12.5** for the molecular case

$$e(k = 0) = \frac{1}{N} [N(\alpha + 2\beta)] = \alpha + 2\beta \tag{13.9}$$

where

$$\alpha = \langle \chi(r - R_p) | H^{\text{eff}} | \chi(r - R_p) \rangle$$

and

$$\beta = \langle \chi(r - R_p) | H^{\text{eff}} | \chi(r - R_{p+1}) \rangle$$

This, of course, is the result from equation 13.3 too. Notice that N, although included in the expression for the wavefunction, has neatly dropped out of the expression for the energy. At $k = \pi/a$ the wavefunction $\phi(k)$ becomes

$$\phi(k = \pi/a) = \frac{1}{\sqrt{N}} [\cdots \chi(r) \exp(i\pi 0) + \chi(r - a) \exp(i\pi)$$

$$+ \chi(r - 2a) \exp(2i\pi) + \cdots]$$

$$= \frac{1}{\sqrt{N}} [\cdots \chi(r) - \chi(r - a) + \chi(r - 2a) - \cdots] \qquad (13.10)$$

which is shown in **13.13**. The energy of this function can be readily seen to be equal to $e(k = \pi/a) = \alpha - 2\beta$. Since $\beta < 0$ the maximum bonding (and therefore maximum stabilization) is found at the zone center $(k = 0)$ and the maximum antibonding character at the zone edge $(k = \pi/a)$. This is in keeping with the form of the band dispersion of **13.11** and the qualitative picture of **13.3**. For a general value of k equation 13.5 may be rewritten as

$$\phi(k) = \frac{1}{\sqrt{N}} [\cdots \chi(r) + \chi(r - a) \exp(ika) + \chi(r - 2a) \exp(2ika) + \cdots]$$

$$(13.11)$$

which leads to a general expression for the energy

$$e(k) = \langle \phi(k) | H^{\text{eff}} | \phi(k) \rangle$$

$$= \frac{1}{N} \cdot N \{ \alpha + [\exp(ika) + \exp(-ika)] \beta \}$$

$$= \alpha + 2\beta \cos ka \qquad (13.12)$$

With a given number of electrons in the solid the levels, doubly occupied, will be filled to a certain energy e_F, called the Fermi level, which corresponds to a specific value of k, k_F. The total one electron energy per unit cell, E/N, is then obtained by integrating equation 13.13

$$\frac{E}{N} = \frac{a}{2\pi} \int_{-k_F}^{k_F} 2e(k)\, dk = \frac{2a}{\pi} \int_0^{k_F} e(k)\, dk \qquad (13.13)$$

This is an exactly analogous equation to the energy *sum* over a discrete collection of levels in the molecular case. In the solid there are, however, a very large number of levels and electrons to occupy them. The total energy of equation 13.13 is, therefore, referred to the contents of one unit cell. In many cases there will be a nonintegral number of electrons per cell as a result of this choice. There will always be the same number of energy bands as there are atomic orbitals in the unit cell. Sometimes, however, the collection of bands arising from the three p orbitals on an atom are referred to collectively as "the p band" or the levels derived from the five d orbitals as "the d band."

One of the important quantities in describing the electronic structure of a molecule or an extended system is the so-called density of states. In a molecule there is a set of discrete levels as shown in **13.14** for the π orbitals of benzene. Thus the

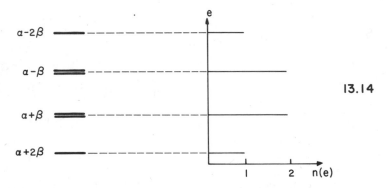

number of allowed orbital levels with energy e [i.e., density of states $n(e)$] is two if e refers to the doubly degenerate levels, one if e refers to the nondegenerate ones, and zero otherwise. Similarly, the density of states $n(e)$ in an extended system is the number of allowed band orbital levels having an energy e. For the one-dimensional case $n(e)$ is known to be inversely proportional to the slope of the e versus k curve (equation 13.14) as shown in Figure 13.2b. At $k = 0$ and π/a Figure 13.2a shows that the slope of this curve is zero and so $n(e) \to \infty$. Such features in $n(e)$ at

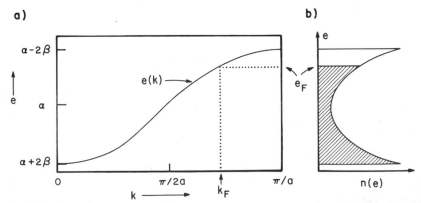

FIGURE 13.2. (a) Dispersion $[e(k)]$ of a one-dimensional energy band formed by overlap of adjacent $p\pi$ orbitals. The orbitals are filled up to the dashed line, the Fermi level (e_F). The corresponding k value is called k_F. (b) A density of states diagram appropriate to Figure 13.2a.

these points are called van Hove singularities. In two and three dimensions the densities of states are invariably more complex, but do not usually display such singularities in $n(e)$.

$$n(e) \propto \left[\frac{\partial e(k)}{\partial k}\right]^{-1} \tag{13.14}$$

The methodology we have just described is a natural extension of the molecular ideas discussed in earlier chapters. This LCAO approach is called the tight-binding method by solid-state physicists. It exists in several different forms, each of which has an analog in the molecular area. We have used simple Hückel theory to derive the results in this section but more sophisticated ones and many-electron approaches are available. Much of the work in this area comes from solid-state physics. In Table 13.1 and **13.15**, we compare some of the jargon used with its nearest molecular equivalent.

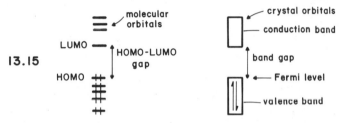

Equations 13.3 and 13.5 give the expressions for the simplest possible case, that of a one-dimensional chain containing a single orbital per unit cell. Most systems are more complex. Suppose that there are a set of atomic orbitals $\{\chi_1, \chi_2, \ldots, \chi_n\}$ contained in each unit cell. Then one can form a set of Bloch functions $\{\phi_1(k), \phi_2(k), \ldots, \phi_n(k)\}$, given in general by

$$\phi_\mu(k) = \frac{1}{\sqrt{N}} \sum_{p=1}^{N} [\exp(ikR_p)] \, \chi_\mu(r - R_p) \tag{13.15}$$

where $\mu = 1, 2, \ldots, n$. In such a case the band orbitals $\psi_j(k)$ $(j = 1, 2, \ldots, n)$ are

TABLE 13.1.   Approximate Analogs Between Molecular and Solid-State Terminology

Molecular	Solid-State
LCAO–MO	Tight-binding
Molecular orbital	Crystal orbital (band orbital)
HOMO	Valence band
LUMO	Conduction band
HOMO–LUMO gap	Band gap
Jahn–Teller distortion	Peierls distortion
High or intermediate spin	Magnetic
Low spin	Nonmagnetic

given by linear combinations of the Bloch functions as

$$\psi_j(k) = \sum_{\mu=1}^{n} c_{\mu j}(k)\, \phi_\mu(k) \qquad (13.16)$$

The energy of such a band orbital $e_j(k)$ is given by the usual expression

$$e_j(k) = \frac{\langle \psi_j(k) | H^{\text{eff}} | \psi_j(k)\rangle}{\langle \psi_j(k) | \psi_j(k)\rangle} \qquad (13.17)$$

The variational theorem, when applied to this problem allows determination of the optimum values of the $c_{\mu j}(k)$ and the generation of a secular determinant

$$\begin{vmatrix} H_{11}(k) - S_{11}(k)\,e(k) & H_{12}(k) - S_{12}(k)\,e(k) & \cdots & H_{1n}(k) - S_{1n}(k)\,e(k) \\ H_{21}(k) - S_{21}(k)\,e(k) & H_{22}(k) - S_{22}(k)\,e(k) & \cdots & H_{2n}(k) - S_{2n}(k)\,e(k) \\ \vdots & \vdots & & \\ H_{n1}(k) - S_{n1}(k)\,e(k) & H_{n2}(k) - S_{n2}(k)\,e(k) & \cdots & H_{nn}(k) - S_{nn}(k)\,e(k) \end{vmatrix} = 0$$

$$(13.18)$$

where the interaction element $H_{\mu\nu}(k)$ and the overlap integral $S_{\mu\nu}(k)$ are defined in terms of the Bloch functions

$$H_{\mu\nu}(k) = \langle \phi_\mu(k) | H^{\text{eff}} | \phi_\nu(k)\rangle$$

$$= N^{-1} \sum_p \sum_q \{\exp\,[ik(R_q - R_p)]\}\langle \chi_\mu(r - R_p) | H^{\text{eff}} | \chi_\nu(r - R_q)\rangle$$

$$(13.19)$$

and

$$S_{\mu\nu}(k) = \langle \phi_\mu(k) | \phi_\nu(k)\rangle$$

$$= N^{-1} \sum_p \sum_q \{\exp\,[ik(R_q - R_p)]\}\langle \chi_\mu(r - R_p) | \chi_\nu(r - R_q)\rangle \qquad (13.20)$$

Equation 13.18 may be written in a shorthand way as

$$|H_{\mu\nu}(k) - S_{\mu\nu}(k)\,e(k)| = 0 \qquad (13.21)$$

This is a very similar equation indeed to the one of equation 1.31 derived for the molecular case. There the basis orbitals used were single atomic orbitals; here they are Bloch functions. In order to derive the energy levels of a molecule, equation 1.31 needs to be solved just once (in principle). For an extended solid-state system equation 13.21 needs to be solved at several "k points" in order to map out the energetic dispersion of the bands. Sometimes we will be able to derive simple algebraic solutions for $e(k)$, as shown above for a particularly simple case. Most often, as is the case too for almost all the molecules we have studied, we will have to rely on a machine solution.

In 13.8 we chose a repeat unit for our calculation which contained a single orbital. If we choose a two-atom repeat unit as in **13.16** where $a' = 2a$, how does the

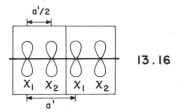

13.16

result change? Any observable property will, of course, have the same calculated value. The $e(k)$ versus k diagram will, however, be different since at each value of k there will be two energy levels, a direct result of the fact that there are now two orbitals per unit cell. To tackle this problem a secular determinant is set up, just as for the ethylene molecule of Section 12.2 but where from equation 13.19 the values of H_{ij} now depend upon k. As before, a considerable simplification of the problem can be made by using the Hückel approximation. First, we need to develop Bloch functions for each of the two orbitals in the unit cell. As shown in **13.16** we shall locate χ_1 and χ_2 on the left- and right-hand side atoms of any given unit cell. Starting with $\chi_1(r)$, this orbital is sent to $\chi_1(r - a')$ by a translation a' and to $\chi_1(r + a')$ by a translation $-a'$ **(13.17)**. Then using equation 13.5

$$\phi_1(k) = \frac{1}{\sqrt{N}} [\cdots \chi_1(r + a') \exp(-ika') + \chi_1(r) + \chi_1(r - a') \exp(ika') + \cdots]$$

$$(13.22)$$

Note that χ_2 and χ_1 are translationally separated by $a'/2$. Since the orbitals $\chi_1(r)$, $\chi_1(r + a')$, and $\chi_1(r - a')$ are not nearest neighbors in the chain, all interaction in-

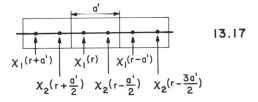

13.17

tegrals between orbitals located on them are zero in the Hückel approximation. This means that the energy of $\phi_1(k)$, $H_{11}(k)$, evaluated as $\langle \phi_1(k) | H^{\text{eff}} | \phi_1(k) \rangle$ is simply equal to α. A similar expansion occurs for $\chi_2(r)$

$$\phi_2(k) = \frac{1}{\sqrt{2}} \left[\cdots \chi_2\left(r + \frac{u'}{2}\right) \exp\left(\frac{ika'}{2}\right) + \chi_2\left(r - \frac{a'}{2}\right) \exp\left(\frac{ika'}{2}\right) \right.$$

$$\left. + \chi_2\left(r - \frac{3a'}{2}\right) \exp\left(\frac{3ika'}{2}\right) + \cdots \right]$$

$$(13.23)$$

Just as before, $H_{22}(k) = \alpha$ within the Hückel approximation. Unlike the diagonal elements, $H_{12}(k)$ does contain nearest neighbor interactions and exhibits a k dependence

$$H_{12}(k) = \langle \phi_1(k) \left| H^{\text{eff}} \right| \phi_2(k) \rangle$$

$$= N \times \frac{1}{N} \times \left[\exp\left(\frac{ika'}{2}\right) + \exp\left(\frac{-ika'}{2}\right) \right] \beta$$

$$= 2\beta \cos\left(\frac{ka'}{2}\right) \tag{13.24}$$

According to the Hückel approximation $S_{11}(k) = S_{22}(k) = 1$ and $S_{12}(k) = 0$. Consequently from equation 13.18 the secular determinant becomes

$$\begin{vmatrix} \alpha - e(k) & 2\beta \cos\dfrac{ka'}{2} \\ 2\beta \cos\dfrac{ka'}{2} & \alpha - e(k) \end{vmatrix} = 0 \tag{13.25}$$

The lower energy root is $e_1(k) = \alpha + 2\beta \cos (ka'/2)$ and the higher energy root is $e_2(k) = \alpha - 2\beta \cos (ka'/2)$. These results are shown graphically in Figure 13.3. Remembering that a' in **13.16** is twice the a of **13.8** the relationship between Figures 13.3 and 13.2 is straightforward. The $e(k)$ vs. k diagram of the two-atom cell is just that of the one-atom cell but the levels have been folded back along $k = \pi/2a$ (Figure 13.4). Now there are two orbitals for each value of k.

The orbitals at various values of the energy in Figure 13.3 are exactly those shown in **13.3**. At the zone center are found the most bonding and most antibonding levels and at the zone edge the nonbonding levels. Figure 13.3 also shows another way of generating these energy bands by starting off from the π and π^* levels of a diatomic unit (located at $e = \alpha + \beta$ and $\alpha - \beta$, respectively). First we write

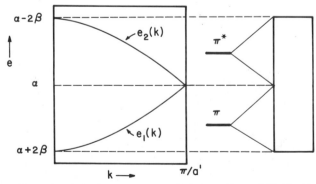

FIGURE 13.3.   Dispersion behavior of the two-orbital problem of **13.16**. Also shown is the identification of the lower and upper halves of this digram with the energetic behavior of the π and π^* levels of the diatomic unit contained in the unit cell of **13.16**.

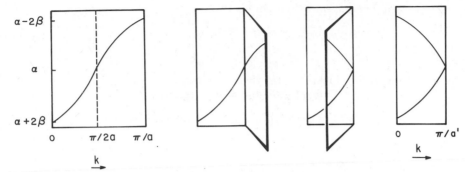

$a-2\beta$

a

$a+2\beta$

0 $\pi/2a$ π/a

0 π/a'

k

k

FIGURE 13.4. The "folding back" of the dispersion curve for the one orbital cell to give the dispersion curve for the two orbital cell.

$$\text{for } \pi \qquad \xi_1(r) = \frac{1}{\sqrt{2}}\left[\chi_1(r) + \chi_2\left(r - \frac{a'}{2}\right)\right]$$

$$\text{for } \pi^* \qquad \xi_2(r) = \frac{1}{\sqrt{2}}\left[\chi_1(r) - \chi_2\left(r - \frac{a'}{2}\right)\right] \qquad (13.26)$$

Using the same notation as before we can construct Bloch functions as

$$\phi_1'(k) = \frac{1}{\sqrt{N}}\ [\cdots \xi_1(r+a')\exp(-ika') + \xi_1(r) + \xi_1(r-a')\exp(ika') + \cdots]$$

$$\phi_2'(k) = \frac{1}{\sqrt{N}}\ [\cdots \xi_2(r+a')\exp(-ika') + \xi_2(r) + \xi_2(r-a')\exp(ika') + \cdots]$$

$$(13.27)$$

Using the Hückel approximation we can readily evaluate $H_{11}(k)$ and $H_{22}(k)$ as

$$H_{11}(k) = \langle \phi_1'(k) \,|\, H^{\text{eff}} \,|\, \phi_1'(k)\rangle$$

$$= \frac{1}{N}\cdot N\left\{\alpha + \beta + \frac{1}{2}\beta\ [\exp(ika') + \exp(-ika')]\right\} = \alpha + \beta + \beta\cos ka'$$

$$H_{22}(k) = \langle \phi_2'(k) \,|\, H^{\text{eff}} \,|\, \phi_2'(k)\rangle$$

$$= \frac{1}{N}\cdot N\left\{\alpha - \beta - \frac{1}{2}\beta\ [\exp(ika') + \exp(-ika')]\right\} = \alpha - \beta - \beta\cos ka'$$

$$(13.29)$$

$H_{12}(k)$ may be evaluated analogously as

$$H_{12}(k) = \langle \phi_1'(k) \,|\, H^{\text{eff}} \,|\, \phi_2'(k)\rangle$$

$$= \frac{1}{N}\cdot N\left\{\frac{1}{2}\beta\ [\exp(ika') - \exp(-ika')]\right\} = i\beta\sin ka' \qquad (13.30)$$

Similar evaluation of $H_{21}(k)$ leads to $-i\beta\sin ka'$, that is, $H_{21}(k) = H_{12}^*(k)$.

The secular determinant then becomes

$$\begin{vmatrix} \alpha + \beta + \beta \cos ka' - e(k) & i\beta \sin ka' \\ -i\beta \sin ka' & \alpha - \beta - \beta \cos ka' - e(k) \end{vmatrix} = 0 \qquad (13.31)$$

Notice that it is $H_{12}(k)$ and $H_{12}^*(k)$ that go into the off-diagonal positions of this equation. Solution of the secular determinant leads to

$$[\alpha - e(k)]^2 - \beta^2(1 + \cos ka')^2 - \beta^2 \sin^2 ka' = 0 \qquad (13.32)$$

and therefore

$$e(k) = \alpha \pm 2\beta \cos \frac{ka'}{2} \qquad (13.33)$$

which is the same result as before. Notice that the value of H_{12} in equation 13.31 is identically zero at $k = 0$ and also at $k = \pi/a'$. At these points the upper and lower bands are then, respectively, pure $\phi_1'(k)$ and $\phi_2'(k)$ in character since there is no mixing between them. Equation 13.27 requires that, at $k = 0$, the coefficient of $\xi_\mu(r - R_p)$ is $\exp(ikR_p) = +1$. Thus the π orbitals are combined as in **13.18** which is bonding between unit cells. Note that this function is not only intercell bonding but is intracell bonding too. At the same time the π^* orbitals are combined as in **13.19** which is antibonding between cells. As a consequence the function that results is both intracell and intercell antibonding. At $k = \pi/a'$ the coefficient of $\xi_\mu(r - R_p)$ is now $(-1)^p$, and the combination of π orbitals of **13.20** is intracell bonding but intercell antibonding. Similarly the π^* levels combine to give a function **(13.21)**

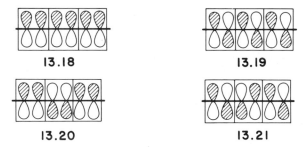

13.18

13.19

13.20

13.21

which is intracell antibonding but intercell bonding. Obviously **13.20** and **13.21** have the same energy and the top of the π band and bottom of the π^* band touch at this point. As noted before, functions that are equally good for this degenerate pair may be obtained by taking a linear combination of **13.20** and **13.21**. The result is shown in **13.22** and **13.23**.

13.22   13.23

13.2.   DISTORTIONS OF ONE-DIMENSIONAL SYSTEMS

The polymeric material of **13.24**, polyacetylene, with one $p\pi$ orbital per atom has a π-band structure which is identical to the one we have spent so much time discussing in Section 13.1. With one $p\pi$ electron per atom this band is half-full and if the electrons are paired (**13.6**) the system should be metallic. Polyacetylene itself does not have the regular structure indicated in **13.24** but is a semiconductor and exhibits the bond alternation shown in **13.25**. Let us see how $e(k)$ varies for this system. Now the distances between one atom and its two neighbors (**13.26**) are not the

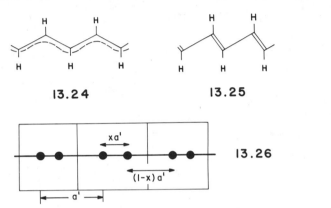

same $[(1 - x)a'$ and xa' where $x < \frac{1}{2}]$. In addition to giving rise to different values of R_p in equation 13.6, different values of the resonance integrals will also be found. β_1 and β_2 may be assigned to the interaction integrals between two neighboring orbitals separated by xa' and $(1 - x)a'$ respectively. Since xa' is smaller than $(1 - x)a'$ we note that $|\beta_1| > |\beta_2|$. Note also that $\beta_1, \beta_2 < 0$ for the case of $p\pi$ orbital overlap. The secular determinant then becomes

$$\begin{vmatrix} \alpha - e(k) & \beta_1 \exp{(ikxa')} + \beta_2 \exp{[-ik(1 - x)a']} \\ \beta_1 \exp{(-ikxa')} + \beta_2 \exp{[ik(1 - x)a']} & \alpha - e(k) \end{vmatrix} = 0$$

(13.34)

Note that as in equation 13.31 one off-diagonal element is the complex conjugate of the other. Solution of this determinant leads to

$$e(k) = \alpha \pm (\beta_1^2 + \beta_2^2 + 2\beta_1\beta_2 \cos ka')^{1/2} \qquad (13.35)$$

Notice that any dependence on x has disappeared from the cosine term. We will take the lower energy level $e_1(k)$ as equation 13.35 with the positive root and the higher energy level $e_2(k)$ as equation 13.35 with the negative root. At $k = 0$, $e_1(k = 0) = \alpha + (\beta_1 + \beta_2)$ and $e_2(k = 0) = \alpha - (\beta_1 + \beta_2)$. At $k = \pi/a'$, $e_1(k = \pi/a') = \alpha + (\beta_1 - \beta_2)$ and $e_2(k = \pi/a') = \alpha - (\beta_1 - \beta_2)$. The $e(k)$ vs. k diagram which results is shown in Figure 13.5a for the case where $|\beta_1| > |\beta_2|$. The corresponding density of states picture is shown in Figure 13.5b. For small distortions $\beta_1 + \beta_2 \simeq 2\beta$. The

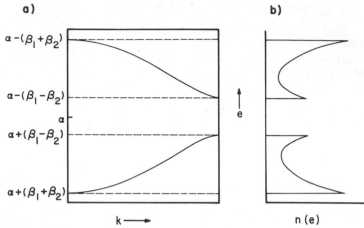

FIGURE 13.5. (a) Dispersion behavior of the two orbitals contained in the unit cell of **13.26** where the internuclear distances along the chain are not uniform. (b) The corresponding density of states.

important result is a splitting of the degeneracy at the zone edge. The form of the new wavefunctions is easy to derive and is given in Figure 13.6. In earlier chapters we have emphasized how symmetrical structures on distortion may either open up a gap or increase an existing energy gap between the HOMO and LUMO. In the case of the nonalternating polyene of **13.24** with one electron per atom the π band of Figure 13.2 is half-full, there is no HOMO–LUMO gap, and the situation is reminiscent of the case of singlet cyclobutadiene of Chapter 12. In the alternating case of **13.25** the lower band of Figure 13.5 is full and the upper empty. Thus the energetic stabilization on distortion of the symmetrical structure to one with bond alternation

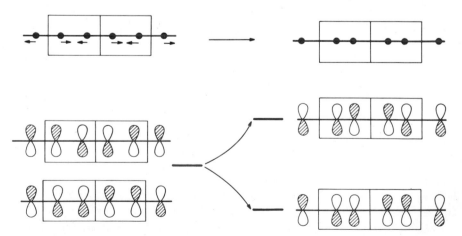

FIGURE 13.6. Generation of the form of the new wavefunctions at $k = \pi/a'$ as a result of the distortion shown in **13.26**.

is really the solid-state analog of a Jahn–Teller distortion.[4–6] It is called a Peierls distortion and the situation is compared with that of cyclobutadiene in **13.27**. How

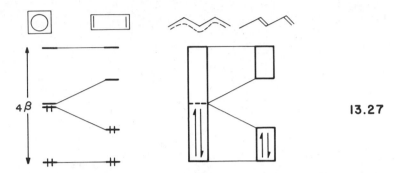

13.27

can we stabilize a system such as **13.24** against such a distortion? There are two obvious strategies that can be borrowed from molecular chemistry. In Chapter 12, we saw that altering the electronegativity of two carbon atoms in cyclobutadiene relative to the other two, split the degeneracy of the nonbonding set and the driving force to the rectangular geometry was lost. **13.28** and **13.29** are two possible related

13.28   **13.29**

examples in the solid state. The first is not known and the second is known but, as yet, poorly characterized. Both are isoelectronic with polyacetylene but the electronegativity difference ensures that the π bands will not touch at the zone edge. (We will examine this case in detail below.) An alternative approach to the problem uses the result that the addition of two electrons to cyclobutadiene completely fills the degenerate nonbonding π set. Consequently, adding two electrons per four-atom unit removes the Jahn–Teller instability so that S_4^{2+}, Se_4^{2+} and Te_4^{2+} are square planar ions. In the present case we need to fill the entire π band with electrons (one extra electron per site). The result is the structure of fibrous sulfur, selenium, and tellurium. These contain chains of atoms in which all the bond lengths are the same. (The chain, however, has distorted so that it is not planar.) Polyacetylene itself may be made conducting by doping either with electron donors or acceptors. The removal of some electron density from the filled band **(13.30)** or the addition of density to the empty band **(13.31)** leads to a conducting (metallic) situation (cf. **13.4**).

Electronically intermediate between the polyacetylene example with one π electron per center and the planar analog of the sulfur chain with two such electrons per center is the $(SN)_x$ polymer with three π electrons per SN atom pair.[6] The band structure of the *trans* isomer **13.32** is shown in Figure 13.7 where we have chosen a

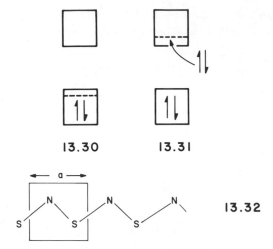

unit cell containing two atoms. It is easy to understand in a qualitative fashion. With two π orbitals per cell there will be two π bands. The splitting at the zone edge in $(SN)_x$, absent in the polyacetylene example of Figure 13.4, is due to the different atomic $p\pi$ (α_S and α_N) energies of the two atoms, sulfur and nitrogen. We show this in the following way. The secular deteminant of equation 13.22 which described a degenerate interaction becomes in $(SN)_x$

$$\begin{vmatrix} \alpha_N - e(k) & 2\beta \cos \dfrac{ka'}{2} \\ 2\beta \cos \dfrac{ka'}{2} & \alpha_S - e(k) \end{vmatrix} = 0 \qquad (13.36)$$

This describes a nondegenerate interaction between the levels α_S and α_N. Using the

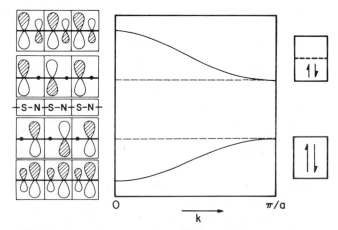

FIGURE 13.7. Dispersion behavior of the π orbitals of polymeric $(SN)_x$. With three π electrons per SN unit the upper band is half-full and so the material is metallic.

ideas of Section 3.2, the energies of the two bands become

$$e_1(k) \simeq \alpha_N + \frac{4\beta^2 \cos^2 ka'/2}{(\alpha_N - \alpha_S)} \tag{13.37}$$

and

$$e_2(k) \simeq \alpha_S - \frac{4\beta^2 \cos^2 ka'/2}{(\alpha_N - \alpha_S)} \tag{13.38}$$

At $k = \pi/a'$ the two energies are simply α_N and α_S, and the form of orbitals just as in **13.22** and **13.23** which are now not degenerate. With three π electrons per cell the upper π band is half-full and the Peierls type of distortion is expected. The actual structure of $(SN)_x$ is in fact the *cis* isomer of **13.33**. Since the repeat unit is

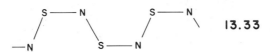

13.33

now four atoms, the band structure is somewhat more complex. (Essentially the levels of Figure 13.7 are folded back as in Figure 13.4). The overall result though is very similar with a similar prediction of a Peierls instability. $(SN)_x$ however, instead of distorting to remove this instability, remains a metal with a half-filled band. It has been suggested that a Peierls distortion is inhibited by interactions between chains of the polymer. This is a very striking material, one composed of sulfur and nitrogen only, which has a copperlike lustre and is metallic. When there are only two π electrons per unit cell (as in **13.28** and **13.29**) then the symmetrical structure is now an insulator and does not suffer from a Peierls instabilty.

13.3.   OTHER ONE-DIMENSIONAL SYSTEMS

Our discussion so far has focused on polyacetylene and related examples. The broad results however, are transferable to many other systems. Algebraically, for example, our discussion applies equally well to the case of a one-dimensional chain of hydrogen atoms bearing $1s$ orbitals. Recall the one-to-one correspondence between the orbitals of finite H_n molecules and their polyene analogs. With one electron per atom, Figure 13.3 indicates a half-filled band for the geometry **13.34**. This will distort (in a Peierls fashion) to a solid composed of H_2 molecules (**13.35**) as chemical

$$-H—H—H—H—H—H- \qquad \longrightarrow \qquad H—H \quad H—H \quad H—H$$

13.34   13.35

intuition predicts. There is, however, considerable interest in generating the three-dimensional analog of **13.34** under high pressure. This species is expected to have a half-filled band and so should be a metal.

For a band describing a chain of *po* orbitals the $e(k)$ vs. k diagram looks a little

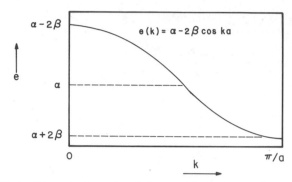

FIGURE 13.8.   Dispersion behavior of the $p\sigma$ orbitals of a one-dimensional chain.

different (Figure 13.8). The phase factor at $k = 0$ requires (equation 13.6) all the atomic coefficients equal to +1 **(13.36)**. This is the point where maximum destabil-

13.36

$k = 0$ $k = \pi/a$

ization occurs. So

$$e(k) = \alpha - 2\beta \cos ka \tag{13.39}$$

At $k = \pi/a$ the phase difference between adjacent orbitals is −1 and maximum bonding results. Apart from this rather simple difference the band structure is identical to our earlier example. Now of course β represents $p\sigma - p\sigma$ rather than $p\pi - p\pi$ interactions. Consequently the band spread in an absolute sense is larger since $p\sigma$ overlap is larger than $p\pi$ overlap, that is, $|\beta_\sigma| > |\beta_\pi|$.

A band made up of z^2 orbitals on each atomic center (in this case a transition metal) in a linear chain will also look very similar to that of Figure 13.2 with β now describing $z^2 - z^2$ interactions **(13.37)**. A well-known example of this situation is that of salts containing the square planar $Pt(CN)_4^{2-}$ unit in $K_2[Pt(CN)_4] \cdot 3H_2O$, which form[7] chains of the type shown in **13.38**. Recalling the level pattern for a square planar species (Figure 16.1) the orbital configuration for a low spin d^8 Pt(II) species is $(b_{2g})^2(e_g)^4(2a_{1g})^2$ or $(xy)^2(xz, yz)^4(z^2)^2$. So with two electrons in the z^2 orbital the band formed by overlap of **13.39** with its neighbors is full. The system is an insulator (conductivity = 5×10^{-7} Ω^{-1} cm^{-1}) as a result, and, in this white salt, equal Pt—Pt distances are found (3.48 Å) which are a little long and reflect the fact that both bonding and antibonding Pt—Pt orbitals are occupied at the zone center and edge, respectively. (One might wonder why the $Pt(CN)_4^{-2}$ chains arrange themselves in a stacked fashion in the first place. A part of the answer lies with the higher energy s and p_z orbitals. In a formal sense these are empty and can mix into the z^2 combinations in such a way as to reduce the net Pt—Pt antibonding between them. Another contribution is via hydrogen bonding involving the water molecules.)

13.37

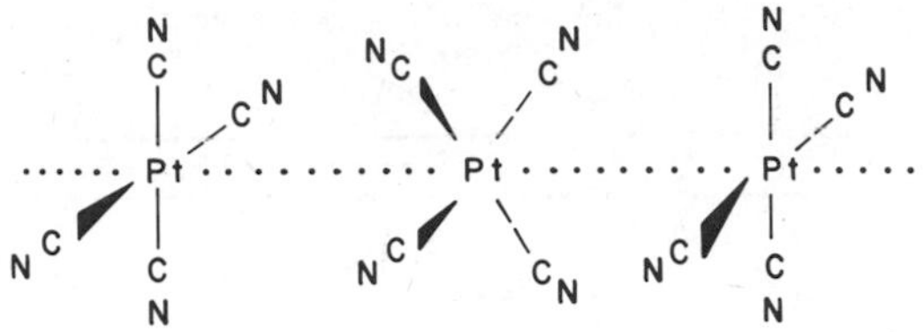

13.38

13.39

These chain systems may, however, be cocrystallized with Br_2, which results in the formation of $K_2[Pt(CN)_4]Br_\delta \cdot 3H_2O$ salts and partial oxidation to $Pt^{(2+\delta)}$. On oxidation, electrons are removed from levels at the top of the band **13.40** and trans-

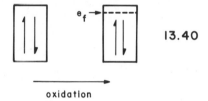

13.40

oxidation

ferred to bromine (as Br^-). These levels are strongly Pt—Pt antibonding and the result is a dramatic shortening of the Pt—Pt distances. In $K_2[Pt(CN)_4]Br_{0.3} \cdot 3H_2O$ the Pt—Pt distance is 2.88 Å. In addition, the presence now of the partially filled band means that the system is metallic, just as in the case of doped polyacetylene. A bronze solid results with a conductivity of 200 Ω^{-1} cm^{-1}.

A more complex example, but one that is understandable along similar lines, is the structure of the niobium and tantalum tetrahalides.[8] They form chains based on the sharing of opposite edges of NbX_6 octahedra with niobium–niobium distances which alternate along the chain **(13.42)**. The structure is an obvious distortion of the ideal edge-sharing chain **(13.41)**. The electronic configuration of the metals is

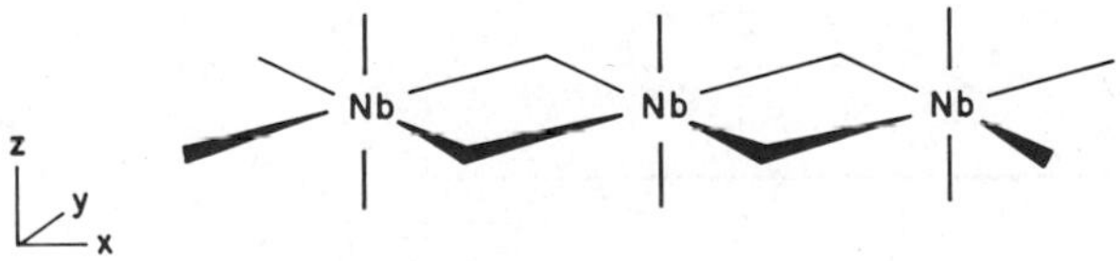

13.41

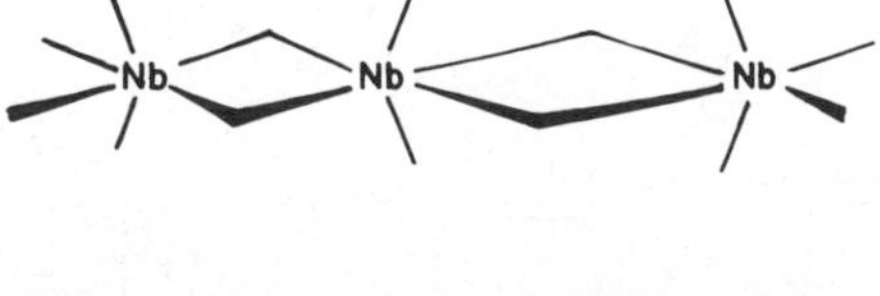

13.42

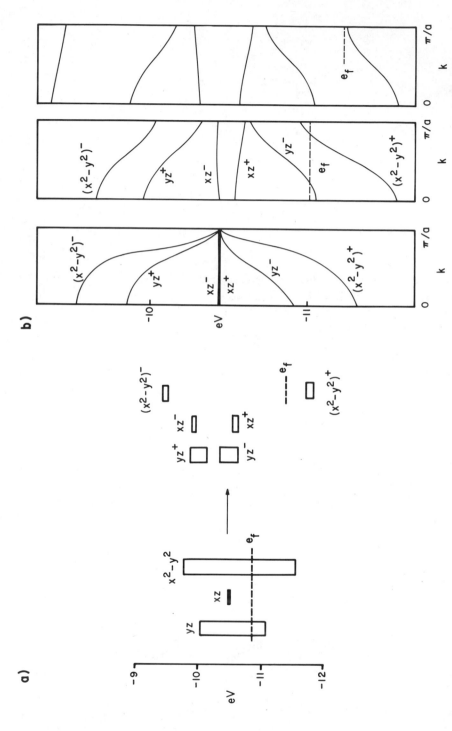

FIGURE 13.9. Behavior of the energy bands of **13.41** on distortion to the structure of **13.42.** (a) Block diagram. (b) $e(k)$ diagrams. The three d bands corresponding to the "t_{2g}" levels of the undistorted geometry split apart into pairs (labeled $+$ and $-$) on distortion. In (b) the middle panel shows how the bands change in energy for a smaller distortion; the right-hand panel shows the situation for a larger distortion.

d^1 which immediately suggests the possibility of a Peierls type of distortion, result-ing in the atoms being linked together in pairs as in the p^1 case of polyacetylene or s^1 case of hydrogen. The band structure associated with the d orbital region of the undistorted structure is shown in Figure 13.9. Its derivation and complete under-standing is beyond the scope of this book. It is somewhat more complex than our previous examples in that the dispersion along the chain direction is controlled not only by direct $d - d$ interactions between the two metal centers but is also mediated by the presence of bridging halide orbitals. The overall picture is one of the two-above-three level structures found for octahedral metal coordination. (We only show the lower energy group of three in Figure 13.9.) Since there are two metal atoms per unit cell there are three pairs of bands as shown in Figure 13.9b. On dis-torting **13.41** to **13.42** it is easy to see how each pair of folded bands splits apart in energy at the zone edge, just as in Figure 13.5. In Figure 13.9 it is clear that if the distortion is large enough then the band labeled $(x^2 - y^2)^+$ will be split off from the rest and, with one d electron per metal atom, will give rise to an insulator. It is interesting that applying pressure to isostructural NbI_4 converts it to a metal. This can be regarded as a result of the pressure-induced reduction of the niobium–niobium bond alternation and the overlap of a filled band with an empty one, as in the inter-mediate case of Figure 13.9.

Many other examples exist,[7] with perhaps very different chemical compositions, but which are understandable in an exactly analogous way.[9] One particularly im-portant series are the organic metals made by stacking planar molecules on top of one another. Tetrathiofulvalene (TTF) shown in **13.43** is one example. The tetra-methylated derivative (TMTTF) and its selenium analog (TMTSF) are two others. Stacked conductors containing these units may be made in an exactly analogous way to the tetracyanoplatinate (TCP) example described above. $(TTF)Br_{\sim 0.73}$, for example, has a conductivity parallel to the chain axis of about $400 \ \Omega^{-1} \ cm^{-1}$. Here the orbital involved in forming the one-dimensional band is not localized on a single atom as in TCP but is delocalized over the organic unit. In $(TTF)Br$ where this band is exactly half-full discrete $(TTF)_2^{+2}$ dimers are found with an exactly analogous explanation to the one for the dimerization of the H atom chain **13.34** to **13.35**. 7,7,9,9-tetracyano-p-quinodimethane (TCNQ) is another example of an organic metal **(13.44)**. The system $[H(CH_3)_3N]^+ \ (I^-)_{1/3} \ (TCNQ)^{-2/3}$ has a conductivity of $\sim 20 \ \Omega^{-1} \ cm^{-1}$. Note that TTF forms a cationic but TCNQ an anionic chain. Al-though this is a simple description of the electronic problem here, these systems are, in fact, somewhat more complex than we have intimated.

13.43   13.44

13.4.   THREE-DIMENSIONAL SYSTEMS

So far we have concentrated on one-dimensional systems, but this approach is readily extended in principle to two and three dimensions. We shall illustrate the three-dimensional case with a simple example. The natural structure to look at is simply the linking together of chains of atoms along the x, y, and z directions.[10] This gives rise to the simple cubic structure of **13.45**. It is easy to generate the **k**

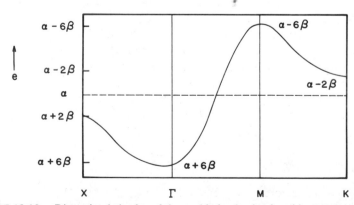

13.45

dependence of the energy of an s orbital located on each atom as equation 13.36 where we specify values of **k** in terms of the three Cartesian directions. The energy dependence is simply the sum of three perpendicular systems each given by equation 13.3. It is problematical to display the energy dependence upon k_x, k_y, and k_z simultaneously, but what is done is to present slices through the $e(\mathbf{k})$ surface as in Figure 13.10. Here the symbols Γ, M, K, and X represent points in the Brillouin zone at $(k_x, k_y, k_z) = (0, 0, 0)$ $(2\pi/a)$, $(\frac{1}{2}, \frac{1}{2}, \frac{1}{2})$ $(2\pi/a)$, $(\frac{1}{2}, \frac{1}{2}, 0)$ $(2\pi/a)$ and $(0, 0, \frac{1}{2})$ $(2\pi/a)$, respectively. Clearly because of the symmetry inherent in equation 13.36, the energy at $(0, 0, \frac{1}{2})$ $(2\pi/a)$, $(0, \frac{1}{2}, 0)$ $(2\pi/a)$, and $(\frac{1}{2}, 0, 0)$ $(2\pi/a)$ are equal. Also $e(\mathbf{k}) = e(-\mathbf{k})$. The points Γ, M, K, and so on, are called the symmetry points of the Brillouin zone.

$$e(\mathbf{k}) = \alpha + 2\beta \left[\cos k_x a + \cos k_y a + \cos k_z a\right] \tag{13.40}$$

For the three p orbitals of the primitive cubic lattice **(13.45)** the result is also simple if π-type overlap and any interaction with the s orbitals are neglected. It is just the sum of three diagrams of the type shown in Figure 13.8 where the energy dependence on **k** is given by three functions of the type of equation 13.39.

$$e(\mathbf{k}) = \alpha - 2\beta \cos k_x a \qquad \text{for } p_x$$
$$e(\mathbf{k}) = \alpha - 2\beta \cos k_y a \qquad \text{for } p_y \tag{13.41}$$
$$e(\mathbf{k}) = \alpha - 2\beta \cos k_z a \qquad \text{for } p_z$$

FIGURE 13.10.   Dispersion behavior of the s orbitals of a simple cubic structure **(13.45)**.

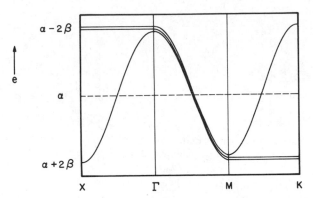

FIGURE 13.11.   Dispersion behavior of the p orbitals of a simple cubic lattice (neglecting π type overlap).

where

$$\mathbf{k} = \hat{\mathbf{i}}k_x + \hat{\mathbf{j}}k_y + \hat{\mathbf{k}}k_z \tag{13.42}$$

$\hat{\mathbf{i}}, \hat{\mathbf{j}}$, and $\hat{\mathbf{k}}$ being three unit vectors along the Cartesian x, y, and z directions, respectively.

A picture analogous to that of Figure 13.10 is shown in Figure 13.11. Note that the triple degeneracy of the three p orbitals only holds at the points Γ and M.

Just as the half-filled one-dimensional chains of the previous section underwent a Peierls distortion, so we expect that the simple cubic lattice with three p electrons would be similarly unstable. A distortion of this type has indeed occurred when we look at the observed structures of elemental black phosphorus (13.46) and arsenic (13.47). Alternate linkages have been broken along all three directions leaving each atom pyramidally three coordinate. The result is in accord with their atomic configuration $s^2 p^3$. At the simple cubic structure p_x, p_y and p_z bands would each be

P   As

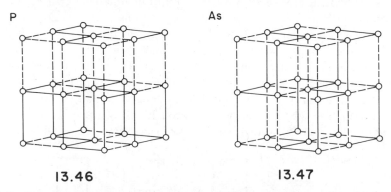

13.46   13.47

half-full. On applying pressure to crystals of black phosphorus the simple cubic structure is produced which is metallic, a process similar to the one described earlier for NbI_4 and the one sought after for hydrogen. Both the arsenic and black phosphorus structures are layer structures with no covalent bonds between the layers.

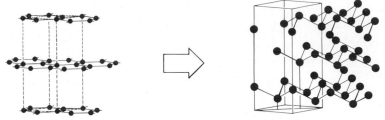

FIGURE 13.12. Relationship between the structures of graphite (left) and arsenic (right). Geometrically, puckering each sheet of graphite and shifting it relative to the one below it leads to the structure of arsenic.

Each atom is in a trigonal pyramidal coordination. Another way of viewing the arsenic structure is as a distortion of another layer structure, namely that of graphite (Figure 13.12). Three of the four electrons from each carbon atom in graphite are used in forming a σ-bonded network. This leaves one electron per carbon atom in a p_z orbital, perpendicular to the graphite plane, which may interact with its neighbors in a π sense. Crudely then, planar graphite is akin to CH_3 or BH_3 with one less valence electron. Both of these species are planar. NH_3, however, with five valence electrons is pyramidal, as described in Chapter 7, for very well-defined reasons. The driving force for pyramidalization in the solid-state analog is similar. In both the molecular example of NH_3 and in the example of arsenic here, lone pairs of electrons occupy the fourth coordination site at each atom.

An important feature of the energy bands described in this chapter is that (with reference to **13.4**) the bottom of the band, which receives the maximum stabilization, is bonding throughout, whereas the most antibonding character is found at the very top of the band. In the middle there are nonbonding orbitals. This is of course just like the pattern of *molecular* orbitals found for the first-row diatomic molecules in Section 6.1. In the molecular case an approximate bond order may be defined as (number of bonding pairs of electrons) – (number of antibonding pairs of electrons). The variation in this function with number of electrons is easy to derive from Figure 6.3 and is shown in Figure 13.13a. It indicates that the strongest bond is found when the atomic p levels are filled to exactly half their capacity (i.e., N_2). For sp solids made of main group elements such as Be, C, O, and so on, then the s and p bands broaden out to give a band of hybrid character as in **13.48**. Similarly the atomic s and d levels of the transition elements broaden into an analogous composite band **(13.49)**. The cohesive energies (equation 13.43) of the main group elements X vary[11] as shown in Figure 13.13b and show in an exactly analogous way the max-

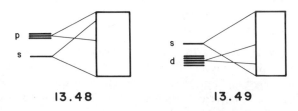

13.48   13.49

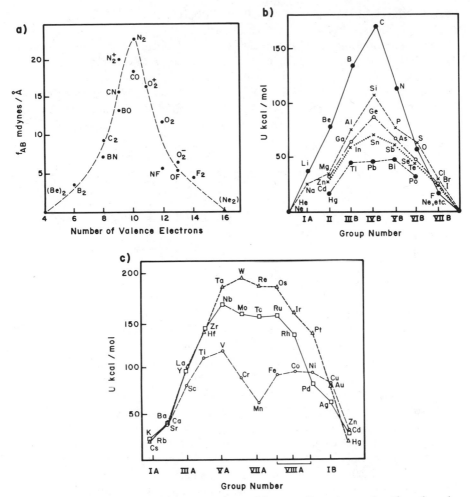

FIGURE 13.13. (a) Vibrational force constants of first-row diatomics as a function of number of electron pairs. (b) Cohesive energy (U) of main group elements as a function of group number (number of valence electrons). (C) Cohesive energy (U) of transition elements as a function of group number.

imum at the half-filled band at the Group IV elements. Similarly the transition metals show a maximum in the cohesive energy for six electrons at the chromium group.[11] For the first-row metals the plot of Figure 13.13c is a little asymmetrical due to the magnetic nature of the elements Mn, Fe, and Co, a topic we discuss below. One interesting point concerning these figures is that the cohesive energy is broadly independent of structure, or phrased alternatively, the energy differences between structural alternatives is small compared to the energy of formation of the crystal.

$$X(\text{monatomic gas}) \longrightarrow X(\text{solid}), \qquad \Delta H^0 = \text{cohesive energy} \qquad (13.43)$$

13.5. HIGH SPIN AND LOW SPIN CONSIDERATIONS

Just as in the molecular case, and discussed in detail in Chapter 8, there is always a choice to be made between filling all the lowest levels with electron pairs and the alternative of allowing some of the higher energy levels to be occupied by electrons with parallel spins. **13.5** and **13.6** showed two extreme cases where all the electrons in a band were either all spin unpaired or all spin paired. An intermediate situation shown in **13.50** is also of importance, where not all of the spins are unpaired. An example of this type, which, in addition to being metallic is magnetic, is found in the body-centered cubic structure of elemental iron. There are about 1.5 unpaired electrons per atom. **13.51** depicts an alternative way of showing this result which

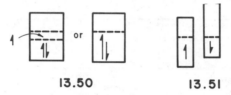

13.50   13.51

emphasizes the lower energy of the up-spin band compared to the down-spin band. This is a result of the larger number of up-spin electrons and a concomitantly larger number of stabilizing exchange integrals between them, compared to the down-spin electrons. As in the molecular case, it is difficult to predict *a priori* whether magnetic or nonmagnetic states will be found in a given instance. One interesting observation which has an exact parallel with molecular chemistry is that a change of spin state is often associated with a change in structure. Just as high and low spin four-co-ordinate d^8 molecules are tetrahedral and square planar, respectively, (Section 16.4) so magnetic iron has the body-centered cubic structure but nonmagnetic iron crystallizes in the hexagonal close-packed arrangement. Note that in Figure 13.13c the magnetic elements had a smaller cohesive energy than expected. This may be understood as a result of the population of some of the higher energy antibonding orbitals as in **13.50** combined with the fact that the atomic configurations of the corresponding gaseous atoms are different.

Another instance where spin state is important and has a direct bearing on structure is in one-dimensional systems. Just as in Section 13.2 where we showed how a half-filled band usually results in a pairing distortion, so similar reasoning suggests

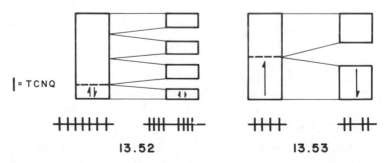

| = TCNQ

13.52   13.53

that a quarter-filled band should result in a tetramerization **13.52**. However, if the distorted arrangement is magnetic then dimerization is the process that is favored **13.53**. Again prediction of the mode of distortion is not at all easy. The state of affairs shown in **13.53** is found for the $(MEM)^+(TCNQ)_2^-$ species [$(MEM)^+$ = methyl ethyl morpholinium cation] where there is one electron per two TCNQ orbitals.[4]

REFERENCES

1. J. C. Slater, *Introduction to Chemical Physics*, McGraw-Hill, New York (1939).
2. B. C. Gerstein, *J. Chem. Educ.*, **50**, 316 (1973).
3. N. W. Ashcroft and N. D. Mermin, *Solid State Physics*, Saunders, Philiadelphia (1976).
4. M-H Whangbo, *Accts. Chem. Res.*, **16**, 95 (1983).
5. M-H Whangbo, in *Extended Linear Chain Compounds*, Vol. II, J. S. Miller, editor, Plenum, New York (1982).
6. M-H Whangbo, and R. Hoffmann, R. B. Woodward, *Proc. Roy. Soc.*, **A366**, 23 (1979).
7. See, for example, J. S. Miller, editor, *Extended Linear Chain Compounds*, Vol. I, Plenum, New York (1982).
8. M-H Whangbo and M. J. Foshee, *Inorg. Chem.*, **20**, 113 (1981).
9. J. K. Burdett, *Progress in Solid State Chemistry* (in press).
10. J. K. Burdett and S. Lee, *J. Amer. Chem. Soc.*, **105**, 1079 (1983).
11. J. Friedel, *J. Physique.*, **39**, 651, 671 (1978).

CHAPTER FOURTEEN

Hypervalent Molecules

14.1. ORBITALS OF OCTAHEDRALLY BASED MOLECULES

In many of the molecules studied so far there were obvious ties between the orbital picture we presented and traditional ideas of electron pair bond formation. But not all molecules are susceptible to the elementary decomposition described in Chapter 7, which showed the correspondence between localized and delocalized bonding viewpoints. For example, the linear H_3^- molecule of Section 3.3 has a single pair of electrons located in a bonding orbital and another pair in a nonbonding orbital. Clearly the two H—H "bonds" cannot be described as two-center-two-electron ones. In this case the best description of the bonding situation is as a three-center orbital arrangement containing two bonding electrons. Such ideas are quite familiar to us in the realm of conjugated organic molecules. In benzene, for example, we consider a π network delocalized over all six carbon atoms of the molecule. With a total of six π electrons located as three bonding electron pairs and six CC linkages, the C—C π bond order is $\frac{1}{2}$, just as in H_3^-. Such considerations are so much a part of the chemist's background that we feel quite comfortable mixing localized (e.g., the benzene C—H linkages) and delocalized (e.g., benzene C—C π linkages) descriptions of bonding within the *same* molecule. In this chapter we will study some main group molecules (such as SF_4 or ClF_3) where, as in benzene, localized and delocalized bonding descriptions may be concurrently used to describe different parts of the molecule. These molecules may also be viewed via a completely delocalized description. For some species, such as SF_6 we have no choice but to use a delocalized description unless higher energy d orbitals are included in the bonding picture. For "penta or hexavalent carbon" (e.g., $C_5H_5^+$ or $C_6H_6^{2+}$ in Section 11.3) where such d orbital participation is unlikely on energy grounds, a delocalized picture is the only one we have.

First we generate the levels of an octahedral AH_6 molecule since it illustrates

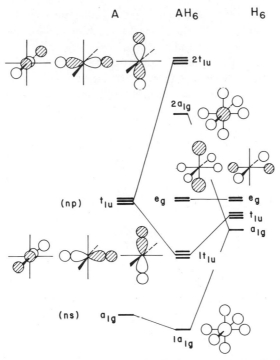

A AH_6 H_6

$2t_{1u}$

$2a_{1g}$

(np) t_{1u} e_g e_g
                                                t_{1u}
                                                a_{1g}

$1t_{1u}$

(ns) a_{1g}

$1a_{1g}$

FIGURE 14.1. Assembly of the molecular orbital diagram of an octahedral AH_6 molecule from the orbitals of A and of H_6. d orbitals are not included on A.

several of the general features associated with these so-called hypervalent molecules —molecules with more than an octet of electrons around the central atom. Group theory plays an important role. Figure 14.1 shows the strikingly simple interaction diagram for AH_6, assembled from an A atom bearing valence s and p orbitals and six hydrogen $1s$ orbitals. The ligand orbitals break down into three sets of a_{1g}, e_g, and t_{1u} symmetry. An important result is that the e_g pair finds no central atom orbitals with which to interact and remains completely ligand located and therefore A—H nonbonding. With a total of six valence electron pairs (e.g., for the hypothetical SH_6 molecule) four occupy A—H bonding orbitals and two are placed in this e_g nonbonding pair. So the molecule has six "bonds" but only four bonding electron pairs. The form of these bonding orbitals is particularly interesting. The lowest energy level arises via the in-phase overlap of the central atom s and ligand s orbitals, but the triply degenerate t_{1u} set are three center-bonding orbitals just like the lowest energy orbital in H_3^- (Figure 3.1). The only difference arises from the fact that the central atom in AH_6 has a p orbital while the central hydrogen in H_3^- has an s orbital. The symmetry of the molecule leads to a single nonbonding, ligand-located orbital for H_3^- but in AH_6 there is a doubly degenerate pair of ligand based orbitals (e_g). Just as in the case of H_3^-, electronic configurations with occupation of the nonbonding e_g pair lead to a buildup of electron density on the ligands. This is difficult to show in a general analytic way, but a simplification of this orbital picture will help. The Rundle–Pimentel scheme[1,2] neglects the involvement of the central atom s orbital, except as a storage location for one pair of electrons. Let us assume that the atomic sulfur $3p$ and hydrogen $1s$ orbitals have the same energy. Then the form of the wavefunctions for the $1t_{1u}$ and e_g orbitals are readily written down as in **14.1**. Using these results leads to a ligand

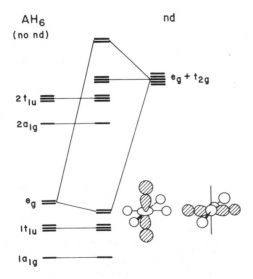

density of $\frac{7}{6}$ electrons (where overlap between the ligand atomic orbitals has been neglected) and a central atom density of five electrons for the configuration $(1a_{1g})^2 (1t_{1u})^6 (e_g)^4$. With specific reference to SH_6 this implies a transfer of one electron from the central atom to the ligands as a result of this electron-rich three-center bonding. The result suggests that the best stabilization will arise when the terminal atoms of such a structure are electronegative ones. This is in general true. For example, SF_6 is known but SH_6 is not. In a qualitative sense we could have anticipated this result by inspection of the interaction diagram of Figure 14.1. If the ligands are electropositive, then the e_g set would lie at high energy and the compound would be expected to undergo oxidation very readily.

One way that has been used to produce a localized bonding picture for molecules of this type is to involve the higher energy valence shell d orbitals in bonding. In the O_h point group these transform as $e_g + t_{2g}$ and the result of their inclusion is shown in Figure 14.2. Now, of course, there are six bonding orbitals and the rules

FIGURE 14.2.   Effect on the AH_6 energy levels of including d orbitals on A. The e_g orbital, nonbonding in Figure 14.1 is lowered in energy.

of Chapter 7 would allow us to generate six localized two-center–two-electron orbitals. The only question concerning such a picture is one of magnitude. Just how important are these orbitals energetically? This is still a controversial question but probably the involvement of, for example, the sulfur $3d$ orbitals in the ground-state wavefunction is small. For "pentavalent" first-row atoms as in the set of compounds[3] in **14.2** then the energy separation between central-atom $2s$, $2p$, and

14.2

$$Ar = p-MeC_6H_5, \ p-FC_6H_5$$
$$R = t-Bu, H$$

$3d$ orbitals is so immense that the higher energy orbitals are of no importance at all. (In any case, any inclusion of d orbitals will slightly stabilize the e_g set and will not alter our conclusions.) We will proceed in this chapter without the use of d orbitals and, therefore, force a delocalized description of the electronic structure in several places.

The hypothetical SH_6 molecule (the arguments for SF_6 will be similar) has one pair of electrons in the lowest energy a_{1g} orbital, three pairs of electrons in three-center bonding orbitals, and two pairs in nonbonding orbitals. With an extra pair of electrons the $2a_{1g}$ orbital is occupied, an orbital that is A—H antibonding. Now there are only three bonding pairs for six bonds. The molecules SbX_6^{n-} ($X = Cl$, Br; $n = 1, 3$) are known which differ in the occupancy of this a_{1g} orbital ($2a_{1g}$ is empty for $n = 1$). In nice verification of our description of the $2a_{1g}$ orbital, the bond lengths are substantially shorter for $n = 3$ compared to $n = 1$. (For $X = Cl$, bond lengths are 2.65, 2.35 Å; for $X = Br$ they are 2.80 and 2.55 Å.)

Figure 14.3 shows the analogous derivation of the level structure of the square planar AH_4 molecule. Once again group theoretical considerations are very useful in its generation. There are now two nonbonding orbitals; one ligand located and the analog of the e_g pair in AH_6 and one purely A located. This second orbital is a pure A p orbital which has a zero overlap with all the ligand s orbitals. The level ordering for square-planar AH_4 in Figure 14.3 is different from that discussed for square planar methane in Section 9.4. Here the central main group atom will be surrounded by electronegative atoms. (We have modeled them by hydrogens for convenience.) For the discussion in this chapter the b_{1g} orbital is placed below a_{2u}, anticipating that a more realistic case would be an AF_4 molecule. Figure 14.4 shows the very interesting energetic correlation between the levels of linear AH_2, square planar AH_4, and octahedral AH_6. The orbitals that directly correlate between AH_2 and AH_4 are connected by a solid line. Two hydrogen

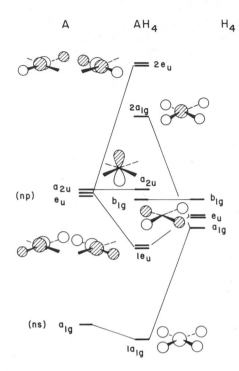

FIGURE 14.3.   Assembly of the molecular orbital diagram of a square AH_4 molecule from the orbitals of A and H_4. d orbitals are not included on A.

atoms are added to AH_2 and consequently there will be two additional valence orbitals in AH_4. One of these is derived from one component of the π_u set in AH_2. Bonding and antibonding combinations to σ_u from the H_2 "fragment" are formed. This is indicated by dashed lines in Figure 14.4. The other orbital that is formed is b_{1g}. This is derived from the σ_g fragment orbital of H_2. Likewise, on going from AH_4 to AH_6 two extra molecular orbitals are created. One is derived from the a_{2u} orbital of AH_4. Combined with the σ_u orbital of H_2 one component of the $1t_{1u}$ and $2t_{1u}$ orbitals result. The other MO is derived from σ_g of H_2, and leads to one component of the e_g set. We can use this diagram to trace the similarities between XeF_2, XeF_4 and XeF_6, the latter assumed to be octahedral. (We will return to its distorted structure below.) The usual assumption is made that the energetics of these molecules are dominated by the σ manifold. These species then have five, six, and seven σ pairs of electrons around the xenon atom, respectively. All three molecules are held together by three center bonds involving the central atom p orbital. The bonding contributed by the occupation of the deep lying orbital involving the central atom s orbital is canceled by occupation of its antibonding counterpart. The σ-bond order in all cases is thus equal to $\frac{1}{2}$. All three species have two nonbonding orbitals. In XeF_6 they are ligand located. In XeF_2 they are central atom located. In XeF_4 there is one orbital of each type. Clearly the nature of the electronic charge distribution is determined by the symmetry of the system in these cases.

A prominent feature of Figure 14.4 is the decreasing HOMO–LUMO gap in AH_n

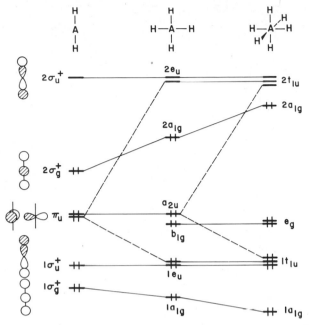

FIGURE 14.4. Correlation diagram for the molecular orbital levels of linear AH_2, square AH_4, and octahedral AH_6. The orbital occupancy is that expected for the XeH_n systems. Each time two hydrogen atoms are added to the system, two new orbitals are created. For example, on adding two H atoms to AH_2, a new ligand-located orbital (b_{1g} is produced). The dashed lines show how one component of π_u, along with one orbital combination from the added H_2 give rise to two new orbitals.

as n increases. In some AX_6 molecules with this electronic configuration the octahedral geometry is unstable and the molecule distorts. Since the HOMO is of a_{1g} symmetry and the LUMO of t_{1u} symmetry, according to the second-order Jahn–Teller recipe of Chapter 7, a t_{1u} symmetry distortion should open up a larger HOMO–LUMO gap as shown in **14.3**. If the driving force is large enough a static distortion will result. As with any degenerate set of orbitals we have a choice of how to write the wavefunctions. To see what happens in the distortion given by **14.3** we will write **14.4** as one component of a new $2t_{1u}$ set by using a linear combination of the old functions. On distortion the $2a_{1g}$ orbital and **14.4** mix together to give a hybrid orbital directed toward one face of the octahedron (**14.5**). As the distortion proceeds this orbital becomes more and more like a lone pair. The result is an interesting one when this orbital behavior is compared with the geometrical predictions of the VSEPR (valence-shell–electron-pair-repulsion) scheme.[4] In this very useful approach for the prediction of the geometries of main group compounds the electron pairs in the valence shell of the central atom are considered to arrange themselves so as to minimize the electrostatic repulsions between them. In methane there are a total of four valence pairs (four electrons from the central carbon atom and one from each coordinated hydrogen atom). These arrange them-

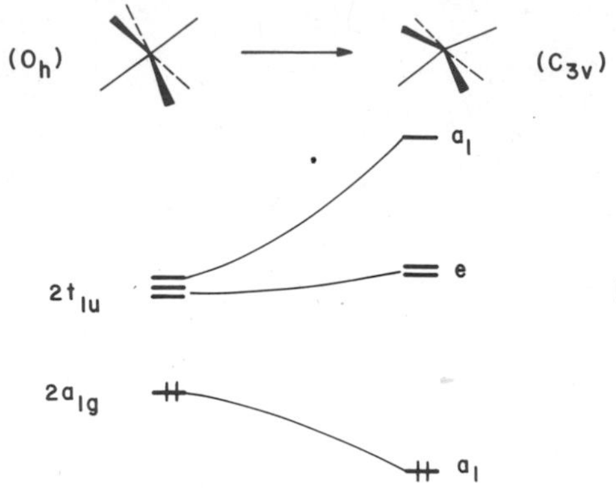

(O_h) (C_{3v})

14.3

$2t_{1u}$ e

$2a_{1g}$ a_1

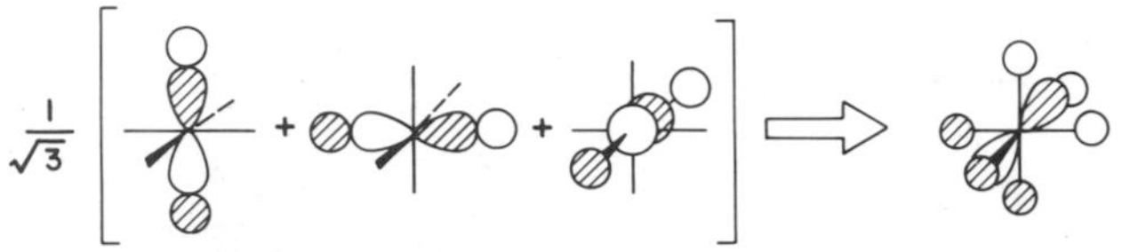

14.4

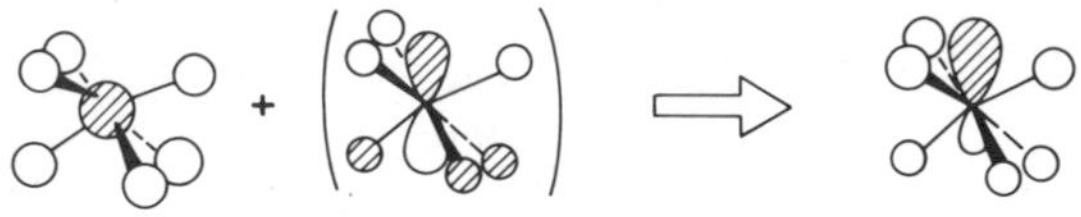

14.5

selves in a tetrahedral geometry. Since each electron pair is a bond pair then the location of these pairs determines the position of the hydrogen atom ligands. The geometry of CH_4 is tetrahedral as a result. In NH_3 there are also a total of four valence electron pairs (five electrons from the central nitrogen atom and one from each coordinated hydrogen atom). Three out of the four are bond pairs and one is, by default, a lone pair. NH_3 as a result has a pyramidal geometry with one lone pair envisaged as pointing out of the top of the pyramid. In XeF_6 with a total of seven pairs we expect to observe an octahedral geometry with a lone pair occupying a seventh site. Some species isoelectronic with XeF_6 have a distorted structure, others have a regular octahedral geometry. In molecules that fall into the second category their geometry is often discussed in terms of an inert pair of electrons. This chemically inert pair of electrons often implied by the structures of complexes containing a heavy atom from the right-hand side of the Periodic Table, has long posed a problem for theorists. At present, the best explanation of the reluctance of the $6s^2$ pair of electrons to enter into bonding is based on a relativistic effect[5], which manifests itself primarily as a contraction of s rather than p orbitals and is expected to be most important for heavy atoms. XeF_2 and XeF_4 present no problems for the VSEPR scheme. With five and six pairs of electrons respectively these molecules should have the structures shown in **14.6**.

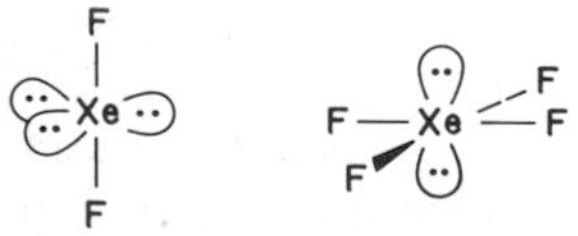

14.6

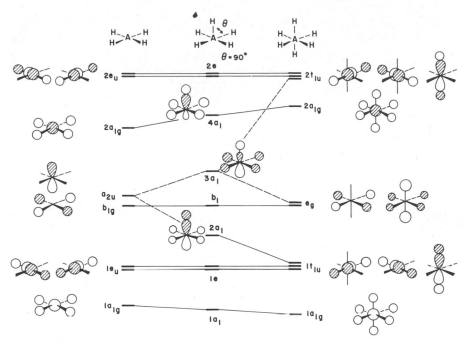

FIGURE 14.5. Correlation diagram for the molecular orbital levels of square AH_4, square pyramidal AH_5 ($\theta = 90°$), and octahedral AH_6. As in Figure 14.4 we show the effect of the new orbitals by the use of dashed tie lines.

The levels of square pyramidal AH_5 are easy to devise either from the square planar AH_4 or octahedral AH_6 units. Figure 14.5 shows a correlation diagram with the orbitals of both these species. Many of the orbitals have descriptions (and therefore energies) identical to those in AH_4 and AH_6 geometries if the axial/basal angle is set at $90°$. The only orbitals that do have different energies are those of a_1 symmetry. In this point group both the p_z and s orbitals transform as a_1 (if we choose the z axis to lie parallel to the fourfold rotation axis of the square pyramid) and will therefore mix together to produce hybrid orbitals. The deepest lying a_1 orbital, $1a_1$, is dominated by central atom s rather than p character and so lies intermediate in character between analogous orbitals in the square and octahedron. The next highest orbital, $2a_1$, although it contains some basal ligand character, is largely a bonding orbital between central atom p_z and the axial ligand s orbital. The $3a_1$ orbital has a large contribution from central atom s and ligand orbitals (just like the $2a_{1g}$ orbital of the AH_6 unit) but also a significant contribution from central atom p and axial ligand orbitals (just like the $2t_{1u}$ orbital of the AH_6 unit). It is an orbital intermediate in character between these two extremes. As a result of the sp mixing in this orbital a lone pair is created pointing toward the vacant sixth site of the square pyramid. The highest energy a_1 orbital, $4a_1$, combines the roles of the $2a_{1g}$ orbital of AH_4, antibonding between central s orbital and basal ligands, and the antibonding partner to the $2a_1$ axial bonding orbital. The level composition thus leads to an approximate description for BrF_5 of three center bonding (as in XeF_4) for the four basal ligands plus a conventional two-center-two-electron bond for the axial ligand. This picture can only be an approximate one of course because of this intermixing between a_1 orbitals. In accord with this picture however, the axial distance (two center bonding) is shorter than the basal ones (three center bonding) in BrF_5 and ClF_5 (**14.7**).

14.7

BrF_5 and ClF_5 do not have the square pyramid geometry with $\theta = 90°$ but exhibit a somewhat smaller value ($\sim84°$). The reason for this is easy to see by considering the energetics of the HOMO as a function of angle. Notice in **14.8**

14.8

$\theta = 90°$ $\theta < 90°$

the phase of the central atom p_z orbital relative to the basal ligand orbitals. At $\theta = 90°$ there is no overlap between these two orbital sets. A distortion to $\theta > 90°$ switches on an antibonding (destabilizing) interaction between these atomic orbitals, but a distortion to $\theta < 90°$ generates a stabilizing, bonding interaction. A distortion too far in this direction leads to repulsive, antibonding interactions between axial and basal ligand orbitals. The VSEPR explanation of this result is the following. In BrF_5 there are six valence pairs which point toward the vertices of an octahedron. Five vertices are occupied by ligands, the sixth is vacant. Into this position a lone pair orbital is directed, **14.9**. One of the VSEPR rules,[4] that

14.9

lone pairs repel bonding pairs more than bonding pairs repel each other, requires the basal ligand's bond orbitals to be pushed away from the lone pair leading to $\theta < 90°$. We see here the orbital explanation of this rule.

The level structure of the T-shaped AH_3 is readily derived from that for the square pyramid by removal of a *trans* pair of ligands. This is shown in Figure 14.6. Indeed the orbitals of AH_5 and AH_3 are very similar. The nonbonding ligand-located b_1 orbital of the square pyramid is replaced by a nonbonding, central-atom-located orbital (b_1). Notice that going from AH_5 to AH_3 two valence orbitals are removed (corresponding to σ_g and σ_u from the H_2 unit which is lost). One of these is b_1 in AH_5, which correlates with σ_g. The other is derived from one component from each of $1e$ and $2e$. The nonbonding b_1 orbital in AH_3 is created (as shown by the dashed lines), and the σ_u orbital of H_2 (not shown), on loss of the two hy-

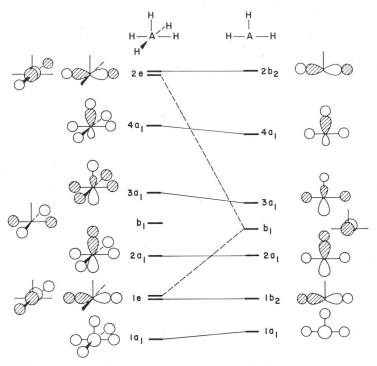

FIGURE 14.6. Correlation diagram for the molecular orbital levels of square pyramidal AH$_5$ and T-shaped AH$_3$. As in earlier figures we show the effect of the new orbitals by the use of dashed tie lines.

drogen atoms. With a total of five pairs of electrons in ClF$_3$ the details of the bonding situation is exactly analogous to that in BrF$_5$. There are two ligands, *trans* to one another, attached to the central atom by three center bonding and one ligand attached by a conventional two-center–two-electron bond. Accordingly in these molecules the unique bond is the shorter one **(14.10)**. Since the *trans* pair of atoms carries the highest charge in the unsubstituted parent (as we showed earlier for electron-rich three-center bonds) in derivatives such as **14.11**, the more electronegative ligands occupy these sites. Just as S$_N$2 processes at tetrahedral carbon proceed through a transition state with a trigonal bipyramid geometry, so the analogous substitution at sulfur in RSX compounds is calculated[6] to proceed through a T-shaped geometry. In equation 14.1, the RSXX$'$ species with the lowest energy is the one with the electronegative halide (X) atoms located in the arms of the T.

$$\theta = 87.5° \quad \overset{\text{F}}{\underset{\text{F}}{\overset{|}{\underset{\diagup}{\text{F}{-}\text{Cl}}}}} \overset{\leftarrow 1.698 \text{ Å}}{} \qquad 86.2° \quad \overset{\text{F}}{\underset{\text{F}}{\overset{|}{\underset{\diagup}{\text{F}{-}\text{Br}}}}} \overset{\leftarrow 1.810 \text{ Å}}{} \qquad \overset{\text{Cl}}{\underset{\text{Cl}}{\overset{|}{\underset{\diagup}{\text{C}_6\text{H}_5{-}\text{I}}}}} \overset{\leftarrow 2.45 \text{ Å}}{}$$

1.598   1.721   2.00

14.10   **14.11**

$$X^- + S\!-\!X' \longrightarrow \left[X\!-\!S\!-\!X' \right]^- \longrightarrow X\!-\!S + X'^-$$
$$\underset{R}{|} \qquad\qquad \underset{R}{|} \qquad\qquad \underset{R}{|} \qquad\qquad (14.1)$$

the electronegative halide (X) atoms located in the arms of the T.

The bending back of the *trans* ligands in the structure **14.10** to give an angle $\theta < 90°$ is explicable along exactly the same lines as the distortion of BrF$_5$ in **14.8**. In VSEPR terms the argument is also similar to the one used for BrF$_5$. ClF$_3$ contains five valence electron pairs which are arranged in the form of a trigonal bipyramid **(14.12)**. Placing two lone pairs in the trigonal plane and using the same

$$\theta < 90° \qquad \overset{F}{\underset{F}{\overset{\displaystyle\nearrow}{\underset{|}{F\!-\!Cl}}}} \quad \text{14.12}$$

argument concerning the relative sizes of the repulsions between the bond pairs and lone pairs leads to the prediction of a $\theta < 90°$. But just where are those two lone pairs drawn in **14.12** from a localized (VSEPR) perspective in a delocalized picture? Linear combinations of them yield two orbitals of b_1 and a_1 symmetries. It is im-

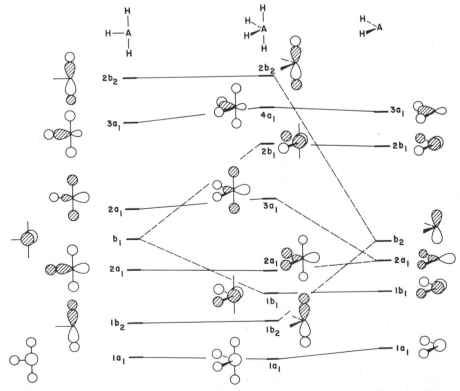

FIGURE 14.7. Correlation diagram for the molecular orbital levels of an AH$_4$ unit with the butterfly (SF$_4$) geometry, with those of T-shaped AH$_3$ and nonlinear AH$_2$. As in earlier figures the effect of the extra orbitals is indicated by the use of dashed tie lines.

mediately apparent that they correspond to the b_1 and $3a_1$ molecular orbitals for AH_3 in Figure 14.6. Figure 14.7 shows an orbital derivation for the butterfly structure of SF_4 from that of ClF_3 by addition of a ligand. The level pattern and description of the molecular orbitals are very similar to those of the square pyramidal AH_4 and T-shaped AH_3 geometries. Also shown in Figure 14.7 is the derivation of the level pattern from the orbitals of a bent AH_2 triatomic unit. As in the cases of ClF_3 and BrF_5 the energetic behavior of the HOMO in SF_4 allows understanding of the angular geometry of the molecule, **14.13**. The distortion away from the ideal struc-

14.13

ture runs, as in all of these molecules, counter to steric reasoning. Also the site preference problem is a similar one to ClF_3 and BrF_5. In SF_4 the electronic description of one pair of *trans* ligands attached by three center bonding, and two other ligands attached by the two-center-two-electron bonds leads to the prediction of the three center sites for electronegative atoms in substituted sulfuranes as found in the examples of **14.14**. The bond lengths of **14.13** are in accord with this picture too.

14.14

The orbital correlation diagrams of Figures 14.4–14.7 were used to highlight the orbital relationships between many AH_n species. As an exercise the reader should work through explicit orbital interaction diagrams for the compounds in terms of interacting an AH_n fragment with one or two hydrogen atoms. One could also consider what happens, in orbital terms, when a hydrogen atom is removed from AH_6 to give a square pyramidal AH_5 molecule. That is a method which will be extensively used for the derivation of the valence orbitals of ML_n fragments in later chapters.

14.2. GEOMETRIES OF HYPERVALENT MOLECULES

We have already mentioned the predictions of the VSEPR approach in the area of molecular geometry. Here we will not exhaustively treat all possible geometric excursions away from a symmetric structure in orbital terms but will show slices through the potential energy surface along some selected distortion coordinates. **14.15** shows the connection between the levels of the planar AH_3 molecule of D_{3h} symmetry and the corresponding levels of the T-shaped C_{2v} planar geometry. With a total of five valence electron pairs the $2a_1'$ orbital of the D_{3h} structure is occupied. This orbital is rapidly stabilized on bending toward the T-shaped structure since it strongly mixes with one component of the LUMO of the D_{3h} structure as

14.15

14.16

shown in **14.16**. Octet molecules have an empty $2a_1'$ orbital and are stable with respect to such a distortion. Molecules with five valence pairs, where this orbital is occupied, should then be unstable at the D_{3h} geometry and distort to a C_{2v} arrangement. The first excited electronic state of NH_3 in planar (D_{3h}) and has the configuration ... $(1e')^4 (1a_2'')^1 (2a_1')^1$. With only one electron in this $2a_1'$ orbital, therefore, the geometry remains trigonal and does not distort to the C_{2v}, T-shaped structure. The distortion of the D_{3h} to C_{2v} geometry for ClF_3 may be envisaged as a second-order Jahn–Teller instability of the trigonal structure with this electronic configuration. A distortion coordinate of $a_1' \times e' = e'$ will allow HOMO and the one component of the LUMO to strongly mix, as shown in **14.16**.

The energy level diagram for the $D_{4h} \rightarrow C_{4v}$ distortion of the square plane, shown in Figure 14.8, is easy to derive. It was briefly discussed in Section 9.4 and has obvious ties with the pyramidalization of planar AH_3 (shown in **9.22**) and the bending of linear AH_2 (shown in **7.19**). A prominent feature is the strong coupling between a_{2u} and $2a_{1g}$ orbitals on bending. With five electron pairs, SF_4 therefore should be unstable at the planar geometry in a second-order Jahn–Teller sense. The situation is reminiscent of that of NH_3, and the pyramidal structure should be stabilized with respect to the planar one. For SF_4 (or hypothetical SH_4) with this configuration the C_{4v} geometry is a possible candidate for the ground-state structure. We have shown that SF_4 actually exists in a C_{2v} geometry **(14.13)**. The relationship between the C_{4v} and C_{2v} structures for these molecules with five pairs of electrons is one that will shortly be explored. XeF_4 with six valence pairs will be more stable at the planar structure since here the HOMO is destabilized on pyramidalization. (The same argument can be used to rationalize the planar rather than pyramidal first excited electronic state of NH_3.) Use of second-order Jahn–Teller

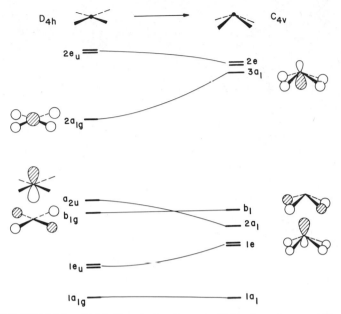

FIGURE 14.8.   Walsh diagram for the pyramidalization of square AH$_4$.

ideas at the tetrahedral geometry leads to a different geometry, the butterfly structure of **14.13**. **14.17** shows how the HOMO ($2a_1$) and LUMO ($2t_2$) may couple

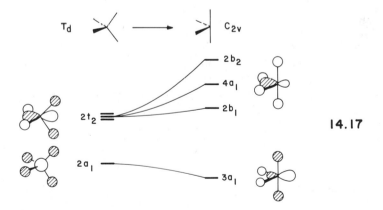

14.17

together during a t_2 distortion which leads to the observed structure of SF$_4$.
Whether the C_{2v} or C_{4v} structure lies lower in energy is very difficult to predict.[7]
Numerical calculations [8,9] suggest that the C_{4v} structure is the lower energy isomer
for the (hypothetical) SH$_4$ molecule but the C_{2v} structure is the lower energy
isomer for SF$_4$ (as observed). The energetic juxtaposition of these two structures
leads to a ready pathway for the isomerization of SF$_4$ (**14.18**). Notice that the initially axial ligands (of the VSEPR trigonal bipyramid) labeled with asterisks become

14.18

the equatorial ligands after rearrangement. This process is just the Berry pseudo-rotation process for five coordinate molecules (which we describe more fully below) but with a lone pair occupying the fifth coordinate position **14.19**. The Berry process (or rather a ligand interchange process consistent with it) has been verified for SF_4 by NMR studies.[10]

14.19

Main group five-coordinate molecules are found either as trigonal bipyramidal molecules (e.g., PF_5) or as square pyramidal species (e.g., BrF_5). Geometrically they are quite close as shown in **14.20**, a projection down the fourfold axis of the

14.20

square pyramid and one of the twofold axes of the trigonal bipyramid. The levels[11] of the AH_5 trigonal bipyramid are built up in Figure 14.9 from an A atom plus five ligands and in Figure 14.10 from the trigonal plane plus a pair of axial ligands. In the five-electron pair molecule the HOMO is a nonbonding orbital. Its origin is best seen in Figure 14.9. Here the ligand orbitals contain two a_1' representations. The central atom s orbital transforms as a_1' too. In AH_5 the a_1' orbitals constructed from these three are an orbital bonding between s and both axial and equatorial ligands, an orbital antibonding between s and both axial and equatorial ligands and a nonbonding orbital. The axial ligands are clearly attached by three-center bonds in this molecule. The equatorial ligands, however, are attached by conventional two-center–two-electron bonds. We could imagine sp hybrid orbitals constructed from the $1a_1'$ and $1e'$ orbitals as shown in **14.21**. As a result the axial linkages are

14.21

longer than the equatorial ones in PF_5, **14.22**. Also in accord with this bonding picture, electronegative substituents preferentially reside in the axial positions as in **14.23**.

Figure 14.11 shows the orbital correlation diagram connecting the square pyramidal and trigonal bipyramidal structures. Comparison of the occupied levels for five valence pairs of electrons shows little energetic preference for either struc-

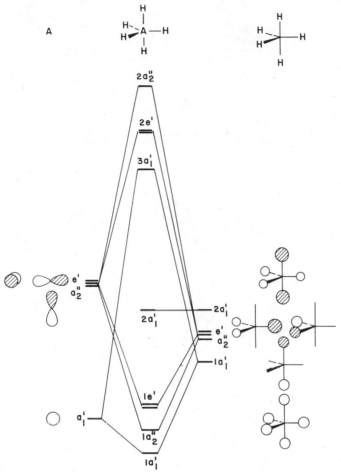

FIGURE 14.9. Assembly of the molecular orbital diagram for trigonal bipyramidal AH_5 from the levels of A and of H_5. d orbitals are not included on A.

14.22   **14.23**

ture. PF_5 itself has the trigonal bipyramidal structure but with a low energy re-arrangement pathway almost certainly via the square pyramidal geometry. With six pairs of electrons the geometric preference for BrF_5 is much more clear-cut. In the square pyramidal geometry there is an accessible low energy orbital $(3a_1)$ to house this sixth pair. In the trigonal bipyramidal geometry the sixth pair has to

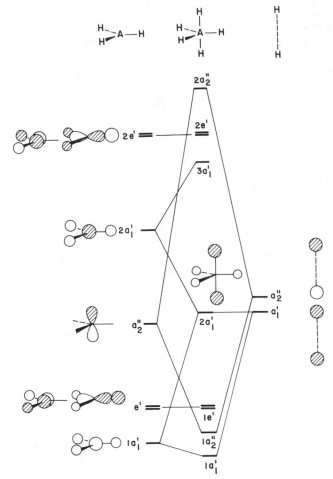

FIGURE 14.10.   Assembly of the molecular orbital diagram for trigonal bipyramidal AH_5 from the levels of A and of trigonal planar AH_3 and those of H_2.

occupy a very high energy orbital and so this geometry is unlikely. It is interesting how the VSEPR approach of **14.9** with six electron pairs but only five ligands captures the essence of this orbital result.

Facile rearrangement of PF_5 may occur via the Berry pseudorotation process **14.24**. After an excursion to the square pyramidal structure and back, axial and

$$+ \longrightarrow \nearrow \longrightarrow \diagup \qquad \textbf{14.24}$$

$$D_{3h} \qquad\qquad C_{3v} \qquad\qquad D_{3h}$$

equatorial sites of the trigonal bipyramid have been exchanged. From the diagram of Figure 14.11 this is an orbitally allowed process for five valence electron pairs.

It is interesting at this point to mention the structures which have been calcu-

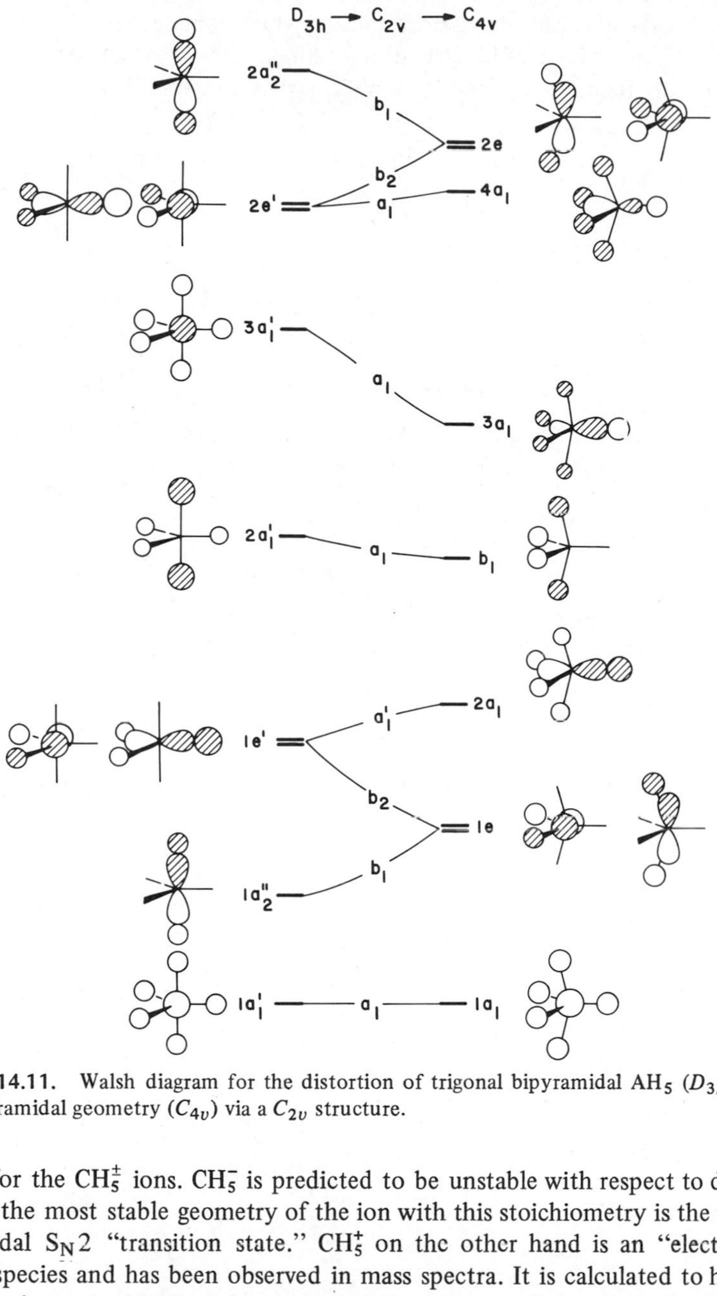

$$D_{3h} \longrightarrow C_{2v} \longrightarrow C_{4v}$$

$2a_2''$

b_1

$\equiv 2e$

$2e' \equiv$

b_2 — $4a_1$

a_1

$3a_1'$

a_1

— $3a_1$

$2a_1'$

a_1 — b_1

— $2a_1$

$1e' \equiv$

a_1'

b_2

$\equiv 1e$

$1a_2''$

b_1

$1a_1'$ — a_1 — $1a_1$

FIGURE 14.11. Walsh diagram for the distortion of trigonal bipyramidal AH_5 (D_{3h}) to the square pyramidal geometry (C_{4v}) via a C_{2v} structure.

lated[12] for the $CH_5^{\pm}$ ions. CH_5^{-} is predicted to be unstable with respect to dissociation but the most stable geometry of the ion with this stoichiometry is the trigonal bipyramidal S_N2 "transition state." CH_5^{+} on the other hand is an "electron deficient" species and has been observed in mass spectra. It is calculated to have the geometry shown in **14.25**, and as indicated may be regarded as being isoelectronic

14.25

with H_3^+. A CH_3 unit provides a single orbital which contains one electron just like one of the H atoms in H_3^+. This structure is estimated[12,13] to be about 11 kcal/mol lower in energy than the trigonal bipyramidal structure.

REFERENCES

1. R. E. Rundle, *J. Amer. Chem. Soc.*, **85**, 112 (1963).
2. G. C. Pimentel, *J. Chem. Phys.*, **19**, 446 (1951).
3. T. R. Forbus and J. C. Martin, *J. Amer. Chem. Soc.*, **101**, 5057 (1979).
4. R. J. Gillespie, *Molecular Geometry*, Von Nostrand-Rheinhold, London (1972).
5. K. S. Pitzer, *Accts. Chem. Res.*, **12**, 271 (1979).
6. V. I. Minkin and R. M. Minyaev, *Zh. Org. Khim.*, **13**, 1129 (1977), Eng. trans., **13**, 1039 (1977).
7. M. M-L Chen and R. Hoffmann, *J. Amer. Chem. Soc.*, **98**, 1647.
8. R. Gleiter and A. Veillard, *Chem. Phys. Lett.*, **37**, 33 (1976).
9. G. M. Schwenzer and H. F. Schaeffer, *J. Amer. Chem. Soc.*, **97**, 1391 (1975).
10. W. G. Klemperer, J. K. Kreiger, M. D. McCreary, E. L. Muetterties, D. D. Traficante, and G. M. Whitesides, *J. Amer. Chem. Soc.*, **97**, 7023 (1975).
11. R. Hoffmann, J. M. Howell, and E. L. Muetterties, *J. Amer. Chem. Soc.*, **94**, 3047 (1972).
12. J. B. Collins, P. v. R. Schleyer, J. S. Binkley, J. A. Pople, and L. Radom, *J. Amer. Chem. Soc.*, **98**, 3436 (1976).
13. K. Raghavachari, R. A. Whiteside, J. A. Pople, and P. v. R. Schleyer, *J. Amer. Chem. Soc.*, **103**, 5649 (1981).

CHAPTER FIFTEEN

Transition Metal Complexes—A Starting Point at the Octahedron

15.1. INTRODUCTION

Apart from a brief digression on hypervalent molecules in Chapter 14 we have really only considered molecules with coordination numbers one through four. The geometries, or more precisely, the angles around the central atom of these AH_n building-block fragments were small in number and fell into rather well-defined classes. We have also utilized a small "basis set" of atomic s and p orbitals to describe their bonding. In the transition metal field coordination numbers of two through eight are common. There is also a richer variety of structural types that are found for these molecules. Many times it is not at all obvious as to whether a compound should be viewed as a member of one class or another. To make matters worse the coordination number of a metal, particularly in the organometallic domain is not always uniquely defined. For example $Cr(CO)_6$, **15.1,** is clearly an octahedron. There are two alternatives for ethylene-$Fe(CO)_4$. One might consider it as a trigonal bipyramid, **15.2,** or as an octahedral complex, **15.3.** Related to this issue is whether one regards the compound as an olefin–metal **(15.2)** or metal-lacyclopropane **(15.3)** complex. In fact there are two basic geometries for an ML_5 complex: the trigonal bipyramid and square pyramid. If we insist that the ethylene ligand in ethylene-$Fe(CO)_4$ occupies one coordination site, then it falls into the trigonal bipyramidal class. But many ML_5 compounds geometrically lie somewhere between the idealized trigonal bipyramid and square pyramid. Ferrocene, **15.4,** is another common example of the coordination number problem. Is it two coordi-

15.1   15.2   15.3

15.4

nate, as shown by the drawing in **15.4**, or ten coordinate? In actual fact it is better described as an octahedron! We shall see in Chapters 20 and 21 that the cyclopentadienyl group effectively utilizes three coordination sites.

Part of this complexity is a result of the fact that the metal utilizes five d as well as its s and p atomic functions to bond with the surrounding ligands, but the reader should not despair. Our focus will naturally be concentrated on the metal-based orbitals. However, all nine atomic s, p, and d functions will rarely be needed. As in the preceding chapters those relationships, and there are many of them, that bridge the worlds of organic/main group chemistry to inorganic/organometallic chemistry will be highlighted. Actually structural diversity is an added bonus. Different vantage points can be exploited when a problem is analyzed. Changes in structure can certainly modify reactivity, so too will oxidations or reductions and fine tuning the electron density at the metal by varying the electronic properties of the ligands. All of this makes life more interesting to the chemist.

This chapter and the next will introduce the use of d orbitals in transition metal complexes. First of all we shall build up the orbitals of octahedral ML_6 and square-planar ML_4 complexes. These molecular levels will be used to develop the orbitals of ML_n fragments which is the topic of Chapters 17–20 so considerable time will be spent on this aspect. How the octahedral splitting pattern and geometry is modified by the numbers of electrons and the electronic naturé of the ligands is also undertaken.

15.2. OCTAHEDRAL ML_6

Let us start with octahedral ML_6. For the moment L will be a simple σ donor ligand. In other words, L has one valence orbital which is pointed toward the metal and there are two electrons in it. Examples are the lone pair of a phosphine, amine, alkyl group, **15.5**, or even the s orbital of a hydride, **15.6**. Some ML_6 examples are $Cr(PMe_3)_6$, **15.7**, or the tris(ethylenediamine)Fe^{2+} complex, **15.8** (here the ethylenediamine group is $H_2NCH_2CH_2NH_2$). What we are initially concerned with are

15.5   15.6

15.7   15.8

the metal–ligand σ orbitals, π bonding is reserved for the next section. In this regard the pattern that is constructed will not change much for $Cr(CO)_6$, **15.1**, $CH_3Ru(CO)_4Cl$, **15.9**, or even the more complicated **15.10**. The π and π^* levels of CO and the phenyl group in **15.10** along with the chlorine lone pairs can be in-

15.9   **15.10**

troduced into the electronic picture at a later stage of the analysis. While the symmetry of **15.9** and **15.10** is low, there is an effective pseudosymmetry that the transition metal experiences in the σ levels which is octahedral. What is important in **15.7–15.10** is that there are six lone pairs directed toward the metal in an octahedral arrangement which brings us back to the ubiquitous L groups. They will be utilized throughout the remaining chapters when we want to present a generalized treatment of a problem. Figure 15.1 illustrates one approach to construction of the molecular orbitals of ML_6. On the left side of the interaction diagram are the nine atomic orbitals of a transition metal. Notice in particular that the d functions are drawn in their familiar form and correspond to the coordinate system at the top center of the figure. The z^2 and $x^2 - y^2$ functions† are of e_g symmetry and xy, yz, xz transform as a t_{2g} set. At higher energies lie the metal s and p levels. Recall that we are concerned only with the valence levels, so that the *inner shells* of s and p electrons on the metal are neglected. On the right side of Figure 15.1 are drawn the symmetry-adapted linear combinations of the ligand σ orbitals. There are six and their relative ordering is set by the number of nodes within each member (see Section 14.1). There are none in a_{1g}, t_{1u} has one, and e_g contains two nodes. The a_{1g} and t_{1u} combinations match with metal s and p so they are stabilized, yielding the molecular levels $1a_{1g}$ and $1t_{1u}$. The e_g ligand set is stabilized by metal z^2 and $x^2 - y^2$ which gives the molecular $1e_g$ levels. The ordering of these M—L bonding orbitals is exactly the same as that for AH_6 (Section 14.1) with the exception that $1e_g$ was left nonbonding in the main group system. There was also not such a large energy gap between the central atom s and p AO set and the "ligand" a_{1g}, t_{1u}, and e_g sets for AH_6. Here the six M—L bonding orbitals are concentrated at the ligands. There are also six corresponding M—L antibonding levels: $2e_g$, $2a_{1g}$, and $2t_{1u}$ which are heavily weighted on the metal atom. Left behind is t_{2g} on the metal. It is nonbonding when L is a σ donor; however, it will play an important role when the ligands have functions that can enter into π bonding with the metal. Inspection of Figure 15.1 shows that an octahedral compound is likely to be stable when t_{2g} is either completely filled or empty. The

---

†Throughout the rest of this book we shall refer to the nd AOs as $z^2, x^2 - y^2, xz, yz$, and xy. The $(n + 1)p$ AOs are given as x, y, and z.

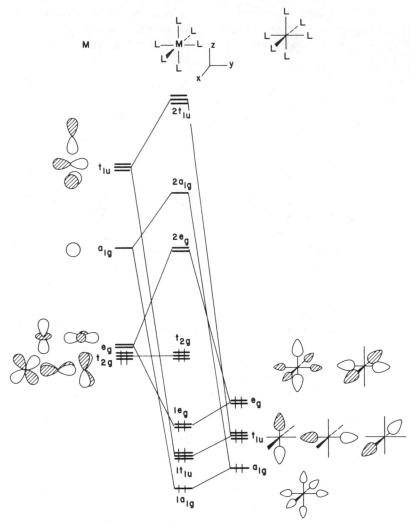

FIGURE 15.1.   Development of the molecular orbitals of an octahedral ML_6 complex where L is an arbitrary σ donor ligand.

former case is more likely for transition metals and one can think of the six electrons in t_{2g} as three sets of lone pairs which are localized on the metal. Together with the 12 electrons from the M—L bonding levels creates a situation where 18 valence electrons are associated with the metal.

It is the HOMO, t_{2g}, and the LUMO, $2e_g$, which will be the focus of our attention throughout the rest of this book. They are shown in **15.11**. The energy gap between t_{2g} and $2e_g$ is a function of the ligand σ donor strength (in the absence of π effects). Raising the energy of the ligand lone pairs causes the energy gap between ligand e_g and the metal d set to diminish. Consequently there is a stronger

$2e_g$ ==

15.11

t_{2g} #

interaction; the antibonding $2e_g$ set is destabilized so the $t_{2g} - 2e_g$ energy gap increases. Overlap between the ligand lone pairs and metal e_g can also play an obvious role. Thus, a strong σ donor set of ligands creates a sizable energy gap between the molecular t_{2g} and $2e_g$ levels. This ensures a singlet ground state (all six electrons reside in t_{2g} for an 18-electron complex) which is called a low spin situation **15.12,** by inorganic chemists. This is the case for nearly all organometallic compounds. Classical coordination complexes, where $L = NH_3$, H_2O, halogen, and so on, sometimes behave differently. The $t_{2g} - 2e_g$ splitting is not so large since the ligands are very electronegative. Therefore, the ground state may be one that contains some spins unpaired—an intermediate spin system, **15.13,** or a maximum of unpaired spins—a high spin system, **15.14.** The energy balance between these spin states can

low spin   intermediate spin   high spin

15.12   **15.13**   **15.14**

be formulated in the same terms as the singlet–triplet situation for organic diradical; see methylene, for example, in Section 8.8 where the $b_2 - 2a_1$ energy gap was small. For coordination compounds the $t_{2g} - e_g$ splitting is in delicate balance with the spin pairing energies, consequently, the high-spin–low-spin energy difference is often very tiny. A change of spin states can even be induced by cooling the sample or application of mechanical stress![1] Putting more than 18 electrons into the molecular orbitals of ML$_6$ in Figure 15.1 will cause problems. The extra electrons will be housed in $2e_g$ which is strongly M—L antibonding. We would expect that the compounds will distort so as to lengthen the M—L distances or perhaps one or two M—L bonds might completely break. An example of this is bombardment of Cr(CO)$_6$ (an 18-electron complex) with electrons in a solid Ar matrix.[2a] Instead of isolating the 19-electron Cr(CO)$_6^-$, the major product was found to be a 17-electron Cr(CO)$_5^-$ complex. But remember for the classic coordination complexes $2e_g$ is not greatly destabilized. Electron counts at the metal which exceed 18 are possible an and population of $2e_g$ with unpaired electrons (**15.13** and **15.14**) is frequent; Ni(OH$_2$)$_6^{2+}$ is a well-known example. This is not to say that there will be no structural changes that accompany the population of $2e_g$. Structural data[2b] exist for low spin $Co(NH_3)_6^{3+}$ and high spin $Co(NH_3)_6^{2+}$. In the former complex the t_{2g} set is filled and e_g is empty. The Co—N bond length is 1.94 Å. In the latter complex the e_g set contains two electrons. The Co—N bond length increases to 2.11 Å. When there are less than 18 electrons associated with the metal the t_{2g} set is partially filled. This signals a Jahn–Teller (Section 7.4) or some other geometrical distortion (for a low spin complex) which lowers the symmetry of the molecule. We will return to this problem in greater depth after substituent effects of the ligands are covered.

What will become obvious in the next chapters is that the primary valence orbitals that we shall need are derived from t_{2g}, $2e_g$, and sometimes $2a_{1g}$ in Figure 15.1. Therefore, our basis set of orbitals will never be larger than five or six. Those orbitals are concentrated on the metal and they lie at intermediate energies. They will be the HOMOs and LUMOs in any transition metal complex. When the octahedral symmetry is perturbed by removing a ligand or distorting the geometry the $2a_{1g}$ and $2t_{1u}$ orbitals may be utilized. For example, metal s or p may mix into the members of e_g or t_{2g}. In other words, $2t_{1u}$ and $2a_{1g}$ provide a mechanism for the hybridization of the valence orbitals.

15.3.   π-EFFECTS IN AN OCTAHEDRON

How does the picture in Figure 15.1 change when π functions are added to the surrounding ligands? Let us start by replacing one of the generalized σ donor ligands in ML_6 by a carbonyl group which yields an ML_5CO complex. CO is isoelectronic to N_2. A detailed discussion of the perturbations encountered on going from N_2 to CO was given in Section 6.3. We shall briefly review the results. The few molecular orbitals of CO that are needed for this analysis are shown in **15.15**. The σ orbital is

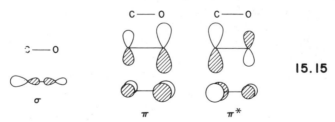

15.15

derived from $2\sigma_g$ in N_2. It is hybridized at carbon and will act as the σ donor function in a transition metal complex. Notice that the hybridization of electron density at carbon makes the CO ligand bind to the metal at the carbon end. There are also two orthogonal π and π^* levels. In N_2 they were π_u and π_g, respectively. The π sets intermix with the perturbation to CO so that the π level becomes more heavily weighted at the electronegative oxygen atom. On the other hand, π^* becomes concentrated at carbon. As mentioned previously the σ donor orbital of CO along with the five σ levels of the L_5 grouping produce a splitting pattern in ML_5CO analogous to that in Figure 15.1. The two members of $2e_g$, for example,

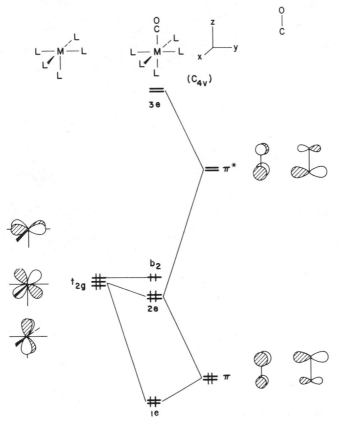

FIGURE 15.2.   An interaction diagram for the π components in a ML_5CO complex where L is an arbitrary σ donor.

will not be at precisely the same energy. That will depend on the relative σ donor strength of CO compared to whatever L is; however, there will be a close correspondence. What does change is the t_{2g} levels as shown in Figure 15.2. Two members of t_{2g}, xz and yz (see the coordinate system at the top of this figure), have the correct symmetry to interact with π and π^* of CO. They become an e set in the reduced symmetry of the complex. The third component, xy, is left nonbonding. What results from this interaction is a typical three-level pattern which is exactly the same as in the linear H_3 (Section 3.3) or allyl (Section 12.1) systems. At low energy, the orbital labeled $1e$ in Figure 15.2, is primarily π with some xz and yz mixed in a bonding fashion. One component of $1e$ is shown in **15.16**. At high energy $3e$ is primarily CO π^*, antibonding to xz and yz; **15.17** shows one component. The middle level, $2e$, is slightly more complicated. It is represented by metal xz and yz perturbed by CO π and π^*. Since π^* and π lie at respectively higher and lower energy than metal t_{2g}, $2e$ contains CO π^* mixed in a bonding way to xz and yz while CO π mixes in an antibonding fashion. This is expressed by **15.18** for the component of $2e$ in the yz plane. The net result, **15.19**, shows cancel-

15.16   15.17

15.18   15.19

lation of electron density at the carbon and reinforcement at oxygen. In actual fact the interaction of CO π^* to metal xz, yz is larger than that to CO π. CO is overall a π acceptor in that there is a net drift of electron density from metal t_{2g} to the carbonyl group. In essence we are saying that the interaction of xz and yz to CO π^* is larger than that to CO π. This is due to overlap factors. Recall that there is a larger AO coefficient at carbon in π^* than there is in π and this creates the larger overlap to the xz, yz set. It is also important to realize that energetically it is the $1e$ levels in Figure 15.2 which are stabilized the most. The stabilization in $2e$ with respect to t_{2g} will be relatively small. The M—CO bonding is synergistic; that is, electron density from the filled σ level (15.15) is transferred to empty metal $s, p,$ and two of the five d orbitals. Likewise, electron density from the filled metal t_{2g} set is transferred to the empty CO π^*.

When the ligand has only one π acceptor function, then one component of t_{2g} is stabilized. A case in point is the carbene ligand, 15.20; it has a filled σ donor and empty π acceptor function (Section 8.8) which is available for bonding to one member of the t_{2g} set. One example is provided by 15.21. When the energy gap and

15.20   15.21

overlap to t_{2g} is favorable there is a strong interaction, 15.22. A good bit of electron density is transferred from metal t_{2g} to the carbene. The carbene carbon becomes nucleophilic (alternatively one could imagine that those two electrons originally came from a carbanionic group and are partially donated to an empty member of t_{2g}). When the interaction is not so strong, 15.23, the carbene carbon remains electrophilic.[3] Related to the carbenes are carbyne ligands, 15.24. Now there are two orthogonal π acceptor functions and one hybrid σ donor. Taking the

15.22

15.23

15.24

15.25

ligand to be positively charged stresses the analogy to the carbene case. The two t_{2g} derived metal orbitals interact with the empty p AOs on the carbyne. Consequently there is a sizable splitting between the stabilized t_{2g} members and the one that is left nonbonding.[4] A good π donor ligand will contain a high lying filled π orbital. It will destabilize one or two components of t_{2g}. Examples of π donor ligands are amido groups, 15.25, or the halogens where there are two filled p orbitals. Many ligands have both π acceptor and π donor functions, for example, the π^* and π orbitals in CO, N_2, NO, and RNC. A detailed theoretical investigation of π acceptor and donor effects for several ligands has been given by Ziegler and Rauk.[5] Trends can be established by using energy gap and overlap considerations.

The substitution of two or more π acceptor ligands at the metal will stabilize two or all three members of the metal t_{2g} set. For example, in $Cr(CO)_6$, symmetry-adapted linear combinations of the 12 π^* levels yields one of t_{2g} symmetry. Therefore, metal xz, yz, and xy are stabilized.

An interesting problem arises when there are two π acceptor ligands, say *trans* to one another and each ligand has only one acceptor function. A hypothetical case is a *trans* $(R_2C)_2ML_4$ complex.[6] The particular question to be addressed is that for an 18-electron complex, is the D_{2h} conformation, 15.26, more stable than the D_{2d} form, 15.27? The σ levels in 15.26 and 15.27 will be at identical energies. The splitting pattern will strongly resemble that for an octahedron in Figure 15.1. The difference lies in the way the empty p functions of the carbene backbond to the metal t_{2g} set. In the D_{2h} conformation of 15.26 both carbenes lie in the yz plane so both p orbitals will interact with metal xz. In the D_{2d} geometry, 15.27, the carbenes are orthogonal. One p function will overlap with xz and the other with

15.26   **15.27**

yz. Figure 15.3 illustrates these differences in π bonding by means of interaction diagrams. Let us start with the D_{2h} geometry. In-phase and out-of-phase combinations of the carbene p functions are taken on the left side of the figure. They are of b_{1u} and b_{2g} symmetry, respectively. They will be nearly degenerate in energy since the carbenes are far from each other. The b_{2g} combination has the same symmetry and will overlap with metal xz. That metal orbital will then be stabilized greatly with respect to xy and yz, which are left nonbonding. Notice also that the in-phase combination, b_{1u}, of carbene p orbitals is left nonbonding. In the D_{2d} geometry of **15.27** the carbene functions transform as an e set—see the right side of Figure 15.3. They will stabilize metal xz *and* yz. It is clear that the interaction in the b_{2g} combination of the D_{2h} geometry will be greater than that in e for the D_{2d} case. But will it be twice as large? In that case the energy difference between the two conformations will be zero. It turns out, and can be proven by perturbation

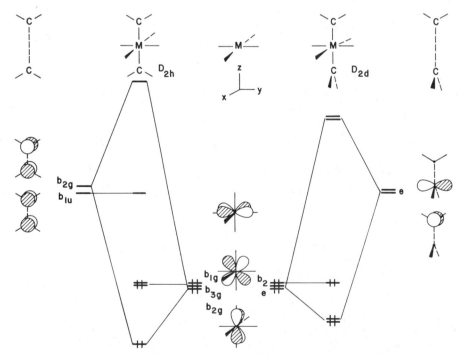

FIGURE 15.3.   Interaction diagrams for two possible conformations in a *trans*-$(R_2C)_2ML_4$ complex. Only the π interactions are illustrated.

theory arguments,[6-8] that if the energy difference between the carbene p and metal d levels is large, then the stabilization in b_{2g} is twice as large as that in the e set. However, as the energy difference becomes smaller, b_{2g} is stabilized by less than twice as much as e. Therefore, the D_{2d} conformation becomes more stable than the D_{2h} one. This result is a general one: if two good acceptors have a choice, they will choose to interact with orthogonal donor functions. A "real" case occurs in *trans*-bis-ethylene-ML_4 complexes. There are two 18-electron complexes which have been shown[9] to have structure **15.28**. There is also an isoelectronic dioxygen complex which has an equivalent geometry.[10] Exactly the same story occurs here. The ethylene ligand has a filled π and empty π^* orbital pointed at the metal. This is topologically equivalent to the situation in a carbene, as indicated in **15.29**. In

15.28

15.29

15.28 the two acceptor π^* functions backbond to orthogonal members of the metal t_{2g} set and, therefore, this geometry is more stable than one with the ethylenes oriented parallel to each other. The actual mechanism of rotation about the ethylene–metal bond in **15.28** is complicated by another electronic factor;[11,12] it is easier to see what happens in the bis-carbene complex and so we will describe what happens in that situation. Starting from the D_{2d} geometry, **15.27**, rotation of both CR_2 units in the same direction will cause little change in the energy. This is a consequence of the fact that the xz and yz donor functions are degenerate. At **15.27** or the rotomer where both methylene groups are twisted by $90°$, the two p orbitals interact with xz and yz. At intermediate geometries they will interact with linear combinations of xz and yz. Recall that any linear combination of a degenerate set yields an equivalent set, so the two stabilized metal orbitals will stay at constant energy. From another point of reference one could say that the ML_4 group was freely rotating with the CR_2 units fixed in space, orthogonal to each other. This is a little bit of an oversimplification which depends on the size of the R groups and L. In **15.26** the carbenes eclipse the M—L bonds, so there may be a steric preference for rotation by $45°$ to a staggered geometry.

There are many other systems where two acceptors interact with orthogonal donors. One way to view allene is by the union of two carbenes with a central carbon atom, **15.30**. The four electrons in the central carbon will artificially be placed in the two p orbitals. The D_{2d} geometry then maximizes π bonding if the two carbene functions are orthogonal. In the di-ylid **15.31** the two carbene p orbitals are filled. Stabilization could then occur via the empty P—H antibonding e' set (see Section 14.2). Again an orthogonal arrangement maximizes bonding this time from the filled methylene p orbitals to orthogonal members of the e' P—H σ^* set (alternatively one could use d orbitals on phosphorus as the acceptor set).

15.30   15.31

An interesting situation arises with two electrons less in **15.31**. The neutral phosphorane is then a diradicaloid species. In the orthogonal structure of **15.31** the stabilized methylene p orbitals come as an e set—see the right side of Figure 15.3. The ground state is then a triplet. Rotation of one methylene group by 90° so that it lies in the plane of other stabilizes one methylene p combination; the other is left nonbonding—see the left side of Figure 15.3. If there is substantial P—H σ^* (or d) involvement, then a sizable splitting would create a singlet ground state at this geometry.[13] The reader should note that removing two electrons from the $(R_2C)_2ML_4$ complex will produce a stable 16-electron system. However, as shown in Figure 15.3 those two electrons would come from the xy orbital in the D_{2d} conformation. The e set is probably stabilized enough to give a ground-state singlet. In the D_{2h} geometry, xy and yz are degenerate so a triplet state is predicted.

This phenomena of acceptor orbitals maximizing their interaction to orthogonal donors need not be restricted to π bonding. Arguments can be constructed for the orientation of σ bonds, as well. Take F_2O_2 as being divided into two F^+ atoms with empty hybrids pointed at O_2^{2-}. Now O_2^{2-} has four more electrons than N_2 (see Section 6.3). Therefore, π_g is totally filled and provides an orthogonal donor set to bond to F^+ (see **15.32**). Both members of π_g in the O_2^{2-} core are then stabilized when the F—O—O—F dihedral angle is 90°. That is a more stabilizing arrangement than when the dihedral angle is 0° or 180° (a *cis* or *trans* structure). In that case only one π_g function is stabilized. Likewise, for the transition metal complexes in **15.33** and **15.34** consider L′ to be a better σ donor than L. In **15.33** the donor functions interact with z^2 *and* $x^2 - y^2$, that is, both members of e_g (see Figure 15.1). In **15.34**, only z^2 will stabilize them. It is easy to see that **15.33** will energetically be preferred. For the reverse situation, when L′ is a weaker σ donor than L, the same preference is predicted. The stronger σ donors interact only with the $x^2 - y^2$ component of e_g in **15.34**. In **15.33** they interact with both. One should be cautious in pushing this argument too far in predicting *cis* and *trans* energy differences for octahedral complexes. Steric effects and other electronic factors (most notably of the π type) can overrule the arguments that have been constructed here.

15.32   15.33   15.34

15.4. DISTORTIONS FROM AN OCTAHEDRAL GEOMETRY

Before we leave octahedral complexes it is worthwhile to consider some of the alternative geometries that are conceivable for ML_6 complexes. The octahedron is a special geometry for an 18-electron count. Referring back to Figure 15.1 it is seen that there are a total of 12 electrons housed in six strong M—L bonding orbitals—$1a_{1g}$, $1e_g$, and $1t_{1u}$. The remaining six electrons are localized at the metal (barring π effects) and are nonbonding. That is, they are directed in space away from the M—L bonding regions (see **15.11**). As we shall see in the next chapter most transition metal complexes are of the 18-electron type and so it is not unusual to find that this geometry is so pervasive in nature. Of course, another way to view this is from a valence-shell–electron-repulsion model.[14] Consider the ligands again as Lewis bases. The optimal way to position six bases around a sphere (the transition metal) is in an octahedral arrangement. This minimizes steric inter-actions, as well as lone-pair repulsions (or repulsions between the electron density in the M—L bonds) between the ligands. But what happens when there are fewer than 18 electrons around the transition metal complex? As mentioned in Section 15.1, the t_{2g} set (see Figure 15.1) will be partially filled. The complex will either adopt a higher spin state or distort in such a way as to lower the energy of the filled components of t_{2g} and raise the energy of the empty ones. Low spin ML_6 complexes with electron counts of 14–16 valence electrons in the levels of Figure 15.1 are then the most likely cases where distortions from an octahedron will be found. We shall examine in this section two typical distortions.[15-18] The first is a decrease of one *trans* L—M—L angle from 180°, as shown by **15.35**. In the second, the *cis* L—M—L angle is varied, **15.36**, in either direction from 90°. A combina-tion of both types of deformation, as shown in **15.37**, takes an octahedron to a bicapped tetrahedron, **15.38**.

| 15.35 | 15.36 | 15.37 | 15.38 |

Let us start with **15.35** where all ligands are solely σ donating. The symmetry of the complex is lowered along this distortion coordinate, as it is in any dis-tortion from octahedral symmetry. Orbitals that were formerly orthogonal now mix. This is a necessary and complicating feature that is to be analyzed in some detail. Let us sidestep that issue for a moment and see what will happen to the splitting pattern of our octahedron in Figure 15.1 solely on the basis of over-lap changes. It is easy to see that those orbitals that contain metal z and z^2 will be perturbed. The metal s orbital is spherical. Therefore, any ligand angular dis-tortion will keep a constant overlap with it. The net result is that the z component of $1t_{1u}$ in Figure 15.1 and z^2 in $1e_g$ will rise in energy since these are bonding orbitals and overlap to the lone pairs of the ligands is lost—see **15.39** and **15.40**.

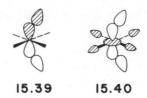

15.39   15.40

Likewise, the corresponding member of $2t_{1u}$ and $2e_g$ will fall in energy since they are antibonding orbitals. For convenience we have rotated our coordinate system for the octahedron by $45°$ to that in **15.35**. The components of the important t_{2g} set will now be $x^2 - y^2$, xz, and yz. Let us concentrate on yz. At the octahedron it was orthogonal, of course, to the z component of $1t_{1u}$ and $2t_{1u}$. However, the symmetry of the molecule is reduced to C_{2v} with the distortion in **15.35**. Both yz and **15.39** are of b_1 symmetry so they will intermix along with the antibonding analog of **15.39**. There are a number of ways within the framework of perturbation theory to see how they intermix; let us look at two briefly. First of all, we start with the elements of these orbitals—yz, z, and the antisymmetric combination of the two lone pair orbitals, σ_a. They are drawn explicitly on the far left and right of **15.41**. Again when the L—M—L angle is $180°$ σ_a and z form bond-

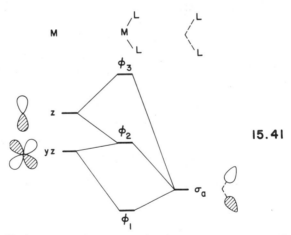

15.41

ing and antibonding combinations; yz is left nonbonding. However, when the L—M—L angle is less than $180°$ the three orbitals intermix. The lowest molecular level, ϕ_1, will be heavily concentrated on σ_a, the fragment orbital closest to it in energy. Some yz and z character will also mix into it in a bonding fashion, **15.42** (z and yz are at higher energies than σ_a), to produce **15.43**. The actual magnitudes of the mixing coefficients, λ_1 and λ_2, will be set by the $\sigma_a - yz$ and $\sigma_a - z$ energy

$$\phi_1 = \Bigg\{ + \lambda_1 \Bigg(\Bigg) + \lambda_2 \Bigg(\Bigg) \Longrightarrow$$

15.42   15.43

gaps. But more importantly they will be sensitive to the L—M—L angle. By symmetry λ_1 is zero and λ_2 is at a maximum at $180°$. When the L—M——L angle decreases, λ_2 will decrease since the overlap between σ_a and z decreases. The magnitude of λ_1 increases to a maximum value at $90°$ and then also decreases. This is again a reflection of overlap changes. The highest molecular level, ϕ_3, in **15.41** is the antibonding analog of **15.43**. It is mostly z with σ_a and yz mixed in an antibonding fashion, **15.44**. The middle level, ϕ_2, is always the most complicated in any three orbital pattern, but the net result is a familiar one that we first encountered in Section 3.3. It is primarily yz in character. Since yz lies at higher energy than σ_a, σ_a will mix into yz in first order in an antibonding fashion. Some z will also enter into this orbital, bonding with respect to σ_a. The net result is given by **15.45**. The reader should refer back to the exercise in Section 3.3 to see that z does indeed mix into ϕ_2 with the sign shown via a second-order perturbation. Note that the mixing of yz into $\sigma_a + z$ (via first order) makes ϕ_1 more bonding, that is, it will be stabilized. Conversely, the mixing of yz in $\sigma_a - z$ makes ϕ_3 more antibonding. The middle level, ϕ_2, becomes hybridized outward, away from the ligands. As the L—M—L angle decreases more z is mixed in and the vertical nodal plane in **15.45** moves to the right. In other words, the nodal plane follows the movement of the ligands, see **15.46**. It is difficult to tell whether ϕ_2 is stabilized or destabilized as the L—M—L angle decreases. We will come back to this point later. For now note σ_a lies quite close in energy to yz in **15.41**. As overlap is turned on between them, the interaction will be very strong and ϕ_2 is destabilized. That is the normal situation for transition metal complexes. But the reader should be fully aware that this three orbital perturbation problem is conceptually the same as bending AH_2 (Section 7.3) and pyramidalization of AH_3 (Section 9.2B). In those two cases the nonbonding level (which started as an atomic p orbital) went down in energy upon bending.

$$\phi_3 = \qquad + \qquad + \qquad \Rightarrow \qquad \qquad 15.44$$

$$\phi_2 = \qquad + \qquad + \qquad \Rightarrow \qquad \qquad 15.45$$

15.46

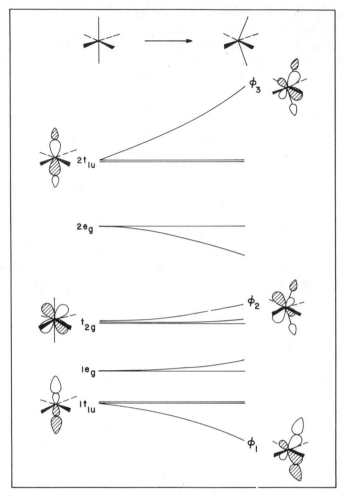

FIGURE 15.4.   A Walsh diagram for bending one *trans* L–M–L angle in a ML$_6$ complex.

The alternative way to develop the orbitals starts with the molecular orbitals of octahedral ML$_6$. On the left side of Figure 15.4 are shown the three orbitals of interest to us. Their intermixing can also be derived from perturbations of the initial molecular levels. As an exercise the reader should do this by recalling the AH$_2$ bending (Section 7.3) and AH$_3$ pyramidalization (Section 9.2) examples. Let us work through ϕ_2 in some detail. As shown in **15.47** $1t_{1u}$ mixes into yz (since it is at lower energy) in an antibonding fashion. Recall that the nonvanishing matrix element that determines the mixing sign will be derived from the overlap of yz with the lone pair hybrids in **15.47**. Likewise, $2t_{1u}$ mixes into yz in-phase **(15.48)**. The resultant orbital is **15.49**. Here the mixing coefficients will be in the order $\lambda_1 > \lambda_2$. In **15.47** the orbital is concentrated on the lone-pair hybrids and **15.48** is more heavily weighted on metal z. Irrespective of this detail ϕ_2 becomes hybridized by the mix-

$$\phi_2 = \quad + \lambda_1 \left(\quad \right) + \lambda_2 \left(\quad \right) \Longrightarrow$$

15.47   15.48   15.49

ing of both orbitals in a direction away from the two ligands. Since **15.47** mixes more strongly into yz than does **15.48**, ϕ_2 will be destabilized along the ordinate of Fig. 15.4. ϕ_1 is also strongly stabilized and ϕ_3 destabilized. It was previously pointed out that overlap factors cause the z^2 component of $1e_g$ to rise in energy and $2e_g$ to fall. A low spin 16-electron ML_6 complex would then be stabilized by this distortion. At the octahedral geometry all levels are doubly occupied up to t_{2g}. There will be 4 electrons in t_{2g}. The downward slope of ϕ_1 overrules the one component of $1e_g$ that rises in energy. Notice also that since ϕ_2 goes up in energy a HOMO–LUMO gap is created for a distorted 16-electron complex.

An analysis of bending for two *cis* ligands in **15.36** can be constructed along the same lines. A convenient top view of this distortion is presented in **15.50**. Here the y component of $1t_{1u}$ and $2t_{1u}$ mix with the $x^2 - y^2$ component of t_{2g}. It is apparent that when the L—M—L angle changes from $90°$ some $x^2 - y^2$ character will mix into $1t_{1u}$ in a way to increase the M—L bonding. In the $x^2 - y^2$ component of t_{2g} there will be a small amount of ligand lone pair and metal y character mixed in. This is shown in **15.51** when the L—M—L angle becomes less than $90°$ and in **15.52** when it is larger than $90°$. Energetically, the distortion in both directions will stabilize one member of $1t_{1u}$ and destabilize $x^2 - y^2$ in t_{2g}. The reader should realize that this is a considerably simplified treatment, especially the analysis of the *cis* angle distortion. The symmetry of the molecule is lowered from O_h to C_{2v}. Metal $x^2 - y^2$ and y are of a_1 symmetry. So too are s and z^2. Therefore, $1a_{1g}$, $2a_{1g}$, and the z^2 components $1e_g$ and $2e_g$ (see Figure 15.1) also become a_1 functions. They will mix into **15.51** and **15.52**. That is certainly a complicating factor. We have neglected it since mixing metal s and z^2 into **15.51** and **15.52** will not change their shapes in the xy plane.

15.50

15.51

15.52

These *cis* and *trans* L—M—L angle distortions split the degeneracy of the t_{2g} set. In both cases one component is destabilized. Recall that π acceptors or π donors will also create an energy difference between the members of t_{2g}. What happens to the π overlap as a function of angular changes can easily be established.[15,16] Normally, a 16-electron ML_6 molecule will utilize both π effects and angular changes together so that one member of t_{2g} lies appreciably higher in energy than the other two. Thus, while low spin 16-electron complexes are unusual, their stability is understandable and there is a growing body of them in the literature.[15-17]

REFERENCES

1. R. L. Martin and A. H. White, *Trans. Metal Chem.* (Monograph Series), **4**, 113 (1968); H. A. Goodwin, *Coord. Chem. Rev.*, **18**, 293 (1976); J. H. Ammeter, L. Zoller, J. Bachmann, P. Baltzer, E. Gamp, R. Bucker, and E. Deiss, *Helv. Chim. Acta*, **64**, 1063 (1981).
2. (a) P. A. Breeze, J. K. Burdett, and J. J. Turner, *Inorg. Chem.*, **20**, 3369 (1981), and references therein. (b) N. E. Kime and J. A. Ibers, *Acta Crystallogr.*, **B25**, 168 (1969); T. Barnet, B. M. Craven, N. E. Kime, and J. A. Ibers, *Chem. Commun.*, 307 (1966).
3. R. J. Goddard, R. Hoffmann, and E. D. Jemmis, *J. Am. Chem. Soc.*, **102**, 7667 (1980), see also W. A. Nugent, R. J. McKinney, R. V. Kasowski, and R. A. Van-Catledge, *Inorg. Chim. Acta*, **65**, L91 (1982).
4. N. M. Kostić and R. F. Fenske, *Organometallics*, **1**, 489 (1982); U. Schubert, D. Neugebauer, P. Hofmann, B. E. R. Schilling, H. Fischer, and A. Motsch, *Chem. Ber.*, **114**, 3349 (1981).
5. T. Ziegler and A. Rauk, *Inorg. Chem.*, **18**, 1755 (1979).
6. M. M. L. Chen, Ph.D. Dissertation, Cornell University (1976).
7. N. Rösch and R. Hoffmann, *Inorg. Chem.*, **13**, 2656 (1974).
8. J. K. Burdett and T. A. Albright, *Inorg. Chem.*, **18**, 2112 (1979).
9. E. Cormona, J. M. Marin, M. L. Poveda, J. L. Atwood, R. D. Rogers, and G. Wilkinson, *Angew. Chem.*, **94**, 467 (1982); J. W. Byrne, H. V. Blaser, and J. A. Osborn, *J. Am. Chem. Soc.*, **97**, 3871 (1975).
10. B. Shevrier, Th. Diebold, and R. Weiss, *Inorg. Chim. Acta*, **19**, L57 (1976).
11. C. Bachmann, J. Demuynck, and A. Veillard, *J. Am. Chem. Soc.*, **100**, 2366 (1978).
12. T. A. Albright, R. Hoffmann, J. C. Thibeault, and D. L. Thorn, *J. Am. Chem. Soc.*, **101**, 3801 (1979).
13. R. Hoffmann, D. B. Boyd, and S. Z. Goldberg, *J. Am. Chem. Soc.*, **92**, 3929 (1970).
14. R. J. Gillespie, *Molecular Geometry*, Van Nostrand–Reinhold, London (1972).
15. P. Kubacek and R. Hoffmann, *J. Am. Chem. Soc.*, **103**, 4320 (1981).
16. J. L. Templeton, P. B. Winston, and B. C. Ward, *J. Am. Chem. Soc.*, **103**, 7713 (1981).
17. M. Kamata, K. Hirotsu, T. Higuchi, K. Tatsumi, R. Hoffmann, T. Yoshida, and S. Otsuka, *J. Am. Chem. Soc.*, **103**, 5772 (1981).
18. R. Hoffmann, J. M. Howell, and A. R. Rossi, *J. Am. Chem. Soc.*, **98**, 2484 (1976).

CHAPTER SIXTEEN

Square Planar, Tetrahedral ML₄ Complexes and Electron Counting

16.1. INTRODUCTION

This chapter is a continuation of the last in that the orbitals of our other molecular building block, a square planar ML_4 complex, are developed. This is a little more complicated than the octahedral case; however, we shall need to use the orbitals of both extensively in subsequent chapters. From the octahedral and square planar splitting patterns a generalized bonding model can be constructed for transition metal complexes. This, in turn, leads to the topic of electron counting. Finally, one distortion that takes a square planar molecule to a tetrahedron is discussed.

16.2. THE SQUARE PLANAR ML₄ MOLECULE

We shall again develop the molecular orbitals for a D_{4h} ML_4 complex in a generalized way where the ligands, L, as before represent two-electron, σ donors. Figure 16.1 constructs the molecular orbitals for this system. On the left side of the figure are the metal s, p, and d levels. On the right side are presented the symmetry-adapted combinations of the four ligand lone pairs. These were developed in some depth for the D_{4h} H_4 system (Section 5.4). Basically the b_{1g} lone-pair combination is stabilized by metal $x^2 - y^2$ and e_u by metal x and y (see the coordinate system at

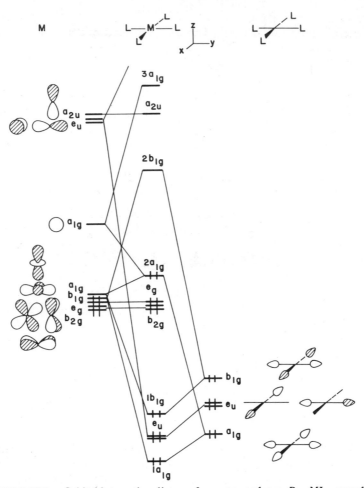

FIGURE 16.1.   Orbital interaction diagram for a square planar, D_{4h} ML₄ complex.

the top center of Figure 16.1). The a_{1g} combination overlaps with and is stabilized by metal z^2 and s. Here again is another three orbital pattern. The molecular level $1a_{1g}$ is mainly lone-pair a_{1g} mixed in a bonding way with metal z^2 and s. There is a fully antibonding analog, labeled $3a_{1g}$, which consists primarily of metal s character. The middle level, $2a_{1g}$, is chiefly z^2 antibonding to the lone-pair a_{1g} combination. Metal s is also mixed into $2a_{1g}$ in a bonding fashion to the lone pairs. The net result is sketched in **16.1**. Metal $x^2 - y^2$ is destabilized by the b_{1g} combination

16.1

yielding molecular $2b_{1g}$ and likewise metal x and y are destabilized by e_u. What is left as nonbonding in Figure 16.1 is metal xy, b_{2g}; metal xz and yz, e_g; and metal z, a_{2u}. While the resultant level splitting pattern looks complicated at first glance, it is quite simple to construct. Notice that there are four levels, $b_{2g} + e_g + 2a_{1g}$, which are primarily metal in character and lie at moderate energy. We have called $2a_{1g}$ a nonbonding orbital because of the pattern in **16.1**. The bonding of metal s to the lone pairs counterbalances the z^2 antibonding and keeps $2a_{1g}$ at low energy. The reader should recall that the orbital is predominantly z^2 in character and, therefore, is often termed z^2. Along with these four nonbonding levels are four M—L bonding ones, $1a_{1g} + e_u + 1b_{1g}$. So there will be a total of eight molecular orbitals which lie at low to moderate energies and are well separated from the antibonding combinations. In other words, a stable complex will be one wherein these eight bonding and nonbonding levels are filled for a total of 16 electrons. This is a different pattern from that in the octahedral system where 18 electrons represented a stable species.

It is instructive to see what is behind this 16–18 electron difference in the two types of complexes. First of all there are two less ligand-based orbitals for ML$_4$ compared to ML$_6$. Compare Figure 15.1 with Figure 16.1. One of the e_g and one of the t_{1u} lone-pair contributions are lost in ML$_4$. Secondly, when the two *trans* ligands in ML$_6$ are removed, the z^2 component of $2e_g$ is stabilized considerably, yielding $2a_{1g}$. Pictorially this is shown in **16.2** and **16.3**. So the ML$_4$ complex gains

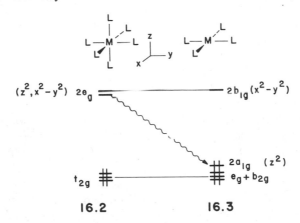

one metal nonbonding orbital over that in ML$_6$ and loses two M—L bonding orbitals. The net result is that there is one less valence level or two less electrons in the stable square planar complex. Notice also in Figure 16.1 that there is one high lying orbital of a_{2u} symmetry which is also left nonbonding. This orbital, primarily metal z, is clearly too high in energy to be filled. In the octahedral ML$_6$ system it was one member of the $2t_{1u}$ set. Our chief concern will be with the metal-based orbitals at moderate energy. As **16.2** and **16.3** indicate, there is a close correspondence between the splitting patterns in O$_h$ ML$_6$ and D$_{4h}$ ML$_4$. Four orbitals are identical in the two systems. It is only the z^2 component of $2e_g$ in ML$_6$ which becomes $2a_{1g}$ in ML$_4$ which is modified. The way in which π acceptors or π donors modify **16.3** can be followed in a way which is identical to that for ML$_6$ in Section

15.3. Therefore, we will not spend time on the issue. The correspondence suggests that there may be a general pattern for any ML_n complex. This, along with electron counting, is the topic of concern in the next section.

16.3. ELECTRON COUNTING

Most stable, diamagnetic transition metal complexes possess a total of 18 valence electrons. We covered some exceptions to this in Chapter 15 and the square planar ML_4 situation presents possibilities for another. But apart from these the overwhelming majority of compounds are of the 18 valence electron type. In other words, the number of nonbonding electrons at the metal plus the number of electrons in the M—L bonds which we have formally assigned to the ligands should total 18. Yet another way of putting this is that there are 18 electrons associated with the metal. The derivation of this rule can be constructed in a number of ways. A transition metal will have five nd (where n is the principal quantum number), three $(n + 1)p$, and one $(n + 1)s$ AO's which form bonding combinations to the surrounding ligands or remain nonbonding. These nine AOs will then house 18 electrons. This is true for most geometries, but it can be seen that when all of the ligands lie in a plane containing the transition metal one p AO (perpendicular to this plane) cannot take part in σ interactions.

The 18-electron rule is therefore nothing but a restatement of the Lewis octet rule. The extra 10 electrons are associated with the five d orbitals. A more elaborate way to express this is shown in Figure 16.2. The orbitals of any ML_n complex can be developed in a way that is analogous to what we have done for ML_6, ML_4, and the AH_n series. Figure 16.2 does so for a generalized transition metal system. There are n ligand-based lone pairs illustrated at the lower right of this figure. Symmetry-adapted linear combinations of the n ligand orbitals will normally find matches with n of the nine metal-based AOs. This produces n M—L bonding and n M—L antibonding MOs. Left behind are $9 - n$ nonbonding orbitals which are localized at the metal. These $9 - n$ nonbonding levels will be primarily metal d in character since the metal AOs start out with the d set lower in energy than s and p. Furthermore, d AOs are more noded than the s- or p-type functions so that it is more likely for the ligand set to lie on or near the nodal plane of d-based functions. (The glaring exception to this generalization occurs when the metal and ligands lie in a common plane.) There are, therefore, a total of $n + (9 - n) = 9$ valence levels at low to moderate energies which constitute bonding and nonbonding interactions and 18 electrons can be housed in them.

The square planar system was different (Figure 16.1) in that one p orbital at the metal, **16.4**, found no symmetry match. There are four metal orbitals primarily of d character at moderate energies, *and* **16.4**, which lies at an appreciably higher energy. It is unreasonable to expect that two electrons should be placed in **16.4** and therefore, stable square planar ML_4 complexes have 16 valence electrons. A trigonal ML_3 complex will also have one empty metal p orbital, **16.5**, and a stable complex will thus be of the 16-electron type. Linear ML_2 compounds have two nonbonding p AOs, **16.6**, so here a 14-electron complex will be stable.

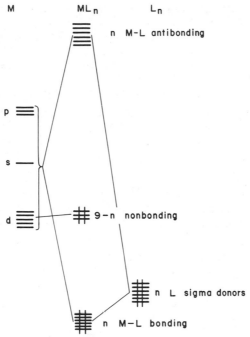

FIGURE 16.2.   A generalized orbital interaction diagram for a ML_n complex where the ligands are arranged in a spherical manner around the transition metal.

16.4   **16.5**   **16.6**

This sort of generalized interaction diagram in Figure 16.2 can be extended to main group compounds where the d orbitals on the central atom have been neglected. The 6-electron, trigonal BR_3 and other electron deficient compounds are then related in an obvious way to 16-electron square planar ML_4 and trigonal ML_3 complexes. In the hypervalent AH_6 and AH_4 molecules (Section 14.1) there are two and one "ligand" combinations, respectively, which do not match in symmetry the central atom's s and p set. For example, the electron count at the central atom in SH_6 is still 8, although the total number of electrons is 12 (there are, however, violations of this rule, see discussion in Chapter 14). This can also happen, albeit with much less frequency, for transition metal complexes.[1] One example that will be discussed shortly is tris(acetylene)W(CO) which appears to be a 20-electron complex. However, here one occupied acetylene π combination does not find a symmetry match with the metal AOs and so the compound is in reality an 18-electron system.

This brings up the mechanics of electron counting. The convention that we shall use is to treat all ligands as Lewis bases. Listed in **16.7** are some typical two-electron σ donor groups. In **16.8** are listed some two-electron σ donors which also have one

$$NR_3 , \quad PR_3 , \quad CH_3^- , \quad SiR_3^- , \quad H^- \qquad \textbf{16.7}$$

$$CO , \quad CNR , \quad SO_2 , \quad CR_2 , \quad NO^+, \quad CR^+ \qquad \textbf{16.8}$$

$$NR_2^- , \quad Cl^- , \quad OR^- , \quad SR_2 \qquad \textbf{16.9}$$

or two π acceptor functions. We covered the CO and CR$_2$ cases explicitly in Section 15.2. For the purposes of electron counting, it does not matter what the strength of π bonding really is between the metal and ligand. The ligands are counted only in so far as their σ donating numbers. The ligands in **16.9** are two-electron σ donors with π donor functions. The ways that electrons are assigned to the ligands in **16.7**–**16.9** are only conventions. One could just as well have had alkyl groups and hydrides as *one*-electron, neutral ligands. Exactly the same electron count at the metal will be obtained. What changes is the oxidation state at the metal—the number of electrons that are *formally* assigned to the metal. The nitrosyl (NO) group is a particularly difficult case. Counting it as a cationic system stresses the analogy to the isoelectronic CO group. The M—NO coordination mode is then expected to be linear. Indeed there are many examples of this type, but there are also compounds where the M—N—O angle is appreciably less than 180°. A detailed discussion of this distortion is reserved for Section 17.5. The point is that with a bent geometry the nitrosyl can be considered as an anionic four-electron donor. The "extra" two electrons are housed in an NO-based π^* orbital (which causes the nitrosyl group to bend; recall the ammonia inversion problem in Section 9.2) and only partial π back donation occurs to an empty metal orbital.

Polyenes are also considered as Lewis bases. They are counted such that all π bonding and nonbonding levels are occupied. Some representative examples are given in **16.10**. Listed below each structure are the number of electrons donated to

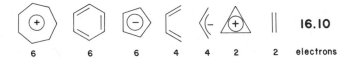

$$\begin{array}{ccccccc} 6 & 6 & 6 & 4 & 4 & 2 & 2 \quad \text{electrons} \end{array}$$

the metal. Some care, however, must be used to establish the connectivity of these polyenes to the metal. The benzene ligand will donate its six π electrons to the metal only if all six carbon atoms are bonded to the metal atom. A short-hand notation to denote this is the hapto number.[2] In this example it would be called an η^6 complex. Bis(benzene)chromium, **16.11**, contains two η^6 benzene ligands; all Cr—C distances are equivalent.[3] In the permethyl derivative of **16.12**[4] one arene ring is η^6 and the other η^4. In other words two Ru–C distances are much longer than that found for the other ten. The hexakis(trifluoromethyl) derivative[5] of **16.13** contains an η^2 benzene ligand. In the η^4 case a total of four π electrons and in the η^2 example two π electrons are donated to the metal. Likewise the cyclopentadienyl ligand will donate six π electrons if it is η^5 as in **16.14**,[6] four electrons at an η^3 geometry, **16.15**,[7] or two electrons at η^1, **16.16**.[8] Thus, the connectivity of the polyene to the metal must be carefully established. Alternatively, one can

16.11

16.12

16.13

16.14

16.15

16.16

assign an 18- or 16-electron count at the metal and this sets the coordination mode of the polyene. The portion of the polyene which is not bonded to the metal bends out of the plane defined by the coordinated carbons in a direction away from the metal.[9] This stereochemical feature has been highlighted in the drawings of **16.11–16.16** along with the actual slippage of the metal over the coordinated portion of the polyene. Thus a fairly detailed geometrical prediction can be made for polyene-metal complexes on the basis of electron counting.

The number of d electrons (the electrons housed in the $9 - n$ nonbonding levels of Figure 16.2) assigned to the metal is determined by adding the number of charges at the ligands and subtracting this sum from the total charge on the mole-cule. This gives the formal charge or oxidation state at the metal. Finally, the num-ber of d electrons is then equal to the number of d electrons of the metal in the zero oxidation state minus the oxidation state (formal charge) assigned to the metal. **16.17** lists the number of d electrons for the transition metals in their zero

d^4	d^5	d^6	d^7	d^8	d^9	d^{10}	$d^{10}s^1$
Ti	V	Cr	Mn	Fe	Co	Ni	Cu
Zr	Nb	Mo	Tc	Ru	Rh	Pd	Ag
Hf	Ta	W	Re	Os	Ir	Pt	Au

16.17

oxidation state. Notice that this is not the atomic electron configuration. In other words, Ti(0) is d^4 rather than the atomic configuration $3d^2 4s^2$. The total number of electrons associated with the metal, that is, the number of electrons in the n ML bonding and $9 - n$ nonbonding orbitals (see Figure 16.2), is equal to the number of d electrons plus the number of electrons that have been donated in a σ fashion by the ligand set.

A few simple examples will make this electron counting rule clearer. In $Cr(CO)_6$ the CO groups donate two σ electrons each for a total of 12 electrons. The charge on the molecule and each CO is zero so Cr is in the zero oxidation state, that is, it is a d^6 complex—see **16.17**. The total number of electrons associated with Cr is $6 + 12 = 18$. In CH$_3$Mn(CO)$_5$ the methyl ligand is taken to be an anionic, 2-electron donor (see **16.7**). Together with the five CO groups this makes for a total of 12 electrons from the ligand set which are donated to the metal. The molecule is neutral and there is one anionic ligand, therefore, the metal is Mn(+1) which is d^6 (note in **16.17** that Mn(0) is d^7). Again there is a total of $6 + 12 = 18$ electrons associated with the metal. The compounds given by **16.18** and **16.19** are also 18-electron systems. In ferrocene, **16.18**, the cyclopentadienyl (Cp) ligands are η^5 and counted as a six-π-electron Cp$^-$ (see **16.10**). The metal is then d^6-Fe(+2). Finding the oxidation state for Mn in **16.19** is a little more complicated. There is an η^1-Cp$^-$, NO$^+$, and CH$_3^-$ ligand. The molecule is neutral, therefore, the oxidation state is Mn(+1) which is d^6. In **16.20** there are three Cl$^-$ and one neutral ethylene ligands,

16.18   16.19   16.20

yielding a total of 8 electrons donated to the metal. The charge on the molecule is -1 so the oxidation state at Pt is given by $-1 - (-3) = +2$. Pt(+2) is d^8 (see **16.17**), therefore, it is an $8 + 8 = 16$ electron complex. Notice that the geometry around **16.20** is, as expected from the previous section, square planar. **16.13** is an example of a 16-electron trigonal complex. The compounds provided by **16.11, 16.12,** and **16.14–16.16** all have 18 electrons. The reader should work through these examples. Special attention should be given to **16.12**. One benzene ring is η^4—a four-π-electron donor and the other is η^6—a six-electron donor. But suppose each benzene was η^6. Then the molecule would be a 20-electron complex. The extra two electrons would enter a M—L *antibonding* orbital in the generalized scheme of Figure 16.2 (in this case a Ru d-benzene π antibonding orbital). Clearly this is expected to be an energetically unfavorable situation. This is not quite true for this *special case* and we shall return to the bonding in metallocenes in Section 20.3.

A practical consideration that must be kept in mind when counting electrons is that the ligand donor orbitals must find a metal function with which to overlap. There are a few "high symmetry" situations where this is not followed. Tris(acetylene)-W(CO), **16.21**, is one example.[1b] The acetylene ligand carries two orthogonal π orbitals. Let us consider that one π orbital at each acetylene is pointed directly at the tungsten atom. This will create an $a_1 + e$ set of "radial" π orbitals. The three orthogonal π orbitals are of $a_2 + e$ symmetry—a tangential set (see the Walsh model for cyclopropane in Section 11.2). The a_1 and two e set of π donor orbitals find overlap with tungsten s, p, and d AOs; however, no function on tungsten matches the a_2 combination **16.22** (an f AO would overlap with **16.22**). Therefore, the three acetylenes donate a total of 10 electrons, making **16.21** an 18-electron complex.

16.21   16.22

Transition metal complexes that have metal–metal bonds pose special problems in electron counting. A couple of examples will suffice to show the general principles. For $Mn_2(CO)_{10}$, **16.23**, each Mn is formally d^7. There are 10 electrons donated by the five CO ligands which would give a 17-electron count at each metal. However, one electron from *each* Mn is shared with the other. In other words, there is a two-center–two-electron, σ bond formed between the two metal atoms, **16.24**. Obviously the two electrons are shared equally so each metal attains an 18-electron configuration. In a formal way each Mn contributes one electron to the Mn—Mn σ bond which leaves six nonbonding electrons just as in any other 18-electron ML_6 complex. The octahedral environment for each Mn is clear from the drawing in **16.23** and one might expect a splitting pattern very similar to that presented for ML_6 in Figure 15.1. This is a point which we shall return to in the next chapter. Each Rh atom in **16.25** is formally d^8. Discounting the Rh—Rh bonding, there are a total of 16 electrons associated with each metal. Sharing two electrons from the neighboring metal will bring each Rh atom up to an 18-electron count. A Rh—Rh double bond is then postulated for this molecule.

16.23   16.24   16.25

Bridging groups in transition metal dimers and clusters are often times a source of confusion and controversy. The problem stems from an ambiguity of how to partition the electrons between the metals and bridging groups. One example is presented in **16.26**. Counting the bridging carbonyl as a neutral, two-electron donor, **16.27**, just as we have done for terminal carbonyls implies that each iron is Fe(+1)-d^7. There will be a total of 10 electrons supplied by the ligands so the total electron count is 17 and the formation of a single Fe—Fe bond is required to attain an 18-electron configuration. This way of counting the bridging carbonyls

16.26   16.27   16.28

implies that they are three-center-two-electron bonds. An alternative and certainly reasonable way to handle the bridging carbonyls is to insist that they are "ketonic." That is, they make two-center-two-electron bonds to each iron. This implies the dianionic formulation, **16.28**, and each iron is then $Fe(+3)$-d^5. The ligand set now donates 12 electrons so there is an electron count of 17 at each iron and again an 18-electron system will be created with an Fe—Fe single bond. Both methods of electron counting lead to the same conclusion, namely an 18-electron configuration is attained by the formation of a metal–metal bond. This is certainly an over-simplification. There are actually a number of occupied metal–metal bonding and antibonding orbitals in **16.26**. The number of occupied bonding levels will exceed the number of occupied antibonding levels by one; however, it is not at all true that each metal–metal orbital carries the same weight toward the total iron–iron overlap population. The bridging carbonyls significantly perturb the electronic environment of the metal-centered orbitals.[10] The situation here is exactly analogous to the one discussed for diborane in Section 10.2B.

The only difference that is created in the two ways of counting electrons for **16.26** is that different oxidation states for iron are obtained and, of course, the number of d electrons formally assigned to the metal changes. That is just a formalism. There is really no right or wrong way to partition the electrons associated with the metal. One hopes that the methods used to assign electrons for the ligands in these complexes will lead to an oxidation state (charge) at the metal that approaches reality. But this is probably an unreasonable expectation. Treating each ligand in **16.7–16.10** as a Lewis base does offer a practical advantage. What we are really saying is that the σ donor orbitals of the ligands lie at lower energy than the metal d levels—see Figure 16.2. This is normally the case. Furthermore, the number of d electrons assigned will then correspond to those contained with the $9 - n$ nonbonding levels of Figure 16.2. For example, in $Cr(CO)_6$ there are three "nonbonding" levels—the t_{2g} set of Figure 15.1. $Cr(CO)_6$ is counted as being d^6 so those six electrons are housed in t_{2g}. A d^4 ML_6 complex will possess four electrons in t_{2g}, and so on.

16.4. THE SQUARE PLANAR–TETRAHEDRAL ML₄ INTERCONVERSION

In the first section of this chapter we built up the orbitals of a square planar ML_4 complex. An alternative geometry would be a tetrahedral species. There are two basic ways to convert a square planar complex into a tetrahedral one. The **16.29** to **16.30** interconversion involves twisting one pair of *cis* ligands about an axis shown in **16.29**. That will conserve D_2 symmetry along all points that interconnect **16.29** with **16.30**. In the other path the two *trans* L—M—L angles are decreased, as shown in **16.31**, ultimately yielding the tetrahedron **16.32**. This conserves D_{2d} symmetry. The elements of this latter pathway have actually been developed in Section 15.4, so we shall briefly explore this distortion. A Walsh

16.29 $\quad$ >M⟨⟨⟨⟨ ⟨ $\quad \xrightarrow[D_2]{}$ $\quad$ >M⟨ $\quad$ 16.30

16.31 $\quad$ ⟨—M⟨⟨ $\quad \xrightarrow[D_{2d}]{}$ $\quad$ ⟩M⟨ $\quad$ 16.32

diagram for a model ML₄ is presented in Figure 16.3. On the left side are the metal–centered, valence orbitals of a square planar ML₄ system which have been taken from Figure 16.1. The $2b_{1g}$ level $(x^2 - y^2)$ is stabilized greatly since this orbital is strongly M—L antibonding and overlap between the ligand lone pairs and $x^2 - y^2$ is reduced upon reduction of the two *trans* L—M—L angles. The e_g set (xz, yz) is destabilized and meets $x^2 - y^2$ at the tetrahedral geometry to form a t_2 set. The rationale for the destabilization is identical to that developed for the distortion in **15.35** (see Section 15.4). The ligands move into a position where overlap to xz and yz is turned on. The ligand lone pairs mix with e_g in an antibonding manner, thus, xz and yz are destabilized. This is abated somewhat by mixing in some metal x and y character. The result is that the metal-centered functions become hybridized away from the ligands. The $2a_{1g}$ level at the square planar geometry is mainly z^2 with some antibonding from the lone pairs (see **16.1**). This distor-

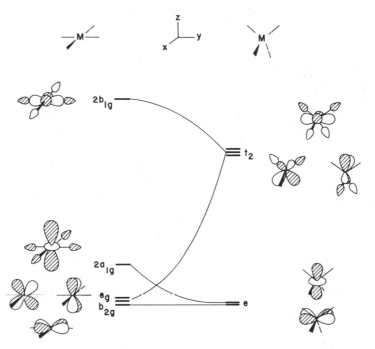

FIGURE 16.3. Distortion of a square planar to tetrahedral ML₄ complex maintaining D_{2d} symmetry.

tion moves the ligands into the nodal planes of z^2. Therefore, $2a_{1g}$ is stabilized; it becomes one partner of an e set, along with xy which is unperturbed along the rearrangement path.

Before we discuss the actual dynamics of the square planar to tetrahedral inter-conversion, it is instructive to compare the two endpoint geometries. In the square planar system the very high lying $2b_{1g}$ level makes it obvious that a singlet d^8 complex (where $2a_{1g}$ is the HOMO) will be a stable species. At the tetrahedral side a d^8 system will have four electrons in t_2. Consequently, a high spin (triplet) situation is required for a stable species. Notice on the right side of Figure 16.3 that the three members of t_2 have M—L antibonding character. At the square planar geometry only $2a_{1g}$ is slightly antibonding. As a result we expect weaker (and therefore longer) M—L distances for high spin d^8 compounds compared to their low spin, square planar counterparts. This is often found to be true. Table 16.1 lists some examples. With two electrons more it is clear from the relative energies of $2b_{1g}$ vs. t_2 in Figure 16.3 that the tetrahedral form will be much more stable. Notice that this is a saturated 18-electron d^{10} ML$_4$ complex. In the tetrahedral geometry the ligands are arranged in a spherical manner around the transition metal. The generalized orbital pattern in Figure 16.2 is appropriate.

A number of d^8 square planar complexes undergo a *cis-trans* isomerization process, **16.33** to **16.34**.[11] Of interest to us in this section is to probe the direct pathway via the tetrahedral structure **16.35**; however, there are at least two alternative paths which we will come back to in the next two chapters. One ligand may dissociate yielding a *cis* T-shaped intermediate **16.36**. It can rearrange to a *trans* T-shaped structure and interception by ligand B gives **16.34**. Another path involves association of an external ligand, L, which yields the five-coordinate intermediate **16.37**. Rearrangement of **16.37** followed by expulsion of L gives **16.34**. So the dynamics of the *cis-trans* interconversion are complicated by several competing pathways. Returning to the direct route for interconversion a schematic illustration for a thermal process is given in **16.38** for a d^8 complex. The correlation of orbitals has been taken from Figure 16.3. Although the symmetry of the two square planar and tetrahedral complexes is lower than D_{4h} and T_d, respectively, the essential details of the splitting patterns will remain very close to the idealized cases. The important feature in **16.38** is that the *cis* to *trans* interconversion via a tetrahedron should be a very high energy process. It is symmetry forbidden under thermal conditions.[12] There is not much difference in a qualitative sense between this

TABLE 16.1.   Mean Nickel–Ligand Distances $(\text{Å})^a$

	Square Planar	Tetrahedral
Ni—N (sp^2)	1.68	1.96
Ni—P	2.14	2.28
Ni—S	2.15	2.28
Ni—Br	2.30	2.36

^{a}Taken from K. W. Muir, *Molecular Structure by Diffraction Methods*, Vol. 1, the Chemical Society, London (1973), p. 580.

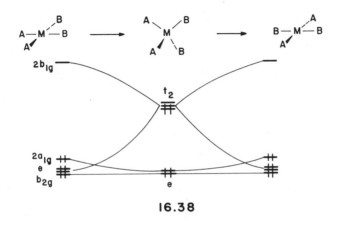

16.38

rearrangement and the $H_2 + D_2$ reaction (Section 5.4) or the dimerization of two olefins (Section 11.3). In all three cases a critical point, a high energy cusp, is reached on the potential energy surface. A path of lower or at least different symmetry will be followed. In this case dissociative or associative paths represent viable alternatives. However, the isomerization can proceed via an excited state species. For example, photochemical excitation of the *cis* compound populates the $2b_{1g}$ orbital. The excited state singlet complex can undergo intersystem crossing to a triplet state. This may then relax to a tetrahedral geometry which decays back to the square planar ground-state singlet with either *cis* or *trans* geometry. Alternatively, the whole rearrangement process may take place within the singlet spin state. This is shown for an arbitrary spin state in **16.39**. The one-electron picture in **16.39** is a very crude representation of the photochemical processes. State correlation diagrams have been constructed which more clearly show the relative energies of the molecule as a function of the electronic and geometrical configurations. Furthermore, the actual details of the spin-state change have been neglected here.[12] In principle, the singlet–triplet interconversion can occur thermally[12] and this accounts for yet another mechanism of the *cis–trans* equilibrium for square-planar complexes.

Another interesting isomerization process can take place between square planar and octahedral systems in coordination compounds. **16.40** shows how the $z^2/x^2 - y^2$ separation changes as two *trans* ligands are brought closer to the square

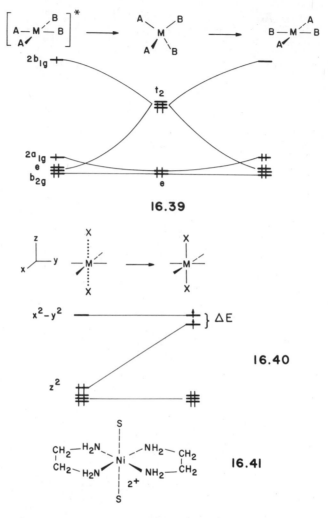

16.39

16.40

16.41

plane. When ΔE is small enough then a high spin d^8 species is formed, as indicated on the right side of **16.40**. This is, of course, just the reverse process, **16.2** to **16.3**, which we have discussed in the first section of this chapter. The Lifschitz salts, Ni(ethylenediamine)$_2^{2+}$ (**16.41**), are either paramagnetic and octahedral with two donor solvent molecules, H_2O for example, occupying the axial sites, or square-planar and diamagnetic (with noncoordinated solvent molecules).[13] Which species is actually found depends critically on crystallization conditions. The diamagnetic d^8 species has z^2 as the HOMO which points toward the fifth and sixth coordination positions of the octahedron. As the solvent molecules are brought closer to Ni, z^2 is greatly destabilized by the symmetric combination of donor lone pairs. One should recall that the symmetric combination of solvent lone pairs is, in turn, stabilized. Furthermore, the antisymmetric combination will be stabilized by metal z. Therefore, there is a delicate balance between the two molecular extremes of

16.40, especially since it involves a spin-state change and two-electron energy terms as well.

REFERENCES

1. (a) R. Hoffmann and J. Lauher, *J. Am. Chem. Soc.*, **98**, 1729 (1976); R. D. Rogers, R. V. Bynum, and J. L. Atwood, ibid., **100**, 5238 (1978); A. Davidson and S. S. Wreford, *Inorg. Chem.*, **14**, 703 (1975); J. Silvestre, T. A. Albright, and B. A. Sosinsky, ibid., **20**, 3937 (1981); S.-Y. Chu and R. Hoffmann, *J. Phys. Chem.*, **86**, 1289 (1982).

    (b) R. M. Laine, R. E. Moriarity, and R. Bau, *J. Am. Chem. Soc.*, **94**, 1402 (1972); R. B. King, *Inorg. Chem.*, **7**, 1044 (1968); J. K. Burdett, *Molecular Shapes*, Wiley, New York (1980), pp. 224–225.

2. F. A. Cotton, *J. Am. Chem. Soc.*, **90**, 6230 (1968).

3. E. Keulen and F. Jellinek, *J. Organomet. Chem.*, **5**, 490 (1966); A. Haaland, *Acta Chem. Scand.*, **19**, 41 (1965).

4. G. Huttner and S. Lange, *Acta Crystallogr.*, **B28**, 2049 (1972).

5. J. Browning and B. R. Penfold, *J. Cryst. Mol. Structure*, 6, 59 (1976).

6. A. F. Berndt and R. E. Marsh, *Acta Crystallogr.*, **16**, 118 (1963).

7. G. Huttner, H. H. Brintzinger, L. G. Bell, P. Friedrich, V. Bejenke, and D. Neugebauer, *J. Organomet. Chem.*, **145**, 329 (1978).

8. M. J. Bennett, F. A. Cotton, A. Davison, J. W. Faller, S. J. Lippard, and S. M. Morehouse, *J. Am. Chem. Soc.*, **88**, 4371 (1966).

9. R. Hoffmann and P. Hofmann, *J. Am. Chem. Soc.*, **98**, 598 (1976).

10. M. Benard, *Inorg. Chem.*, **18**, 2782 (1979).

11. G. K. Anderson and R. J. Cross, *Chem. Rev.*, **9**, 185 (1980).

12. E. A. Halevi and R. Knorr, *Angew. Chem.*, **94**, 307 (1982); *Angew. Chem. Suppl.*, 622 (1982) and references therein; an *ab initio* calculation has given a barrier to 26 kcal/mol for the low spin process; O. Gropen, U. Wahlgren, and L. Petterson, *Chem. Phys.*, **66**, 453 (1982).

13. A. F. Wells, *Structural Inorganic Chemistry*, fourth edition, Clarendon Press, Oxford (1975), p. 964–967.

CHAPTER SEVENTEEN

Five Coordination

17.1. INTRODUCTION

Throughout this book we have stressed one technique for understanding the molecular orbitals of complicated molecules, namely their construction from the valence orbitals of smaller subunits. In the organometallic area this is particularly useful since the molecules consist of an ML_n unit bonded to some organic ligand. For this purpose we shall need to build up a library of valence orbitals for common ML_n fragments, where $n = 2 - 5$ and L is a generalized two-electron σ donor ligand. We could do this by interacting an ensemble of L_n functions with a transition metal, just as was carried through for the octahedron (Section 15.1) and square plane (Section 16.1) cases. However, an easier method[1-3] starts with the valence, metal-centered orbitals of the octahedron and square plane. One or more ligands are then removed. This is illustrated in Chart 17.1. The valence orbitals of a C_{4v} ML_5 fragment, **17.2**, can easily be derived by taking those of ML_6, **17.1**, and considering the perturbation induced by removing one ligand. A C_{2v} ML_4 species, **17.3**, is derived by removing two *cis* ligands from ML_6 and removal of three *fac* ligands will yield the C_{3v} ML_3 fragment, **17.4**. We shall be primarily concerned with the perturbation **17.1** to **17.2** in this chapter. Now those fragments, **17.1-17.3**, can be distorted to give fragments of other types. For example, the C_{2v} ML_4 fragment can easily be distorted to a C_{4v} structure, **17.5**, or a tetrahedron. Likewise, we will find it useful to generate the levels of **17.6** from those of the square pyramid.

Once the orbitals of a trigonal bipyramid have been derived, they can be used in turn to establish the orbitals of a C_{3v} ML_4 fragment like **17.7** which may then be distorted to a tetrahedron, and so on. Thus, the reductive approach illustrated in Chart 17.1 offers many ways to interrelate the orbitals of different systems. The fragments are interesting molecules in their own right and we shall spend some time with their structure and dynamics. Our other starting point is the square plane, **17.8**.

CHART 17.1

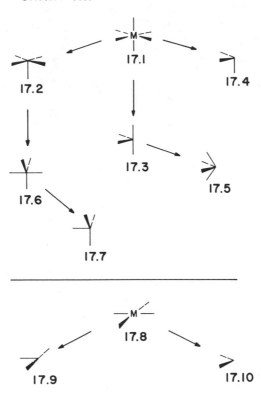

Removal of one ligand gives a C_{2v} ML$_3$ fragment, **17.9**. We shall see in Section 18.1 that the orbital structure of **17.9** is very similar to that of the C_{4v} ML$_5$ fragment, **17.2**. Removing two *cis* ligands from **17.8** gives **17.10**, with orbitals similar to those of **17.3**. This correspondence between different ML$_n$ fragments is an important way to simplify and unify organometallic chemistry, and forms a common thread running through the next three chapters.

17.2.   THE C_{4v} ML$_5$ FRAGMENT

On the left of Figure 17.1 are listed the metal-centered d block of orbitals for octahedral ML$_6$. From Section 15.1 (see Figure 15.1) we established that there is a lower group of three levels, xz, yz, and xy, using the coordinate system at the top of Figure 17.1, which have t_{2g} symmetry. These are filled for a saturated (18-electron) d^6 complex. At much higher energy is the $2e_g$ set. It will be empty in most organometallic examples and consists of $x^2 - y^2$ and z^2 antibonding to the ligand ione pairs. When one ligand is removed from the octahedron,[1] to a first approximation the t_{2g} set is left unaltered. The resultant levels are labeled as $e + b_2$ in the C_{4v} point group of the fragment. No hybridization or energy change is introduced be-

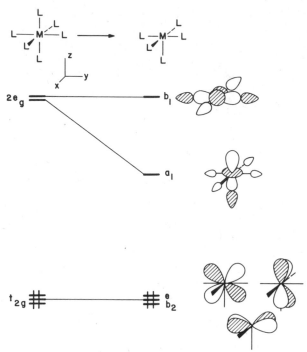

FIGURE 17.1. Orbital correlation diagram for the octahedral to square pyramidal conversion. Only the d orbital part of the diagram is shown. Note the rehybridization of z^2 toward the empty coordination site.

cause the lone pair of the missing ligand is orthogonal to t_{2g}. The same is true for the $x^2 - y^2$ component of $2e_g$. Suppose that the ligand removed from $Cr(CO)_6$ was CO—a π acceptor.[4] Then the xz and yz components of t_{2g} would rise slightly in energy and xy is left untouched. Consequently, a relatively small energy gap will be introduced between e and b_2. The major perturbation occurs with the z^2 component of $2e_g$. That orbital, labeled a_1, will be greatly stabilized. Removing the ligand loses one strong antibonding interaction between metal z^2 and the ligand. The a_1 level also becomes hybridized by mixing some s and z character in a way which reduces the antibonding between the metal and surrounding ligands. The origin of this hybridization in a_1 is not much different from that in the variation of *cis* and *trans* L—M—L angles in ML_6 (Section 15.3). We shall outline one way to view the resultant hybridization. The O_h ML_6 to C_{4v} ML_5 conversion invloves a reduction of symmetry. The $2a_{1g}$ orbital (see Figure 15.1) and the z component of $2t_{1u}$ lie close in energy to $2e_g$. Both orbitals also have a_1 symmetry. Consequently, they mix into the z^2 component of $2e_g$, **17.11** (in first order), in a way that reduces the antibonding between the metal and its surrounding ligands. Recall that $2a_{1g}$ and $2t_{1u}$ lie at higher energy than $2e_g$, thus they mix into **17.11** in a bonding manner. This is diagrammed by **17.12**. Notice that it is the phase relationship shown for the metal s and z in **17.12** to the ligand lone pairs in **17.11** that sets the mixing sign. $2a_{1g}$ and $2t_{1u}$ are, after all, concentrated at the metal. Therefore, the largest

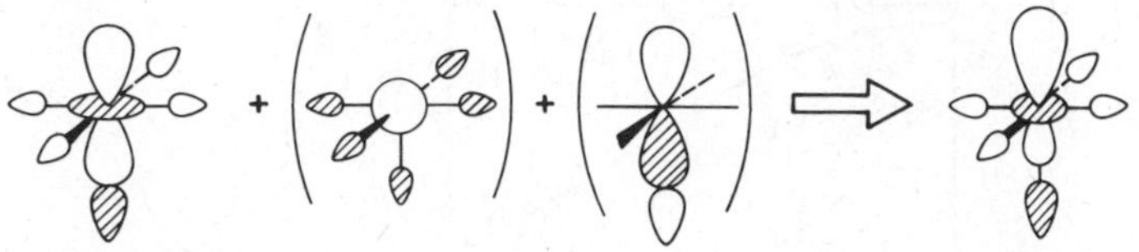

17.11   17.12   17.13

interorbital overlap will occur between the atomic components of **17.12** at the metal and the lone pairs in **17.11**. The resultant orbital, **17.13**, is stabilized further by this mixing process and it becomes hybridized out away from the remaining ligands, toward the missing one. The a_1 orbital is empty for a d^6 fragment. It obviously will play a crucial role when real molecules are constructed from the ML_5 fragment. Its directionality and the fact that it lies at moderate energy makes it a superlative σ-accepting orbital. Below lie a nest of three "t_{2g} like" orbitals which are utilized for π bonding. Before we use the ML_5 as a building block for larger molecules, it is instructive to examine it as a molecule in its own right.

17.3.   FIVE COORDINATION

We have looked at the orbital properties of the main group ML_5 molecules in Chapter 11. Two basic structures are known, the square pyramid **(17.14)** and the trigonal bipyramid **(17.15)**. A whole spectrum of geometries between the two extremes are also found in practice. The interconversion of the two geometries can occur in a simple way via the Berry pseudorotation process **(17.16)**, a geometrical change with an obvious resemblance to the variation of the apical/basal angle, θ **(17.14)** of the square pyramid shown in **17.17**.

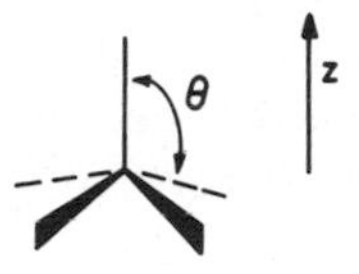

17.14 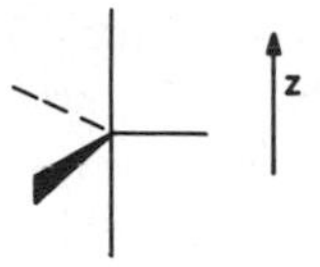   17.15

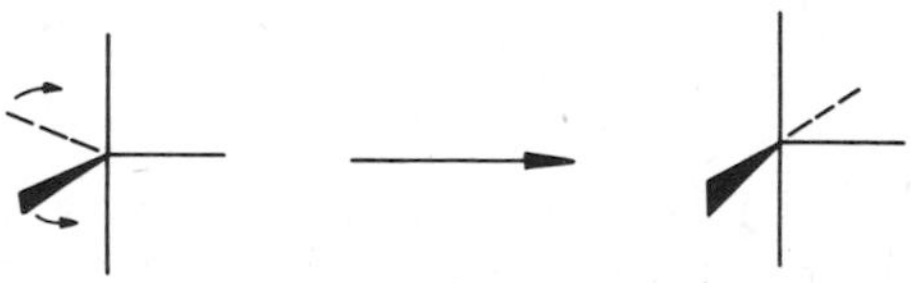

17.16

17.17

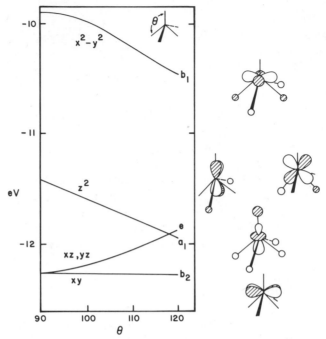

FIGURE 17.2. Orbital correlation diagram for the metal d orbitals on bending a square pyramidal along the coordinate of **17.17**.

The energy levels of the square pyramid[5] with $\theta = 90°$ have been derived in the first section of this chapter. First, we will see how they change in energy as the angle θ varies. As θ increases from $90°$ the σ overlap of the basal ligands with z^2 and $x^2 - y^2$ decreases (Figure 17.2), they becomes less antibonding and drop in energy. Concurrently σ interaction with the xz, yz pair of orbitals is turned on and they are pushed to higher energy. Such a geometry change also changes the shape of these metal d-based orbitals since they become hybridized with the $(n + 1)p$ metal orbitals. In an exaggerated way this is shown in **17.18**. We have seen this d–p

   **17.18**

mixing previously in Section 15.4 for a related angular geometry change. The resulting hybridization out away from the ligands is entirely analogous to this previous case.

Since the d orbital energies of the ML_5 square pyramid change significantly with the angle θ, the details of the geometry of such species will depend upon the number of d electrons and how the orbitals are occupied. Low spin d^6 species are expected to have a flat pyramid ($\theta \sim 90°$) since xz and yz (filled for a d^6 system) rise in energy as θ increases but low spin d^8 species where z^2 is occupied are more distorted

TABLE 17.1.   Some Bond Angles in Square Pyramidal Units as a Function of Electronic Configuration

Electronic Configuration	Compound	Apical–M–Basal Angle (degrees)
d^{10}	Cu^1 macro · CO^a	117
hs $d^{8\ b}$	$Ni(5\text{-}ClsalenNEt_2)_2$	100.4
	$[Ni(dmp)Cl_2]_2 · 2CHCl_3$	100.9
	$Ni(bddae)(NCS)_2$	99.9
	$[Ni(tpen)](ClO_4)_2 · MeNO_2$	98.8
ls d^8	$Ni(bda)Br_2$	93.9
	$Ni(DSP)I_2$	92.4
	$Ni(CN)_5^{3-}$	101.0
	$Co(CNC_6H_5)_5^{2+}$	101.8
	$Co(CN)_5^{3-}$	97.6
hs d^6	Deoxyhemoglobin (FeII)	110
ls d^6	Oxyhemoglobin (FeII)	~90
hs d^5	Chlorohemin (FeIII)	93
ls d^5	Cyanomethemoglobin (FeIII)	90–92

[a] Macro = difluoro-3,3'-(trimethylenedinitrito) bis(2-butanone oxamate).
[b] hs and ls refer to high spin and low spin arrangements, respectively.

$(\theta > 90°)$. This is a trend found in general for the examples of Table 17.1. The iron atom in a heme unit, **17.19** lies in a site of square pyramidal coordination. There are four such heme units, connected to peptide chains, in hemoglobin.[6] Commensurate with the high spin d^6 electronic configuration, θ is larger than 90°. On coordination of O_2 the iron atom becomes six coordinate and the spin state changes to low spin. Both of these factors lead to a θ of about 90° in oxyhemoglobin. Thus

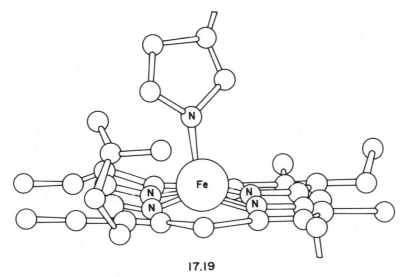

17.19

the stereochemical change on oxygenation leads to a considerable movement of the iron atom shown in **17.20** and, of course, the imidazole ring attached to it in the apical position of the square pyramid. Connected to the imidazole ring is the organic peptide part of the molecule. The deformations induced in this framework by the movement shown in **17.20** have been suggested to be important for the triggering of the important cooperative peptide reorganization process upon oxygen binding of one heme unit. Such movement exposes the other heme groups so that attack by further O_2 molecules is facilitated.

17.20

Just as the electronic configuration is very important in determining the geometry along the deformation coordinate **17.17**, so too is it important in influencing the relative stabilities of the square pyramid and trigonal bipyramid along the related coordinate **17.16**. A minor complication arises in that the obvious axis choice in the two molecules is different (**17.14, 17.15**) so that the z^2 orbital of the trigonal bipyramid becomes the $x^2 - y^2$ orbital of the square pyramid. The molecule, of course, does not know about x, y, z axes; these labels are there to identify orbitals. Figure 17.3 shows the orbital correlation diagram which relates the orbitals for the two geometries. On the far right the orbitals of a square pyramid are listed for a geometry with $\theta > 90°$. The basic motion that is followed in Figure 17.3 takes the square pyramid (**17.21**) to a trigonal bipyramid (**17.22**), by decreasing one *trans*

17.21   **17.22**

L—M—L angle in the xz plane and increasing the other in the yz plane. The xy level for the square pyramid is unchanged along this pathway. It becomes one member of the e'' set at the trigonal bipyramidal geometry. The other member of e'' is derived from yz. As the one *trans* L—M—L angle is increased, the ligands move into the node of yz, causing this orbital to be stabilized. (This also results in the loss of hybridization with metal y.) The xz orbital of the square pyramid is destabilized. As the *trans* L—M—L angle in the xz plane is decreased the lone pair on the ligand increases its antibonding interaction with xz. This is reduced somewhat by increased mixing of metal x character. Ultimately at the trigonal bipyramidal geometry this orbital lies at moderate energy and is substantially hybridized out away from the ligands in what now is the xy plane. What happens to the two highest

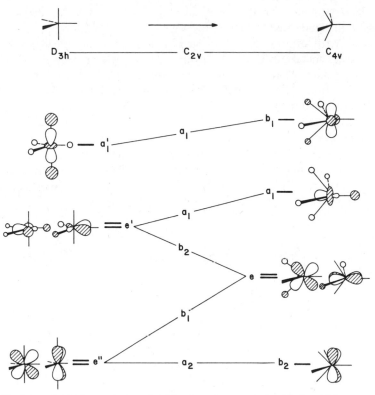

FIGURE 17.3. Orbital correlation diagram for the metal d orbitals which connect square pyramidal and trigonal bipyramidal geometries.

levels of the square pyramidal and trigonal bipyramidal geometries of Figure 17.3 is more difficult to describe. The z^2 and $x^2 - y^2$ "character" of these two levels switch. One way we can trace this intermixing is by noting that the symmetry of the molecule is C_{2v} at a geometry intermediate between the two extremes. The three orbitals which we have just examined are of $a_1 + b_1 + b_2$ symmetry. The two higher orbitals are both of a_1 symmetry. They can, and will, intermix along the reaction path. Starting from the square pyramidal side the z^2 orbital will mix some $x^2 - y^2$ character into it until at the trigonal bipyramidal structure it is predominantly $x^2 - y^2$. (Remember that we have changed the coordinate system. It would become an $x^2 - z^2$ orbital if we had stayed with the axis system in **17.21**.) Two ligands in the yz plane move into the nodal plane of this $x^2 - y^2$ function. Furthermore, metal y mixes into the orbital in a bonding way to the three ligands in the xy plane. Therefore, this level is stabilized and it becomes the other member of the e' set at the trigonal bipyramidal geometry. The $x^2 - y^2$ orbital at the square pyramidal side mixes some z^2 character into itself. At the trigonal bipyramidal geometry it is primarily z^2, antibonding to the surrounding ligands. There is some metal s character in this orbital which reduces the antibonding interactions with the ligands in the zy plane.

The level structure for the valence levels of the trigonal bipyramid is worth studying with some care. At low energy there is an e'' orbital, a pure metal d combination which is orthogonal to the ligand lone pairs. At intermediate energy are two hybridized metal functions of e' symmetry. At higher energy a_1' is fully metal–ligand antibonding. What has been left off this diagram are the five metal–ligand bonding orbitals (17.23). Except for $2a_1'$ (and the z^2 orbital in Figure 17.3), these are exactly analogous to the orbitals of the AH_5 main group compound (Chapter 14). We have introduced a strong mixing with metal d orbitals so that e' in 17.23 is a mixture of $x^2 - y^2$ and xy character as well as x and y at the metal. The e' set displayed in Figure 17.3 are the nonbonding components of this three orbital pattern. Likewise the nonbonding $2a_1'$ orbital of AH_5 will now find a perfect symmetry match with metal z^2. The $2a_1'$ level in 17.23 is the bonding component and a_1' shown at the upper left of Figure 17.3 is the antibonding partner.

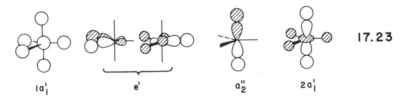

$1a_1'$ e' a_2'' $2a_1'$ 17.23

From Figure 17.3 we can comment on the preferred geometries of ML_5 compounds as a function of d electron configuration. Recall that there is a slight favoring of the D_{3h} trigonal bipyramidal geometry for the d^0 configuration from our discussion of main group stereochemistry. Figure 17.3 indicates that for d^3, d^4 the trigonal bipyramid should be favored even more since the yz component of e'' rises in energy on distortion away from this structure. For d^5, d^6 a square pyramid (with $\theta \sim 90°$ from Figure 17.2) is expected. For d^7 we need to weigh a two-electron stabilization along the $D_{3h} \rightarrow C_{4v}$ coordinate against a one-electron destabilization. The D_{3h} geometry, however, is Jahn–Teller unstable. In low temperature matrices where low spin d^5, d^6, and d^7 pentacarbonyls have been made,[7] these compounds have square pyramidal geometries. The d^6 case is particularly interesting since the level pattern for the D_{3h} and C_{4v} structures suggests the singlet and triplet states might have different geometries. The situation therefore is very similar to that for cyclobutadiene in Chapter 12 and just like the tetrahedral/square planar problem discussed for four-coordinate d^8 molecules in Chapter 16. The d^6 singlet state is unstable at the D_{3h} geometry since the e' orbital would be half-full but is stabilized on distortion to a C_{4v} or C_{2v} geometry. A triplet trigonal bipyramid appears from Figure 17.3 to be stable at this geometry. A singlet square pyramidal but triplet trigonal bipyramidal structure is expected. A closely related geometrical problem is that of the photochemical rearrangement processes in the low spin square pyramidal d^6 ML_5 molecule. Experimentally, irradiation of $W(CO)_4 CS$ leads[8] to the exchange of apical and basal CS groups via a mechanism that does not involve photodissociation (17.24). We can make use of Figure 17.3 to understand this. Promotion of an electron from the e to a_1 orbital of the square pyramidal structure leads to a geometrical instability and the molecule distorts to the trigonal bipyra-

$$17.24$$

midal geometry (17.25) which is a way point on the overall route of returning the high energy, excited electron to the e pair of orbitals from where it originated. (Notice how the labeled ligand has changed places). This can be regarded as a photochemical Berry process.[9]

17.25

Thermal interconversion of these d^6-ML_5 systems will be quite complicated. It is clear that going from the far right in 17.25 to the D_{3h} isomer will be energetically difficult. Not only do the two electrons in the b_2 level rise to high energy, but the D_{3h} geometry is Jahn–Teller unstable. A more favorable way for the thermal interconversion is illustrated in 17.26–17.28. One of the four CO ligands is bent toward

the missing sixth coordination site. The transition state, 17.27, does *not* have D_{3h} symmetry. Three SC—W—CO angles[9] are ~90° and the other is ~135°. The bending motion is continued from 17.27 to 17.28 so that overall it appears that the CS group has migrated from a position *trans* to the sixth coordination site to a *cis* position. The virtue of this pathway is that the $x^2 - y^2/xy$ set is *not* degenerate in 17.27. In fact, 17.27 maximally has C_{2v} symmetry. The center of Figure 17.3 shows the level splitting pattern for this species. The path from **17.26** to **17.27** maintains only C_s symmetry and there is a good bit of intermixing within the orbital set which we will not cover here. The potential surface for the thermal rearrangement of this molecule is similar in form to that for H_3^- and $C_5H_5^+$ described earlier and shown in Figure 7.5. The point E is the trigonal bipyramidal energy maximum, the points A-C correspond to the square pyramidal structures and the saddle points D to the geometry of **17.27**.

In addition to details concerning the angular geometry choices in ML_5, the orbital diagrams also allow us to understand the relative metal–ligand distances and site preferences in these molecules.[5] Within the same molecule there are two symmetry inequivalent linkages (and therefore sites); axial and equatorial in the trigonal bipyramid and apical and basal in the square pyramid (**17.29**). For the trigonal bi-

pyramid, recall the weaker axial than equatorial bonding for the main group (d^0) examples (Section 14.2). This will also be found for d^0–d^4 transition metal systems since the e'' pair of d orbitals (Figure 17.3) are not involved in M—L σ interactions. For the low spin d^8 complex however, four electrons reside in d orbitals (e') which are net metal–equatorial ligand antibonding. The result is now the prediction of a stronger axial than equatorial M—L linkage—a reversal of the main group situation. With 10 d electrons we expect a return to the d^0 situation, although as we will see later this is not in fact found in practice. For the square pyramid the result is a little more complicated since the molecule has an angular degree of freedom, and the e pair of orbitals, σ nonbonding at $\theta = 90°$, become σ antibonding for $\theta > 90°$. The result is a weaker basal than apical M—L linkage for d^0 through d^6 when these e orbitals are occupied. This effect reinforces the d^0 result which was described in Chapter 14. The a_1 orbital is strongly antibonding between the metal and apical ligand. For the low spin d^8 configuration it is the apical linkage which should be weaker. These results are summarized in **17.30** and **17.31**. Table 17.2 shows some bond lengths for a selection of square pyramidal molecules which are understandable along these lines. For the high and low spin d^8 complexes note the tremendous difference between the $(z^2)^2$ and $(z^2)^1(x^2 - y^2)^1$ configurations. In order to be able to understand the trigonal bipyramid problem we need to look at π bonding.

TABLE 17.2.   Some Representative Bond Lengths (Å) in Square Pyramidal Molecules

Electronic Configuration	Molecule	Apical MY Distance	Basal MY Distance	Bond
d^9	$Cu(pyNO)_2(NO_3)_2$	2.44	1.96	Cu—O, Cu—N
	$Cu(dmg)_2$	2.30	1.95	Cu—O, Cu—N
hs d^8	$Ni(5\text{-}Clsalen NEt_2)_2$	1.98	2.00	Ni—N
	$[Ni(dmp)Cl_2]\ 2CHCl_3$	2.06	2.07	Ni—N
	$Ni(bddae)(NCS)_2$	1.97	1.95	Ni—N
	$[Ni(tpen)]\ (ClO_4)_2 \cdot MeNO_2$	2.10	2.10	Ni—N
ls d^8	$Ni(bda)Br_2$	2.70	2.33	Ni—Br
	$Ni(CN)_5^{3-}$	2.17	1.85	Ni—C
	$Ni(DSP)I_2$	2.79	2.19	Ni—S
ls d^7	$Co(CN)_5^{3-}$	2.01	1.90	Co—C
	$Co(CNC_6H_5)_5^{3-}$	1.95	1.84	Co—C
ls d^6	$Ru(CO)(PPh_3)_2((CF_3)_2C_2S_2)$	2.27	2.35	Ru—P
hs d^4	$MnCl_5^{2-a}$	2.58	2.30	Mn—Cl
	$MnCl_5^{2-b}$	2.46	2.27	

[a] Bipyridinium counterion.
[b] Phenanthrolium counterion.

For the trigonal bipyramid there are four symmetry-allowed interactions shown in **17.32–17.35**. Three involve interaction with the e'' orbitals and one interaction with the e' orbitals. **17.34** and **17.35** are equivalent by symmetry. Since the e' orbitals are hybridized away from the ligands as described above, the π-type overlap of a ligand orbital with e' in **17.32** is significantly larger than any of the other interac-

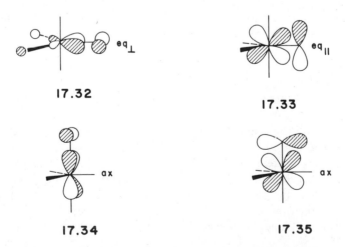

17.32   17.33

17.34   17.35

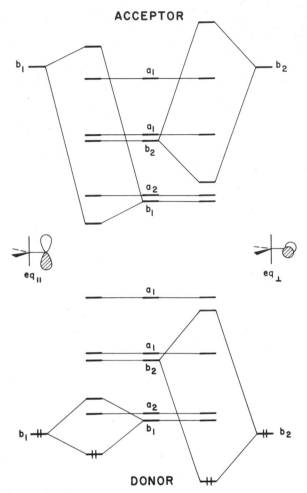

FIGURE 17.4. Interaction diagrams for a single-faced π acceptor or donor in eq$_\parallel$ and eq$_\perp$ orientations. A larger interaction has been shown for the eq$_\perp$ case commensurate with the better overlap via d–p mixing shown in **17.32**.

tions, that is eq$_\perp$ > eq$_\parallel$ ≃ ax. However, it is important to realize that just because the eq$_\perp$ interaction is larger than ax, π-bearing ligands will not always prefer the eq$_\perp$ site. The site preferences depend on the number of electrons and on whether the ligand is a π acceptor or donor. Figure 17.4 shows interaction diagrams for a weak equatorial acceptor or donor which we have constructed by ensuring a larger interaction energy for eq$_\perp$ compared to eq$_\parallel$. Clearly for a d^4 system with an equatorial π donor eq$_\perp$ is preferred over eq$_\parallel$ since this results in maximum stabilization of the occupied b_2 ligand donor orbital compared to b_1 in the eq$_\parallel$ case. For a d^8 system, however, at the eq$_\perp$ geometry the occupied higher energy b_2 orbital is destabilized more than the corresponding orbital in the eq$_\parallel$ case. Here the favored arrangement is the eq$_\parallel$ one. For a π acceptor ligand Figure 17.4 shows that a d^6 or

d^8 system will prefer the eq$_\perp$, and d^2 or d^4 systems, the eq$_\parallel$ arrangement. An example of the d^8 case is provided by the molecule Fe(CO)$_4$C$_2$H$_4$ (17.36) where the ethylene ligand is a π acceptor via its π^* orbital. The reverse orientation is found for the d^9 molecule 17.37 where the π donor orbital on the equatorial imidazole ring is oriented eq$_\parallel$. Presumably this is set by steric interactions between the imidazole ligands.

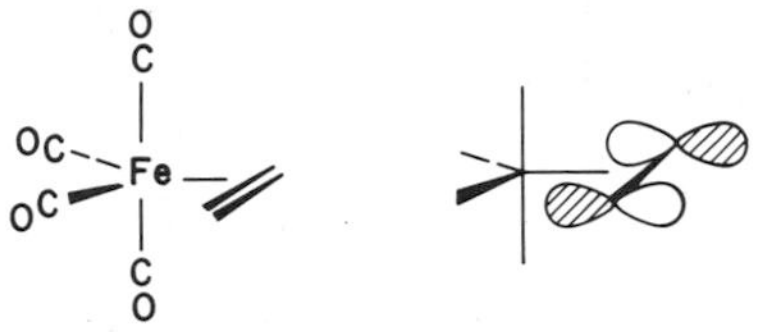

17.36

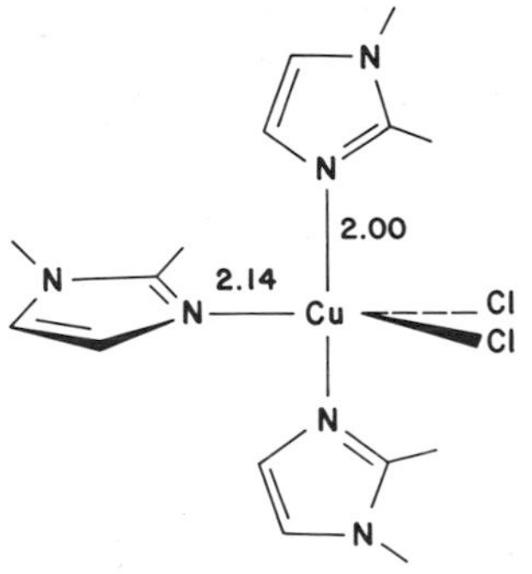

17.37

The structure of a highly active olefin metathesis catalyst has been established[10] by NMR spectroscopy to be that shown in 17.38. These are formally W(IV)-d^2 systems, if the CHR$'$ carbene group is treated as a neutral two-electron donor with an empty p orbital—a superlative π acceptor; see 15.20. Clearly the eq$_\parallel$ conformation on the upper left of Figure 17.4 is the only way to stabilize these highly electron-deficient compounds. The structure shown in 17.38 is a nice confirmation of the theory. By way of contrast the species Fe(CO)$_4$CR$_2$ with a d^8 configuration has the eq$_\perp$ orientation, perhaps the orientation least favored on steric grounds. Another example is provided[11] by 17.39. One could view the CAl$_2$Me$_4$Cl unit as an anionic carbene species to help out in the electron counting. The molecule is then d^4. Notice that the olefin, as well as the carbene ligands, are oriented in the eq$_\parallel$ direction. Both members of the e'' set (a_2 and b_1 in Figure 17.4) are stabilized.

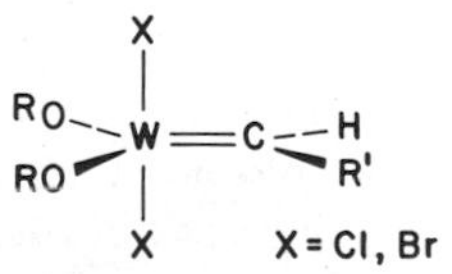

17.38

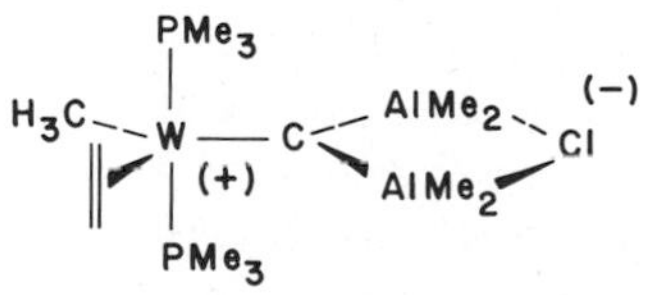

17.39

For all those cases where eq$_\parallel$ is suggested to be favored ax substitution will also be close in energy from the size of the interactions 17.32–17.35. For those systems

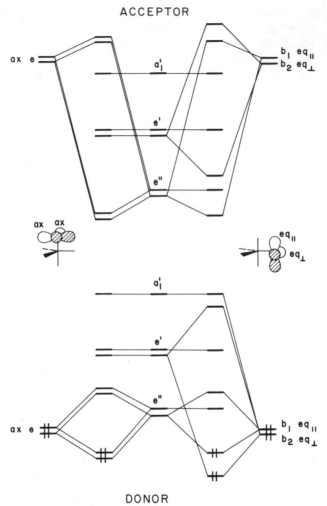

FIGURE 17.5.   Axial and equatorial interaction diagrams for the case where the π donor or acceptor carries a pair of π orbitals.

where the ligand carries a pair of π orbitals (e.g., a donor such as Cl$^-$ or an acceptor such as CO) then Figure 17.5 predicts an equatorial site for a d^8–d^{10} molecule containing an acceptor, and an axial site for a donor. In a molecule with one of these electronic configurations but where all the ligands are the same, this argument then says that on π-bonding grounds, axial bonds should be stronger than equatorial for π donors and the converse for π acceptors. If we combine both of these results and those associated with the σ framework, described earlier, the general conclusion for a d^8 system (the most common electronic configuration for these molecules) is that σ and π effects reinforce each other for π acceptor ligands (eq stronger than ax) but for π donors the two effects, σ (eq stronger than ax) and π (ax stronger than eq), work in opposite directions.

Although it is hazardous to try and weigh competing σ and π effects several of these theoretical trends are borne out experimentally. Strong π acceptor ligands tend to occupy the equatorial sites of the trigonal bipyramid as shown in the selection of molecules **17.40**. In **17.40b** the methyl group, a strong σ donor, occupies

17.40

the axial site, but in **17.40c** a phosphine ligand occupies this position in a complex where there is a weaker σ donor present (Cl_3Sn). Table 17.3 shows some bond lengths in trigonal bipyramidal molecules. The axial linkages are either shorter than the

TABLE 17.3.   M-Y Bond Lengths in Some Trigonal Bipyramidal MY_5 Structures

Molecule	Configuration	M–Y (Å)	
		Axial	Equatorial
$Fe(N_3)_5^{2-}$	d^5 (hs)	2.04	2.00
$Co(C_6H_7NO)_5^{2+}$	d^7 (hs)	2.10	1.98
$Ni(CN)_5^{3-\ a}$	d^8	1.84	1.99, 1.91
$NiP_5^{2+\ b}$	d^8	2.14	2.19
$Fe(CO)_5$	d^8	1.81	1.83
$Co(CNCH_3)_5^+$	d^8	1.84	1.88
$Pt(SnCl_3)_5^{3-}$	d^8	2.54	2.54
$Pt(GeCl_2)_5^{3-\ a}$	d^8	2.40	2.43
$Mn(CO)_5^-$	d^8	1.82	1.80
$CuCl_5^{3-}$	d^9	2.30	2.39
$CuBr_5^{3-}$	d^9	2.45	2.52
$CdCl_5^{3-}$	d^{10}	2.53	2.56
$HgCl_5^{-3}$	d^{10}	2.52	2.64
AsF_5	d^{10}	1.71	1.66

[a] C_{2v} structure intermediate between D_{3h} and C_{4v}.
[b] The ligand is 2,8,9-trioxa-l-phosphaadamantane.

equatorial ones or about the same length for d^8 and d^9 complexes containing π donor ligands. For systems containing π acceptors there seems to be no universal trend, perhaps in agreement with our comments above concerning competing σ and π effects. Notably for d^{10} systems the longer equatorial than axial distances is certainly not in keeping with the idea that the d^0 preferences (cf. AsF$_5$ in Table 17.3) should hold here. It has been suggested that strong s-d mixing will lead to enhanced bond strength to the two axial ligands in these d^{10} complexes (see $2a_1'$ in **17.23**). This will give a linear Cl—Hg—Cl unit with short Hg—Cl distances. Such a structural unit, often associated with other weakly coordinated ligands, is a prominent feature of Hg(II) chemistry and of the coinage metals in their +1 oxidation state. HgCl$_2$ itself contains isolated triatomic molecules of this type; the mineral cinnabar (HgS) contains spirals made up of linear —S—Hg—S— units. MHgCl$_3$ and M$_2$HgCl$_4$ complexes (M = alkali metal) contain linear HgCl$_2$ moieties with other much longer Hg—Cl contacts. Clearly here our simple orbital ideas are at present unable to convincingly rationalize these results.

17.4.   MOLECULES BUILT UP FROM ML$_5$ FRAGMENTS

In this section these valence C_{4v} ML$_5$ fragment orbitals are used to build up the orbitals of more complex units. First, we will look at the level structure[12,13] of a simple dimer M$_2$L$_{10}$ (**17.41**). The ML$_5$ d orbitals neatly partition into $\sigma(z^2)$ $\pi(xz,$

17.41

yz) and $\delta(x^2 - y^2, xy)$ types in this geometry. The details of the resulting orbital diagrams however, depend crucially on the identity of the ligands L. Let us look at the two cases, L = Cl and L = CO, typical simple π donor and acceptor ligands, respectively. Recalling that π donors destabilize and acceptors stabilize the "t_{2g}" orbitals (Chapter 15) and that although xy may interact with four ligand π orbitals, xz and yz may only interact with three, we end up with a two above one level arrangement for M(CO)$_5$ and a one above two arrangement for MCl$_5$. These are shown at the middle of Figure 17.6. The $x^2 - y^2$ level is at very high energy being destabilized by the four basal ligands and is not shown in this figure. Since xy, xz, and yz are destabilized by the lone pairs on Cl, these levels are energetically closer to the z^2 hybrid orbital for ReCl$_5^{2-}$ than in Re(CO)$_5$. These factors are important in understanding the differences in the orbital pictures which result when two MCl$_5$ or two M(CO)$_5$ units are brought together. The metal–metal distance in Re$_2$(CO)$_{10}$ of 3.04 Å is much longer than the corresponding distance (2.22 Å) in Re$_2$Cl$_8$X$_2^{2-}$ (X = H$_2$O). As a result, all of the metal–metal interactions are larger in the halide. Because of this fact and the other points we have just noted, d^7 Re$_2$(CO)$_{10}$ has a single σ bond between the two metal atoms but Re$_2$Cl$_8$X$_2^{2-}$ has a quadruple bond made up of one δ, one σ, and two π components (**17.42**) as shown in Figure 17.6.

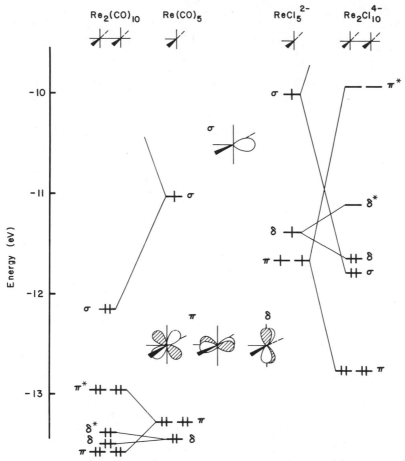

FIGURE 17.6. Interaction diagrams for two M_2L_{10} systems. Notice how the π levels in $Re(CO)_5$ lie to lower energy than in $ReCl_5^{2-}$, a direct result of the π acceptor and donor nature of the ligands respectively. Combined with a shorter metal–metal distance in the halide the final level diagrams are quite different.

17.42

How can we increase the bond order between the two ML$_5$ fragments for the case of L = acceptor? By shortening the MM distance the relevant orbitals change in energy in the obvious way shown in **17.43**. For the case of 10 electrons (a d^5 metal) a formal triple bond is predicted ($\pi^4\delta^2\delta^{*2}\sigma^2$). Indeed $Cp_2M_2(CO)_4$ species (M = Cr, Mo, W), isoelectronic with the unknown $V_2(CO)_{10}$ molecule, with this electron configuration have very short metal–metal distances. As we will see later Cp is equivalent to three coordinated ligands.

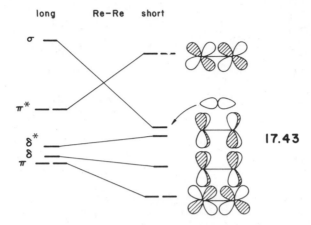

long   Re–Re   short

σ

π^*

δ^*
δ
π

17.43

The energy levels of an $H(ML_5)_2$ complex (**17.44**) may be derived in a similar way by adding the hydrogen $1s$ orbital to the orbital picture produced by two ML_5 units set at the metal–metal distance expected in a molecule of this type (Figure 17.7). Note the small interaction between orbitals of δ type at this distance. We would expect a stable molecule to result for the electron configuration shown at

$$
\underset{OC}{\overset{OC}{}}\!\!Cr\!-\!H\!-\!Cr\!\!\underset{}{\overset{}{}}\quad (-)
$$

17.44

the right-hand side of Figure 17.7 and an example is $HCr_2(CO)_{10}^-$ (**17.44**). The structures of the $HM_2(CO)_{10}^-$ ions are actually a little more complex than that shown in **17.44**. In each case the hydrogen atom lies off the metal–metal axis (**17.45**). The simplest way to view this distortion is to gradually move the metal atoms closer together (and thereby increase their interaction) and at the same time move the hydrogen atom off the M—M axis. The result is shown in **17.46** for the pertinent

$$(CO)_5 M\!-\!H\!-\!M(CO)_5^- \qquad (CO)_5 M\overset{H}{\diagup}\!\!\diagdown M(CO)_5^-$$

σ^*

n

17.45

σ

17.46

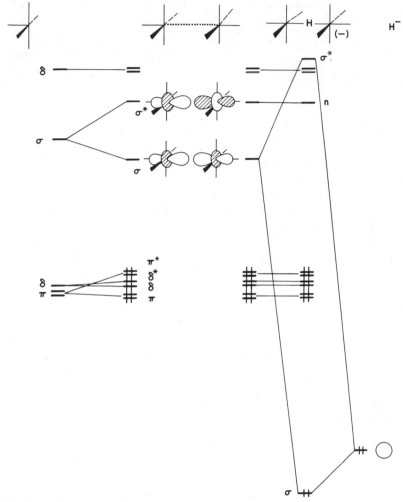

FIGURE 17.7. Generation of the level diagram for an HM_2L_{10} species by allowing the $1s$ hydrogen orbital to interact with the orbitals of the M_2L_{10} unit.

orbitals. The unoccupied out-of-phase z^2 orbital combination (metal–hydrogen nonbonding) goes to higher energy as the metal atoms increase their overlap, and the corresponding bonding combination experiences a stabilization. At the same time however, the hydrogen $1s$ orbital moves toward a node in the ML_5 z^2 hybrid orbital and overlap is reduced. These two factors operate energetically in opposite directions. This means that the bending motion is rather soft and a variety of geometries are observed. If the distortion **17.45** proceeds further, the orbital pattern and bonding picture becomes very similar to that of triangular H_3^+ (Section 5.2).

Another problem which may be tackled in the same way as the bridging hydride case of **17.43** is that of a bridging halide which contains s and p orbitals. Figure 17.8 shows a diagram, analogous to Figure 17.7 for this particular case. Now both symmetric and antisymmetric z^2 hybrid combinations of the two ML_5 units find suitable partners on the bridging halide. The diagram has been constructed to emphasize the larger σ than π type interactions in this unit. Stable M_2L_{11} species of this type are found for many transition metal halides (e.g., Nb_2F_{11}) and the tetrameric unit M_4L_{20} (**17.47**) is a common structure for many transition metal fluorides MF_5.

17.47

The scheme shown in Figure 17.8 gives rise to a collection of six closely spaced orbitals derived from weak π overlap of the "t_{2g}" orbital sets of the two square pyramids with the bridging ligand orbitals. Two d^4 metals, with a total of eight electrons occupying this collection of six orbitals, are then expected to lead to a paramagnetic situation. If the π interaction between the ML_5 units and the bridging

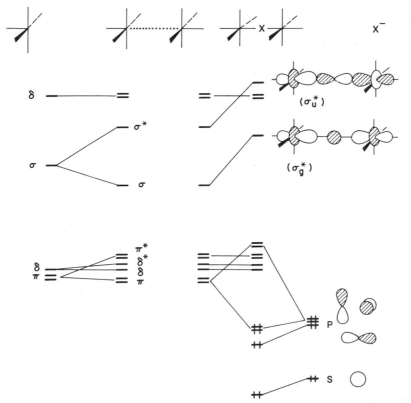

FIGURE 17.8. Generation of the level diagram for an XM_2L_{10} species by allowing the valence s and p orbitals of X to interact with the orbitals of the M_2L_{10} unit.

ligand is large then situation changes. The result is a much stronger destabilization of the $(ML_5)_2$ level labeled π than shown in the middle of Figure 17.8. With a total of eight d electrons a sizable HOMO—LUMO gap opens up and a diamagnetic species is formed. This is the case[14] for the molecule $Cp(CO)_2Cr—S—Cr(CO)_2Cp$, isoelectronic with $(CO)_5Cr—S—Cr(CO)_5^{2+}$. The good π contribution to the Cr—S linkages suggests the description $Cp(CO)_2Cr=S=Cr(CO)_2Cp$ for this molecule.

Sometimes in these units the M—X—M bridge is linear, otherwise it is bent. We will be particularly interested in a different type of distortion, that shown in **17.48**,

$$\text{17.48}$$

the distortion of the symmetric structure to an asymmetric one. A clue to understanding this particular motion lies in the energetic behavior of the σ_g^* and σ_u^* orbitals of Figure 17.8. It is difficult to predict *a priori* whether σ_g^* or σ_u^* lies higher in energy, but we will see that for our purposes it is not important. As the distortion **17.48** proceeds the center of symmetry is lost and σ_g^* and σ_u^* orbitals mix together. The top orbital goes up in energy and the bottom orbital drops in energy as a result (Figure 17.9). The change in the description of the σ_g^* and σ_u^* orbitals on distortion is an interesting one.[15] The higher energy orbital at the symmetrical structure ends up as a σ antibonding orbital (one of the e_g pair) on the now approximately octahedral unit, and the lower energy orbital becomes a pure z^2 hybrid orbital located on the ML$_5$ square pyramidal fragment. Figure 17.9 shows

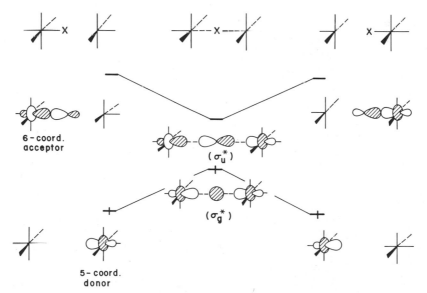

FIGURE 17.9. Orbital correlation diagram for the σ_u^* and σ_g^* orbitals of the symmetric bridge structure, as the X atom is moved off center. Energetically the picture is symmetrical about the center but the orbital descriptions at the left- and right-hand sides of the diagram are different.

this pictorially for both the left and right distortions of the bridging atom or alternatively as the bridging atom is moved from one side of the bridge to the other.

Let us work with the example on the left side of Figure 17.9. As the bridging halide moves toward the metal atom σ_u^* mixes (17.49) into σ_g^* in a way to reduce the antibonding interaction between the metal atom on the left and the bridging atom (z^2 in σ_u^* is bonding to the halide s orbital in σ_g^* and the halide z in σ_u^* is bonding to metal z^2 in σ_g^*). The resultant orbital cancels amplitude on the left ML_5 unit and reinforces it on the right ML_5 unit. Now σ_g^* must mix into σ_u^* with the opposite phase relationships. The result, shown in **17.50**, has reinforced amplitude at the

17.49

17.50

left ML_5 fragment and canceled amplitude at the right ML_5. If σ_u^* lies below σ_g^* in Figure 17.9, exactly the same results are obtained. We shall continue with the ordering of σ_g^* below σ_u^*. With one or two electrons in the σ_g^* orbital this simple result indicates that such species will be unstable at the symmetrical structure and should distort to the asymmetric arrangement. This is a typical example of a second-order Jahn–Teller distortion. For the case of two electrons the electronic ground state is $^1\Sigma_g^+$ and the lowest excited singlet state is of symmetry $^1\Sigma_u^+$. The distortion mode which will lower the energy of the system via a second-order Jahn–Teller mechanism is of symmetry $\sigma_g \times \sigma_u = \sigma_u$ that is, the asymmetric motion of the the central atom. One example of a static distortion of this type is the CrF_5 species which exists as an extended solid-state system. The local geometry about a pair of chromium atoms, one Cr(II), the other Cr(III), is shown in **17.51** and shows that the bridge is asymmetric. Since Cr^{3+} and Cr^{2+} ions are high spin d^3 and d^4, respectively, at the symmetrical geometry one electron would occupy the σ_g^* orbital. With two electrons in the σ_g^* orbital the classic series of Pt(II)/Pt(IV) mixed valence compounds are found (17.52). These are also extended solid-state arrays and are

17.51

17.52

viewed as such below but their local geometry clearly shows the bridge asymmetry. Both of these examples are mixed valence compounds because, as we can see from Figure 17.9 at the asymmetric structure the σ_g^* electron(s) are located on the five-coordinate unit in CrF$_5$ (and the analogous orbital for the square planar Pt case,.

There are strong links between these mixed valence species and an important class of reactions—namely those arising via electron transfer.[15,16] The inner sphere redox behavior of the Cr(II)/Cr(III) system has been studied in great detail. By using labeled chloride (Cl*) it was cleverly shown that the redox process is associated with atom transfer (17.53) and that this occurs in the opposite direction to electron transfer, perhaps via the inner sphere complex (17.54). In 17.53 we use the terms

$$17.53$$

Cr (III)   Cr (II)   Cr (II)   Cr (III)

inert   labile   labile   inert

$$17.54$$

labile and inert to describe the kinetic stability of these complexes. Ligand substitution at Cr(III) is very slow and so the identity of the CrCl$_5$Cl^{*3-} ion is preserved in solution. By way of contrast ligand substitution at Cr(II) is fast and so the ion is best described as an aquo complex constantly exchanging water molecules with the solvent. After electron transfer the coordination sphere around the old Cr(II) ion (new Cr(III) ion) is frozen since it is now the inert species in solution. The coordination sphere around the old Cr(III) ion (new Cr(II) ion) will rapidly be replaced by water. We can use the scheme of Figure 17.9 to see how this takes place in detail. At the left-hand side of the diagram the electron is totally associated with the square pyramidal five-coordinate reductant Cr(II). As the X atom from the Cr(III) unit moves to the center of the bridge (a transition state from our discussion above) the orbital containing this electron has equal weight from both metal atoms. Technically "half an electron" has been transferred at this point. As the bridging atom moves past the symmetric structure to the right-hand side of the bridge, then the electron transfer is now complete and Cr(II) and Cr(III) species are again produced. Thus the electron transfer has proceeded in a smooth way initiated by the atom transfer. We stress that not all redox processes are this simple (many proceed by the outer sphere route where no species such as 17.54 occurs) but within the context of this electronic model one can think about ways that the other ligands around the metal and the transferred halogen can perturb the rate of the reaction.

As we mentioned above the Pt(II)/Pt(IV) mixed valence compounds are in fact found as infinite chains. So, instead of the two orbitals σ_g^* and σ_u^* of Figure 17.8 we have an energy band[17] shown in **17.55**. The very bottom of the band corresponds to the σ_g^* type of orbital with a phase factor of +1 between adjacent cells, and the top of the band is the corresponding σ_u^* combination. Just as the pair of orbitals of Figure 17.9 increased their separation as the bridge is made asymmetric, so the band **17.55** of the infinite system splits into two on such a distortion **(17.56)**. The result is a Peierls-type distortion of the one-dimensional chain with a half-filled band. Note that the stabilization results in a square planar environment for low spin d^8 Pt(II) and an octahedral environment for the low spin d^6 Pt(IV) species, two typical geometries for these oxidation states.

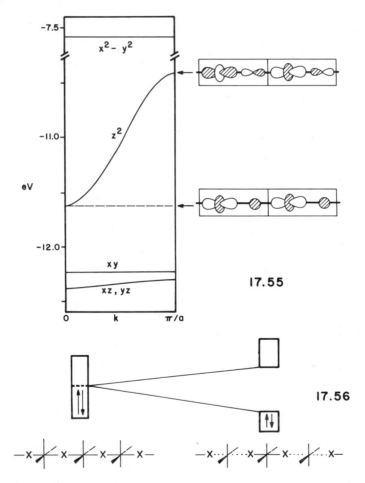

17.55

17.56

17.5.   PENTACOORDINATE NITROSYLS

Coordinated NO is found in two basic geometries in transition metal complexes, linear and bent, exemplified by the molecules **17.57** and **17.58** as well as structures intermediate between the two. We shall concentrate on five-coordinate examples

17.57

17.58

which may have a square pyramidal, trigonal bipyramidal, or an intermediate co-ordination environment. We will be interested in understanding in broad terms when the MNO unit is linear and when it is bent.[18,19] We begin with a square pyramidal ML_4NO complex containing an apical nitrosyl group. Figure 17.10 shows the assembly of such a diagram in the obvious way, using the important frontier orbitals (n, π^*) of the NO group and those of square planar ML_4. On the far left of Figure 17.10 are the orbitals of a square planar ML_4. Making the four ligands pyramidal leaves the xy orbital unchanged in energy. It stays totally nonbonding. The $x^2 - y^2$ orbital is stabilized somewhat since some overlap to the ligands is lost. This also occurs in z^2 except that metal s and z hybridize with z^2 so that the

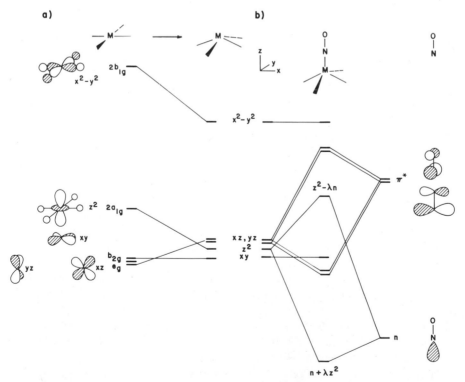

FIGURE 17.10. (a) Energetic behavior of the square planar ML_4 d levels on pyramidalization. (b) Interaction of these levels with those of NO to give a square pyramidal ML_4NO species.

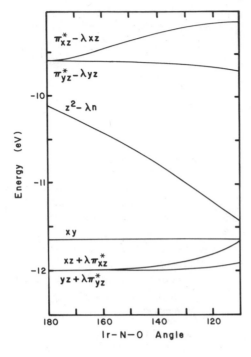

FIGURE 17.11. Energetic behavior of the metal d and nitrosyl π^* levels on bending the M–N–O unit. Adapted from Ref. 18 the calculation refers to an iridium nitrosyl species.

orbital points out away from the ligands. The mechanism for this change is identical to that for pyramidalization in AH_3 (Chapter 6). Finally xz and yz are destabilized and somewhat hybridized. A close comparison of the ML_4 orbitals and those of the C_{4v} ML_5 unit on the right side of Figure 17.3 shows that there is only one difference. Removal of the apical ligand in ML_5 stabilizes the z^2 orbital greatly and rehybridization occurs so that it is pointed toward the missing apical ligand. It finds a strong interaction with the lone-pair orbital of NO which has been labeled n on the right side of Figure 17.10. Likewise xz and yz interact with the π^* levels of NO. When filling this manifold with electrons we need to keep track, not only of the number of d electrons but in addition those which lie in the nitrosyl π^* levels. The sum of the two (m) is given by a new notation $\{MNO\}^m$. Figure 17.11 shows how the energies of these levels change as the MNO angle decreases from $180°$. z^2 is stabilized quite dramatically. As shown in **17.59** there are two effects behind this. The antibonding interaction with the nitrosyl lone pair (n) is reduced on bending since now M, N, and O are not collinear. Concurrently a bonding interaction between z^2 and the nitrosyl π^* orbital is turned on. The interaction of one component of the nitrosyl π^* orbitals (π^*_{xz}) with xz decreases on bending **17.60**, and xz becomes less M—L π bonding and rises in energy. The overall increase in the energy of π^*_{xz} shows the dominating influence of the σ destabilization of the interaction **17.59** compared to the π interaction of **17.60**. In a simple way, then, Figure 17.11 indicates two opposing factors influencing bending. Occupation of xz favors linearity but occupation of z^2 favors bending. For $\{MNO\}^6$ systems where xy is the HOMO the approach definitely predicts a linear geometry. There is a single example, that of the dithio-

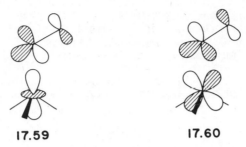

17.59   17.60

carbamate (dtc) complex $Fe(NO)(S_2C_2(CN)_2)_2^-$ which confirms this. For the $\{MNO\}^8$ configuration z^2 is filled and the result of inspection of Figure 17.11 suggests that bending should occur. In confirmation of this there is the Co(NO) complex of tetraphenylporphorin with an MNO angle of 135°. In many $\{MNO\}^7$ compounds an MNO angle of between 150 and 172° is found. Examples include $Fe(NO)(dtc)_2^{2-}$ species with different counterions. Beyond the scope of our discussion here, by changing the nature of the ligands L in the ML_4NO complex the relative slopes of z^2 and xz may be changed and the details of the MNO bending are subtly altered.[18]

One other distortion which may occur as a result of the bending process is the slipping of the nitrosyl off the coordination axis as in 17.61. Eventually, of course, the nitrosyl might attain a sideways geometry. Small distortions of this type are actually found in $IrCl_2(NO)(PPh_3)_2$ and $IrI(CH_3)(NO)(PPh_3)_2$. A detailed account of the slipping motion in 17.61 has been developed for several diatomic ligands.[20]

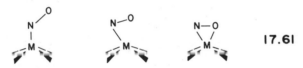

17.61

REFERENCES

1. M. Elian and R. Hoffmann, *Inorg. Chem.*, **14**, 1058 (1975).
2. R. Hoffmann, *Science*, **211**, 995 (1981).
3. T. A. Albright, *Tetrahedron*, **38**, 1339 (1982).
4. D. E. Sherwood and M. B. Hall, *Inorg. Chem.*, **22**, 93 (1983).
5. A. R. Rossi and R. Hoffmann, *Inorg. Chem.*, **14**, 365 (1975).
6. M. F. Perutz, H. Muirhead, J. M. Cox, and L. C. G. Goaman, *Nature*, **219**, 131 (1968).
7. J. K. Burdett, *Coord. Chem. Rev.*, **27**, 1 (1978).
8. M. Poliakoff, *Inorg. Chem.*, **15**, 2022, 2892 (1976).
9. P. J. Hay, *J. Amer. Chem. Soc.*, **100**, 2411 (1978).
10. J. Kress, M. Wesolek, and J. A. Osborn, *J. Chem. Soc., Chem. Comm.*, 514 (1982).
11. M. R. Churchill and H. J. Wasserman, *Inorg. Chem.*, **20**, 4119 (1981).
12. F. A. Cotton and R. A. Walton, *Multiple Bonds Between Metal Atoms*, Wiley, New York (1982).
13. S. Shaik, R. Hoffmann, C. R. Fisel, and R. H. Summerville, *J. Amer. Chem. Soc.*, **102**, 1194 (1980).
14. C. Mealli and L. Sacconi, *Inorg. Chem.*, **21**, 2870 (1982).

15. J. K. Burdett, *Inorg. Chem.*, **17**, 2537 (1978).
16. J. K. Burdett, *J. Amer. Chem. Soc.*, **101**, 5217 (1979).
17. M-H. Whangbo and M. J. Foshee, *Inorg. Chem.*, **20**, 113 (1981).
18. R. Hoffmann, M. M-L Chen, M. Elian, A. R. Rossi, and D. M. P. Mingos, *Inorg. Chem.*, **13**, 2666 (1974).
19. J. H. Enemark and R. D. Feltham, *Coord. Chem. Rev.*, **13**, 339 (1974).
20. R. Hoffmann, M. M-L Chen, and D. L. Thorn, *Inorg. Chem.*, **16**, 503 (1977).

CHAPTER EIGHTEEN

The C_{2v} ML₃ Fragment

18.1. INTRODUCTION

There is a strong electronic resemblance between the C_{4v} ML₅ fragment which was discussed in Section 17.1 and the "T-shaped" ML₃ fragment which is covered here. That relationship will be probed further in the last section of this chapter. In the second section some examples are presented which use the C_{2v} ML₃ fragment.

18.2. THE ORBITALS OF A C_{2v} ML₃ FRAGMENT

The valence orbitals of the T-shaped ML₃ fragment, **18.1**, can be derived by removing one ligand from square planar ML₄, **18.2**. This is shown in Figure 18.1. The five d block and one p orbital of ML₄ (see Section 16.2) are displayed from a top view

```
      L                          L
      |                          |
  L — M — L        ——————→   L — M
      |              - L         |
      L                          L

     18.2                       18.1
```

on the left side of this figure. All of the orbitals are basically unperturbed when one ligand is removed except for $x^2 - y^2$, $2b_{1g}$. This is stabilized greatly because one strongly antibonding interaction with a ligand lone pair is removed. The orbital also becomes hybridized as some metal s and y character are mixed into **18.3** in a way which is bonding to the lone-pair hybrids in **18.3**. This hybridization comes about in a way that is analogous to the a_1 hybrid in the C_{4v} ML₅ fragment (Section 17.2). The resultant orbital, **18.4**, is labeled $2a_1$ in Figure 18.1. The $2a_{1g}$ (z^2) level will

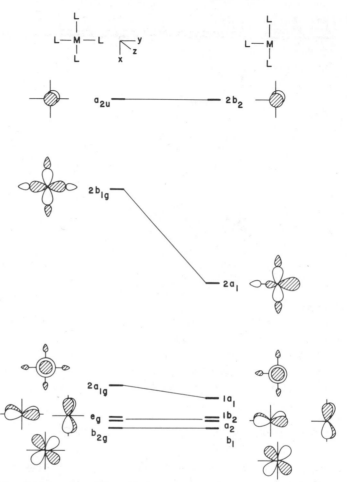

FIGURE 18.1.   Construction of the orbitals of a C_{2v} ML$_3$ fragment from a square planar complex. The L ligands contain only σ donor hybrids.

also be stabilized very slightly by removing one ligand. The reader should note that we have labeled each orbital in the ML$_3$ fragment according to the C_{2v} point group. We want to emphasize, however, that *one* antibonding orbital is shifted to moderate energy and it becomes hybridized out toward the missing ligand. The rest of the levels remain basically unchanged, just as we saw for the square pyramidal ML$_5$ fragment.

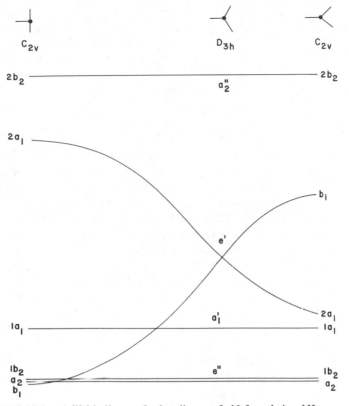

FIGURE 18.2.   A Walsh diagram for bending one L–M–L angle in a ML$_3$ complex.

Next, it is interesting to examine what geometrical options are available to ML$_3$ as a molecule itself. A surface that we explored in the main group area (Section 14.2) is the variation of one angle from a T- through D_{3h} to a Y-shaped structure. This is done for ML$_3$ in Figure 18.2. The orbitals listed on the left correspond to the C_{2v} T-shaped structure derived in Figure 18.1. The a_2, $1b_2$, and $2b_2$ orbitals are orthogonal to the lone-pair functions of the ligands on all points of the distortion coordinate. The $1a_1$ orbital is primarily metal z^2 and s. Varying the L—M—L angle does not change the overlap of the ligand hybrids to them, thus these four levels remain at constant energy. The b_1 orbital, **18.5**, is destabilized as the *trans* L—M—L angle is decreased. Overlap between the ligands and metal xy is turned on and as seen in **18.6** this is an antibonding interaction (see Section 15.4 for a related case). The ligand-based level will be stabilized. The destabilization is somewhat abated because metal x character is also mixed into this level in a way that is bonding to the ligands. Conversely, the $2a_1$ level, **18.7**, is stabilized. As shown in **18.8**, antibonding between the ligand hybrids and $x^2 - y^2$ is diminished while bonding to metal y is turned on. When the three L—M—L angles are 120°, **18.6** and **18.8** meet and become an e' set. At this special point the symmetry of the molecule is D_{3h}. The orbitals have been labeled at the middle of Figure 18.2 to reflect this.

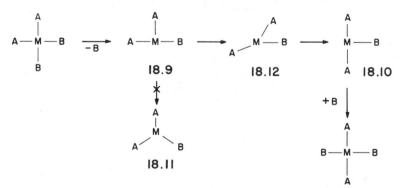

A d^8, 14-electron ML$_3$ complex would have the lowest four levels filled in Figure 18.2. These are $b_1 + a_2 + 1b_2 + 1a_1$ on the T side or $a_2 + 1b_2 + 1a_1 + 2a_1$ at the Y geometry. At a D_{3h} structure there will be a degeneracy. The e' set will be half-filled which signals that either the complex must be high spin (a triplet) or it will undergo a Jahn–Teller distortion to the T or Y geometry. Structural evidence for these very reactive 14-electron complexes is largely lacking; however, the structure of $(Ph_3P)_3Rh^+$ has been determined.[1] It is diamagnetic and approximately T-shaped (one P—Rh—P angle is 159.4°).

There are a number of connections that make these d^8 systems interesting. First of all, in Section 16.4 one pathway for *cis–trans* square planar ML$_4$ interconversion involved a d^8-ML$_3$ species, **18.9**. A rearrangement must take **18.9** to its *trans* isomer **18.10** which then may be intercepted to yield the *trans*-ML$_4$. There are two ways that the rearrangement from **18.9** to **18.10** can occur. A geometrically obvious pathway would be to decrease the *trans* A—M—B angle in **18.9** to a trigonal species, **18.11**, where all angles around the metal are 120°. However, as shown in Figure 18.2 this is energetically prohibitive (provided that there is no change of spin state). Instead the molecule must distort via the Y structure of **18.12** where the

A—M—A angle has opened considerably and the *cis* A—M—B angle somewhat decreases. Relaxation of this Y arrangement yields the *trans*-T intermediate, **18.10**. In Section 19.5 we shall cover a reaction where a d^8 $(CH_3)_3M$ intermediate with a T ground state rearranges to a Y geometry and undergoes reductive elimination to ethane and $(CH_3)M$. There are also interesting connections with problems we have discussed elsewhere. For example, in Section 17.3 the mechanism of the thermal and photochemical rearrangements in $W(CO)_4CS$ was investigated. The reader should carefully compare the electronic details for the rearrangement of this

d^6 ML$_5$ species with d^8 ML$_3$ case here. They are identical! A Mexican-hat surface like that given by Figure 7.5 occurs for each.

With 10 d electrons in the valence levels of Figure 18.2 it is clear that a D_{3h} structure will be preferred. Remember that although b_1 goes up in energy from the T geometry to D_{3h}, there is a lower ligand-based orbital of b_1 symmetry which is stabilized (see, for example the situation in Figure 15.4). Thus, the dominant factor is the stabilization of the HOMO, $2a_1$, which sets the D_{3h} geometry for these d^{10} ML$_3$ compounds. For example, Pt(PPh$_3$)$_3$ and trisethylene nickel adopt this structure. Depending upon the steric constraints of the surrounding ligands, there is some latitude in the L—M—L bond angles that are observed,[2] so distortions toward the T or Y structures are relatively soft for these 16-electron complexes. With two additional electrons the high lying metal z orbital (a_2'') becomes filled. One would anticipate that the molecule will be more stable at a pyramidal geometry. This is exactly the same as in the pyramidalization of AH$_3$ (Section 9.2). The empty metal s orbital mixes into a_2'' which causes it to be stabilized and hybridized out away from the ligands. The pyramidal ML$_3$ fragment orbitals can be constructed in this manner;[3] however, something a little more complicated happens with e' and e'' and we will return to this in Section 20.1.

18.3. ML$_3$-CONTAINING METALLACYCLES

In Section 11.2 we showed that the concerted or least-motion dimerization of two olefins requires excessively high activation energies. This is the classic case of a symmetry-forbidden reaction. A two-step reaction mechanism, or at least a different reaction path has to be followed. In this section a somewhat analogous reaction, the dimerization of two olefins in the presence of Fe(CO)$_5$ and CO[4], is investigated. An initial sequence in this reaction is the photosubstitution of CO by two olefins on Fe(CO)$_5$ which gives the 18-electron intermediate, **18.13**. It rearranges to the 16-electron metallacyclopentane, **18.14**, wherein one C—C and two Fe—C bonds have been formed. Intermediate **18.14** is then trapped by CO, yielding the 18-electron metallacyclopentane, **18.15**. Finally, **18.15** presumably undergoes carbonyl insertion, addition of an olefin, reductive elimination and addition of the second olefin to regenerate **18.13** and cyclopentanone, **18.16**. The oxidative cycloaddition step, **18.13** to **18.14**, is the focus of our interest here. A careful theoretical study of the reaction has been carried out by Stockis and Hoffmann,[5] and the reader should consult the original work for details on alternative reaction sequences, stereoselectivity questions, and so on. There have been a number of other investigations of the coupling of two coordinated olefins to form a metallacyclopentane.[6-11]

18.13   18.14   18.15   18.16

The reader is cautioned that the exact details depend *critically* on the number of ligands, their geometrical disposition around the metal, as well as the number of electrons assigned to the metal. We shall return to the oxidative cycloaddition reaction again in Section 20.4 with a totally different ligand set and we shall see that there are many differences between the two reactions.

We start our analysis by building up the valence orbitals of **18.13** in terms of a C_{2v} Fe(CO)$_3$ fragment and two ethylenes. Notice that the olefins lie in the equatorial plane. This, recall from Section 17.3, is the electronically preferred way to orient olefins in a d^8 trigonal bipyramidal complex. Symmetry-adapted linear combinations of the π and π^* levels of the ethylenes are shown in **18.17** to **18.20**

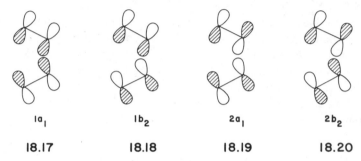

$1a_1$	$1b_2$	$2a_1$	$2b_2$
18.17	18.18	18.19	18.20

from a top view. They are simply the in-phase and out-of-phase combinations and have been redrawn from a side view on the right side of Figure 18.3. The orbitals of a d^8 Fe(CO)$_3$ fragment are illustrated on the left side. Notice that there are two fragment orbitals of a_1 symmetry on Fe(CO)$_3$ and two of b_2 symmetry that interact with **18.17–18.20**. Consequently *four* molecular orbitals of a_1 and four of b_2 symmetry are formed from this union (only the lowest three of each type are explicitly shown in Figure 18.3). The *molecular* $1a_1$ and $1b_2$ orbitals are primarily **18.17** and **18.18** stabilized by the $1a_1$ and $1b_2$ fragment orbitals of Fe(CO)$_3$. Molecular levels $2a_1$ and $2b_1$ are primarily Fe(CO)$_3$ $1a_1$ and $1b_2$. The ethylene π functions mix into these MOs in an antibonding fashion. Molecular $2a_1$ and $2b_2$ are kept at low energy because the $2a_1$ and $2b_2$ fragment orbitals of Fe(CO)$_3$ mix in second order very heavily into them. Molecular $3a_1$ and $3b_2$ consist primarily of the $2a_1$ and $2b_2$ fragment orbitals of Fe(CO)$_3$ antibonding to the ethylene π set and bonding to ethylene π^*. Not shown in Figure 18.3 are two very high lying molecular orbitals which are the ethylene set, **18.19** and **18.20**, mixed with $2a_1$ and $2b_2$ on Fe(CO)$_3$ in an antibonding way. Finally, the b_1 and a_2 orbitals on Fe(CO)$_3$ are left nonbonding. What we want to stress is that there is a distinct resemblance here to the splitting pattern of a trigonal bipyramid (Section 17.3), that is, the ethylene and CO ligands are electronically similar. The a_2 and b_1 molecular levels correspond to e'' in a D_{3h} ML$_5$ molecule. The $2b_2$ and $2a_1$ molecular orbitals are analogs of the e' set.

Two major geometrical parameters can be used to describe a reaction path for the oxidative coupling of **18.13** to the ferracyclopentane tricarbonyl, **18.14**. They are illustrated from a top view of the complex in **18.21**. Decreasing r_1 causes forma-

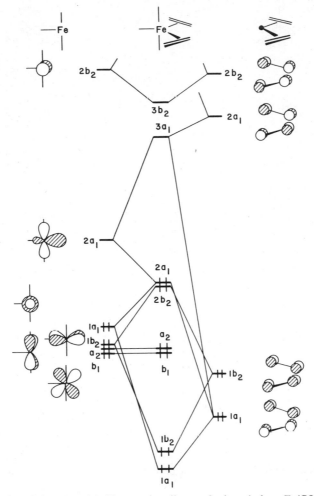

FIGURE 18.3.   An orbital interaction diagram for bis-ethylene–Fe(CO)$_3$.

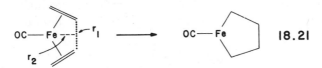

$$18.21$$

tion of the C—C σ and σ^* bonds in the ferracyclopentane. Increasing r_2 maximizes formation of the two Fe—C σ and σ^* bonds. Figure 18.4 shows an idealized Walsh diagram for the orbital energy changes along this reaction path. Since the a_2 and b_1 levels are concentrated at the iron atom and are nonbonding to the two olefins, they remain unperturbed throughout the transformation. Let us carefully look at the other molecular orbitals from the bis-olefin side of the reaction. Molecular $1a_1$ is concentrated on the ethylene π combination **18.17**. It is stabilized as r_1 decreases

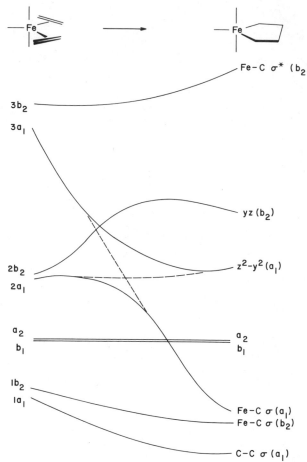

FIGURE 18.4.   An idealized plot of the orbital energies for the oxidative coupling reation.

becoming the C—C σ bond, **18.22**. The $1b_2$ molecular orbital correlates to an Fe—C σ level of b_2 symmetry, **18.23**. Notice that the electron density on the olefinic carbons becomes greatly redistributed. This is a result of the other MOs of

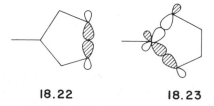

18.22   18.23

a_1 and b_2 symmetry mixing into $1a_1$ and $1b_2$. The exact details are not important to this analysis;[5] however, one can easily see that $1a_1$ is C—C bonding and $1b_2$ is Fe—C bonding. The two very high lying MOs not shown in Figure 18.3 smoothly correlate to the C—C $\sigma^*(b_2)$ and an Fe—C $\sigma^*(a_1)$ combination. The other molecular orbitals behave in a more complicated way. Let us start with $3a_1$. It is primarily $z^2 - y^2$ bonding to ethylene π^*. This is illustrated from a top view in **18.24**. There is substantial density on the olefinic carbons, thus decreasing r_1 stabilizes it. Molecular $2a_1$ is concentrated heavily on the iron atom, **18.25**. Decreasing r_1 and

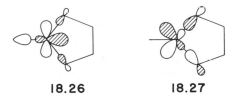

18.24   18.25

increasing r_2 increases the antibonding between the ethylene π and metal d functions, and therefore, $2a_1$ initially rises somewhat in energy. Initially then $3a_1$ attempts to correlate with the C—C σ or Fe—C σ level and $2a_1$ with the nonbonding $z^2 - y^2$ or Fe—C σ^* orbital. But molecular orbitals of the same symmetry can never cross and what actually occurs is an avoided crossing (Section 4.7) between $2a_1$ and $3a_1$. This is indicated by the dashed line in Figure 18.4, so $2a_1$ becomes an Fe—C σ orbital and $3a_1$ correlates to the nonbonding $z^2 - y^2$, **18.26**. Something similar happens to the molecular $2b_2$ orbital. On the bis-olefin side it is primarily metal yz antibonding to the olefin π combination in **18.18**. The antibonding between the two olefinic carbons as r_1 decreases makes this orbital rise in energy. An actual correlation to the σ^* C—C level is avoided by the $3b_2$ molecular orbital and $2b_2$ actually evolves into the yz nonbonding orbital shown in **18.27**.

18.26   18.27

   The important result of this exercise is that given all of the avoided crossings along the reaction path and the complexity that the metal d functions add to the problem, the reaction is still symmetry forbidden. The empty $3a_1$ orbital on the bis-olefin side becomes filled and the filled $2b_2$ level becomes empty. The reader should note that this is true only if $z^2 - y^2$, **18.26**, lies lower in energy than the yz orbital, **18.27**. One expects a trigonal bipyramidal splitting pattern on the ferracyclopentane side of the reaction. **18.26** and **18.27** correspond to the e' set of D_{3h} ML$_5$. However, we have clearly not shown these two molecular orbitals to be degenerate on the right side of Figure 18.4. The reason behind this lies in the relative σ donor strengths of the equatorial ligands. In the ferracyclopentane intermediate $z^2 - y^2$ is antibonding primarily to carbonyl σ; however in yz it is antibonding to two alkyl hybrid functions. The latter interaction is stronger (more destabilizing) for energy gap and overlap reasons. Therefore, we are left with the notion that this reaction path which maintains C_{2v} symmetry cannot be the correct one.

A number of routes can be envisioned which are symmetry allowed.[5] The most plausible one involves twisting the equatorial carbonyl off from the y axis as the cyclization proceeds to form **18.28**. **18.28** is ready to coordinate an additional CO

18.28

ligand at the sixth coordination site to yield **18.15**. Moving the equatorial CO group lowers the symmetry to the molecule to C_s. The $z^2 - y^2$ and yz orbitals at the right of Figure 18.4 have the *same* symmetry (a'). At some point along the reaction path they undergo an avoided crossing, and therefore, the reaction becomes symmetry allowed.

The preparation and study of metallacycles has been a subject of active investigation for organometallic chemists. We have just seen one example where metallacycle formation is a key step in a catalytic process and there are several others most notably, olefin metathesis. The metal acts as a geometrical and electronic template in these reactions. For unsaturated metallacycles there are interesting questions concerning delocalization.[12] Recently, a tungstenacyclobutadiene complex, **18.29**, was prepared.[13] The compound is quite stable and not very reactive (in contrast to cyclobutadienes with similar substituents). Its structure[13] shows relatively short $W-C_1(C_3)$ and $C_1-C_2(C_3)$ bond lengths indicative of substantial delocalization. Furthermore, the $C_1-C_2-C_3$ bond angle is very oblique, $119°$, so that the $W-C_2$ distance is relatively short, 2.12 Å (compared to the $W-C_1(C_2)$ distances of 1.86 Å). Furthermore, the $W-C_1-C_2(W-C_3-C_2)$ angle is $78°$, somewhat less than the idealized value of $90°$. So the compound has distorted from a square to a rhombus. The electronic structure can be developed by interacting a bisdehydroallyl^{3-} fragment, **18.30**, with WCl_3^{3+}. The full interaction diagram is given in Figure

18.29   18.30

18.5 at a "square" geometry with C_{2v} symmetry. The complexity is deceiving; the interaction diagram is easily constructed because σ- and π-type orbitals are orthogonal. On the right side of this figure is the bisdehydroallyl fragment. There are two lone pairs **(18.30)** which are directed toward the two missing hydrogens of an allyl anion. Linear combinations yield two fragment orbitals of a_1 and b_2 symmetry.

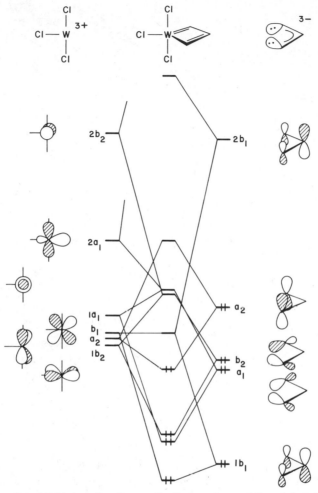

FIGURE 18.5.   An orbital interaction diagram for the tungsten–metallacyclobutadiene complex **18.29** at a square geometry.

There are also three π levels, $1b_1 + a_2 + 2b_1$. We have formally adjusted the electron count so that $1b_1$ and a_2 are filled as in the allyl anion. On the left side of the orbital interaction diagram are displayed the valence orbitals of the Cl_3W^{3+} fragment. Since halogens, like alkyl groups, are treated as anionic two-electron donors, the metal is $W(6+)$. Therefore, the metal is formally d^0. There is a slight difference in the level ordering for this fragment (and this was also true for the $Fe(CO)_3$ case) from that in Figure 18.1. Our earlier treatment of the ML_3 fragment ignored π effects so the b_1, a_2, and $1b_2$ levels were equienergetic. This is not the case for the WCl_3^{3+} fragment since Cl is a π donor. The ordering in Figure 18.5 is a reflection of this. All three chloro ligands destabilize b_1. The two *trans* chlorines destabilize a_2 and only one interacts with $1b_2$ (the π acceptor CO groups stabilize the d set so the level ordering in Figure 18.3 is opposite to that shown here).

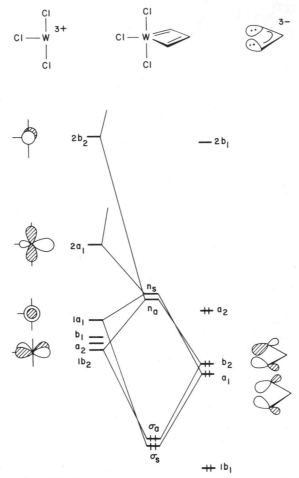

FIGURE 18.6.   An orbital interaction diagram showing only the σ portion in **18.29**.

As previously mentioned the π and σ molecular orbitals remain orthogonal in this molecule. Figure 18.6 shows the σ component. The a_1 and b_2 lone pairs of the bisdehydroallyl unit are stabilized primarily by $1a_1$ and $1b_2$ along with $2a_1$ and $2b_2$ on the metal. The resultant molecular orbitals are listed as σ_s and σ_a for mnemonic purposes in this figure. The molecular levels n_s and n_a are primarily WCl$_3^{3+}$ fragment orbitals $1a_1$ and $1b_2$ antibonding to the lone-pair hybrids on the dehydroallyl fragment. The $2a_1$ and $2b_2$ fragment orbitals of WCl$_3^{3+}$ also mix into these molecular orbitals. n_s and n_a then are hybridized and closely match the e' set in a trigonal bipyramidal ML$_5$ complex.

Overlaying these σ orbitals are the π orbitals of the metallacyclobutadiene. Figure 18.7 shows the pattern which evolves. Three fragment orbitals are of b_1 symmetry. Let us start with the molecular level labeled π_1. It is composed, as shown in **18.31**, by the in-phase combination of the lowest allyl π level and metal b_1. Some allyl $2b_1$ is mixed in second order. There is an obvious correspondence to the low-

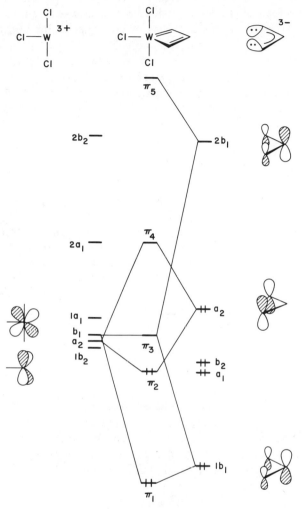

FIGURE 18.7.   An orbital interaction diagram showing only the π portion in **18.29**.

st π level in cyclobutadiene (Section 12.3). Likewise, π_3 which is illustrated in **18.32** corresponds to one of the nonbonding cyclobutadiene set. The other member would be the middle allyl π level, a_2. In this system, and this is part of the reason why the metallacycle is less reactive than cyclobutadiene, the allyl a_2 orbital interacts with the a_2 metal function. The bonding combination, π_2 (see **18.33**), is markedly stabilized and a healthy energy gap between π_2 and π_3 ensues. Finally, π_4 and π_5 are the antibonding analogs of **18.33** and **18.31**, respectively.

The π bonding in the metallacyclobutadiene can be improved still further by the rhomboid distortion indicated by the arrows in **18.34** (see Figure 12.6b for the re-

$$\pi_2$$

18.33	**18.34**

lated distortion in cyclobutadiene). π_1 is stabilized as the W—C$_2$ distance (see numbering system in **18.29**) becomes shorter. Overlap between metal b_1 and allyl $1b_1$ increases. The LUMO, π_3, rises in energy—see **18.32**. As the C$_1$—C$_3$ distance increases, π_2, **18.33**, will also be stabilized because the antibonding between C$_1$ and C$_3$ is diminished and bonding is increased from C$_1$ and C$_3$ to the metal a_2 orbital. Thus, distortion to a rhomboid structure is stabilizing and actually increases π conjugation in the metallacyclobutadiene ring.

18.4.   COMPARISON OF C_{2v} ML$_3$ AND C_{4v} ML$_5$ FRAGMENTS

The valence orbitals of a d^6 ML$_5$ and d^8 ML$_3$ fragment are explicitly shown in Figure 18.8. There is an obvious, direct correspondence between the $e + b_2$ trio in ML$_5$ and $b_1 + 1b_2 + a_2$ in ML$_3$. The z^2 fragment orbital in ML$_5$ also resembles $x^2 - z^2$. Both orbitals are hybridized out away from the ligands and present the same symmetry properties to an incoming probe ligand. (In this case both are of a_1 symmetry, but what is important is that both have hybrid functions which are cylindrically symmetric.) The $x^2 - y^2$ orbital in ML$_5$ does not find a match in the ML$_3$ fragment. However, it is an orbital that is strongly metal–ligand antibonding. It will not overlap to any significant extent with an additional, sixth ligand (that overlap would be of the δ type). In the ML$_3$ fragment y^2 ($1a_1$) and y ($2b_2$) would also be essentially nonbonding to a *fourth* ligand which forms a square planar complex. In Section 16.1 (see **16.2** and **16.3**), we saw that there was a relationship between the molecular orbitals of octahedral and square planar systems. Here we emphasize the correspondence between the orbitals of an ML$_5$ fragment when it is used to combine with a sixth ligand and form an octahedral complex and the ML$_3$ fragment when an analogous square planar complex is formed. In this regard when two *trans* ligands are removed from an ML$_5$ fragment, **18.35**, the antibonding $x^2 - y^2$ level is greatly stabilized. It becomes y^2 in Figure 18.8 and is doubly occupied for a low spin d^8 system. There is certainly some intermixing of $x^2 - y^2$, z^2, and metal s that creates $1a_1$ and $2a_1$. However, it should be clear that the ML$_5$

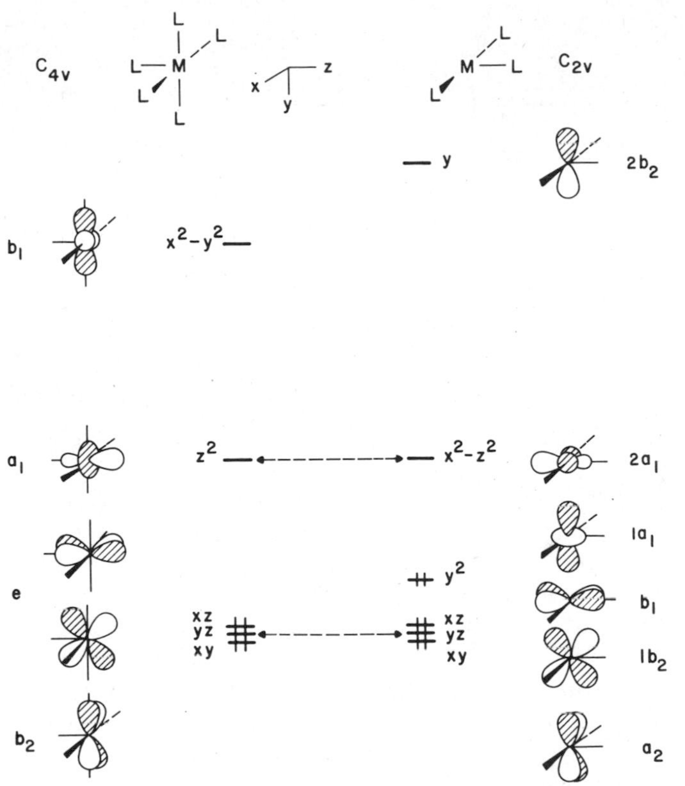

FIGURE 18.8. The valence orbitals of a d^6 C_{4v} ML$_5$ (left) and a d^8 C_{2v} ML$_3$ fragment (right).

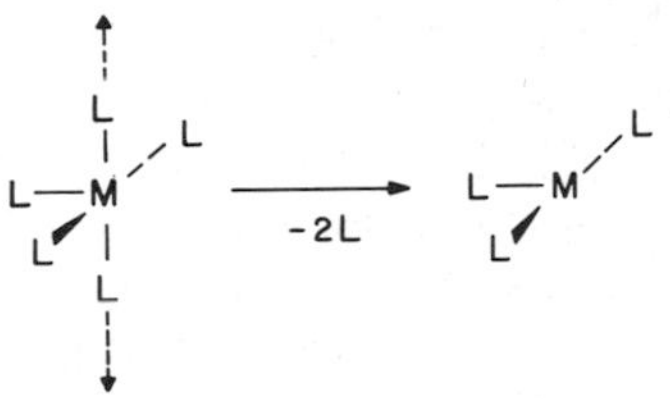

and ML$_3$ fragments have four analogous valence orbitals. In a d^6 ML$_5$ and d^8 ML$_3$ fragment three are filled and one is empty. The extra two electrons in ML$_3$ come from the nonbonding y^2 - $1a_1$. The three filled metal orbitals in each fragment are utilized for π bonding. The low lying, empty fragment orbital in each will form a σ bond with a donor orbital from an extra ligand.

Let us see how this relationship works out in terms of some simple, representative examples. The molecular orbitals of CH$_3$Mn(CO)$_5$ are developed in **18.36**. The a_1 hybrid orbital on Mn(CO)$_5^+$ interacts strongly with the lone pair of the methyl group. σ and σ^* molecular orbitals are formed. The rest of the valence orbitals on Mn(CO)$_5^+$ are left nonbonding to the sixth ligand so we have reconstructed the

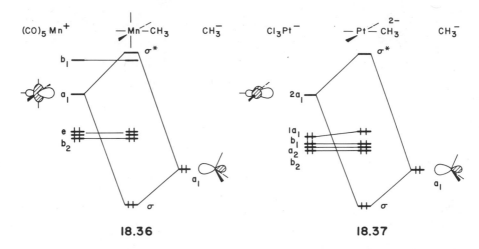

18.36   18.37

octahedral splitting pattern with three (filled) MOs lying below two. An orbital interaction diagram for $Cl_3Pt(CH_3)^{2-}$ is given in **18.37**. A square planar splitting pattern is restored. Notice that again the major perturbation between the fragment orbitals occurs between the hybrid orbital on Cl_3Pt^- and the lone pair of the methyl group that forms the σ and σ^* MOs. To be fair the $1a_1$ Cl_3Pt^- fragment orbital is destabilized slightly by the methyl lone pair, but this is a minor effect (for overlap reasons) and the other three orbitals on Cl_3Pt^- are left rigorously nonbonding. A number of d^9-d^9 M_2L_6 dimers exist; **18.38** is one such example[14] and $Pt_2(CO)_2Cl_4^{2-}$ is another.[15] Each metal atom achieves a 16-electron count by the formation of a metal–metal single bond. It is a straightforward matter to build up the MOs of **18.38** from the union of two d^9 ML_3 fragments. The $2a_1$ hybrids form bonding (σ) and antibonding (σ^*) molecular orbitals. **18.39** shows the orbital that houses the two electrons from each singly occupied $2a_1$ fragment orbital. There will also be eight closely spaced MOs which are filled. They are derived from the in-phase and out-of-phase combinations of $1a_1$, b_1, a_2, and b_2 orbitals of the ML_3 unit. The separation between bonding and antibonding partners will depend on the metal–metal separation. This distance is fairly large for the Pd—Pd single bond in **18.38** and the splitting is therefore small. The orbital pattern is not at all different from that derived for $Mn_2(CO)_{10}$ (see Section 17.4). The σ bond, **18.40**, has an obvious resemblance to **18.39** for the M_2L_6 dimers.

18.38   L= CH_3NC 18.39   18.40

The left side of Figure 18.9 builds up the molecular orbitals for Zeise's salt, ethylene–$PtCl_3^-$. The ethylene π level is stabilized by the $2a_1$ acceptor orbital. One member of the group of nonbonding metal functions, namely the b_2 level, has the

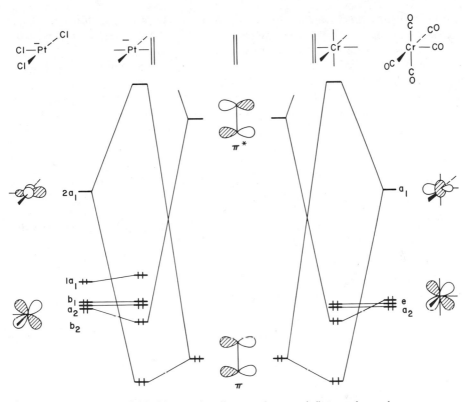

FIGURE 18.9.   Orbital interaction diagrams for two olefin–metal complexes.

right symmetry to find a match with ethylene π^*. Consequently, the metal b_2 orbital is also stabilized. This is the essence of the Dewar–Chatt–Duncanson model[17] for metal–olefin bonding. Charge from the filled ethylene π orbital is transferred to an empty metal hybrid orbital, **18.41**. There is also a backbonding component; charge is transmitted from a filled metal d function to the empty ethylene π^* orbital. This pattern is also readily apparent for ethylene–Cr(CO)$_5$ on the right side of Figure 18.9. The ethylene π orbital is stabilized by a_1 on Cr(CO)$_5$ and one component of the e set is stabilized by ethylene π^*. The amount of forward and back donation in **18.41** and **18.42** is not expected to be precisely the same in both com-

e⁻ flow ⇐   e⁻ flow ⇒

18.41   **18.42**

plexes. We can say with some certainty that both effects will be important.[18] Computationally this is a quantity which is difficult to pin down. It is sensitive to the method, parameters (basis set, etc.), and the exact details for partitioning electron density between the atoms. What we can do is to establish trends. If π acceptor groups are substituted on the ethylene, then the energies of π and π^* drop. This makes the 18.42 ($b_2 + \pi^*$ or $e + \pi^*$) interaction stronger since the energy gap between the fragment orbitals becomes smaller. The 18.41 interaction ($a_1 + \pi$) must necessarily become smaller at the same time since the energy gap between a_1 and π is larger. On the other hand substitution of π donors on the ethylene ligand causes 18.41 to become stronger and 18.42 weaker.

The analogy between the C_{2v} ML$_3$ and C_{4v} ML$_4$ fragments can be carried much further. For example, the electronic features of the olefin insertion reaction,[19] 18.43, are very similar to the catalytic olefin hydrogenation step in 18.44,[20] the carbonyl insertion reaction for $CH_3Mn(CO)_5$,[21] or even the chain propagation step in Ziegler–Natta polymerization.[22] The electronic factors that modify the *trans* M—L bond length for octahedral L'ML$_5$ complexes are identical to those in square planar L'ML$_3$.[23] These studies utilized extended Hückel, CNDO, $X\alpha$, and *ab initio* techniques. The important lesson is that the basic electronic structure of these molecules, as set by their fragment orbitals, is expected to be invariant with respect to the computational technique.

18.43   18.44

REFERENCES

1. Y. W. Yared, S. L. Miles, R. Bau, and C. A. Reed, *J. Am. Chem. Soc.*, 99, 7076 (1977). For calculations on this complex see A. Dedieu and I. Hyla-Kryspin, *J. Organomet. Chem.*, 220, 115 (1981).
2. M. Barrow, H. B. Bürgi, D. K. Johnson, and L. M. Venanzi, *J. Am. Chem. Soc.*, 98, 2356 (1976); N. C. Baeziger, K. M. Dittemore, and J. R. Doyle, *Inorg. Chem.*, 13, 805 (1974).
3. M. Elian and R. Hoffmann, *Inorg. Chem.*, 14, 1058 (1975).
4. E. Weissberger and P. Laszlo, *Accts. Chem. Res.*, 9, 209 (1976).
5. A. Stockis and R. Hoffmann, *J. Am. Chem. Soc.*, 102, 2952 (1980).
6. R. G. Pearson, *Fortschr. Chem. Forsch.*, 41, 75 (1973).
7. R. G. Pearson, *Symmetry Rules for Chemical Reactions*, Wiley, New York (1976).
8. P. S. Braterman, *J. Chem. Soc. Chem. Commun.*, 70 (1979).
9. R. J. McKinney, D. L. Thorn, R. Hoffmann, and A. Stockis, *J. Am. Chem. Soc.*, 103, 2595 (1981).
10. Y. Wakatsuki, O. Nomura, H. Tone, and H. Yamazaki, *J. Chem. Soc., Perkin II*, 1344 (1980).
11. J. W. Lauher and R. Hoffmann, *J. Am. Chem. Soc.*, 98, 1729 (1976).
12. For theoretical studies, see D. L. Thorn and R. Hoffmann, *Nouv. J. Chim*, 3, 39 (1979); M. J. S. Dewar, E. A. C. Lucken, and M. A. Whitehead, *J. Chem. Soc.*, 2423 (1960); D. P. Craig and N. L. Paddock, *ibid.*, 4118 (1962); G. Hafelinger, *Forschr. Chem. Forsch.*, 28, 1 (1972), and references therein.
13. S. F. Pederson, R. R. Schrock, M. R. Churchill, and H. J. Wasserman, *J. Am. Chem. Soc.*, 104, 6808 (1982).

14. J. R. Boehm and A. L. Bach, *Inorg. Chem.*, **16**, 778 (1977); S. Z. Goldberg and R. Eisenberg, ibid., **15**, 53 (1976).
15. A. Modinos and P. Woodward, *J. Chem. Soc., Dalton Trans.*, 1516 (1975).
16. D. M. Hoffman and R. Hoffmann, *Inorg. Chem.*, **20**, 3543 (1981); R. H. Summerville and R. Hoffmann, *J. Am. Chem. Soc.*, **98**, 7240 (1976).
17. J. Chatt and L. A. Duncanson, *J. Chem. Soc.*, 2939 (1953); M. J. S. Dewar, *Bull. Soc. Chim. Fr.*, **18**, C79 (1951). For an amusing historical perspective see M. J. S. Dewar and G. P. Ford, *J. Am. Chem. Soc.*, **101**, 783 (1979).
18. T. A. Albright, R. Hoffmann, J. C. Thibeault, and D. L. Thorn, *J. Am. Chem. Soc.*, **101**, 3801 (1979); P. J. Hay, ibid., **103**, 1390 (1981); T. Ziegler and A. Rauk, *Inorg. Chem.*, **18**, 1558 (1979).
19. D. L. Thorn and R. Hoffmann, *J. Am. Chem. Soc.*, **100**, 2079 (1978).
20. A. Dedieu, *Inorg. Chem.*, **19**, 375 (1980); **20**, 2803 (1981); A. Dedieu and A. Strich, ibid., **18**, 2940 (1979).
21. H. Berke and R. Hoffmann, *J. Am. Chem. Soc.*, **100**, 7224 (1978); M. E. Ruiz, A. Flores-Riveras, and O. Novaro, *J. Catal.*, **64**, 1 (1980).
22. A. C. Balazs and R. H. Johnson, *J. Chem. Phys.*, **77**, 3148 (1982).
23. J. K. Burdett and T. A. Albright, *Inorg. Chem.*, **18**, 2112 (1979).

CHAPTER NINETEEN

The ML_2 and ML_4 Fragments

19.1. DEVELOPMENT OF THE C_{2v} ML_4 FRAGMENT ORBITALS

The removal of two *cis* ligands from an octahedron generates the C_{2v} ML_4 fragment, **19.1**.[1] The lessons we have learned from the ML_5 and ML_3 fragments suggest that removal of these two ligands will create two empty hybrid orbitals that are

19.1

directed toward the missing ligands. Figure 19.1 shows this decomposition. The three members of t_{2g} are not affected by the perturbation, nor basically is one member of the e_g set. The other member of e_g, xz using the coordinate system at the top of this figure, and $2a_{1g}$ are stabilized greatly. In both cases a substantial portion of metal–ligand antibonding is removed. Both orbitals are also hybridized by mixing metal x and z into them since the symmetry of the molecule is lowered from O_h to C_{2v}. For the b_2 case the x component of $2t_{1u}$ mixes in a bonding way with respect to the overlap between the metal function in one orbital and the ligand portion in the other, see **19.2**. The $2a_1$ orbital also is perturbed by metal z from the $2t_{1u}$ set, yielding **19.3**. The two empty hybrids of **19.1** when symmetry adapted generate orbitals of b_2 and a_1 symmetry which correspond to those in **19.2** and **19.3**. The remainder of the octahedral set is basically unperturbed. Notice from Figure 19.1 that there has been some redistribution of d character in $1a_1$ and $3a_1$.

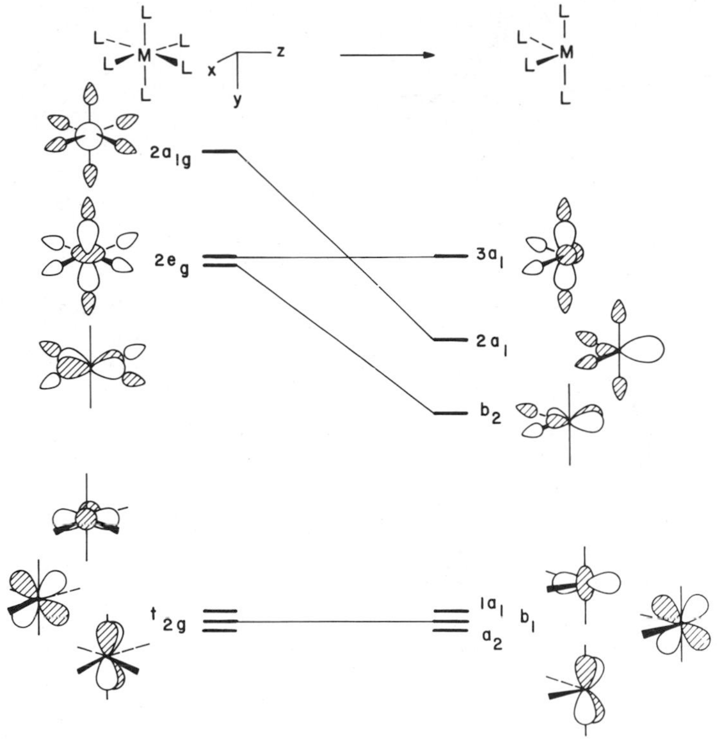

FIGURE 19.1.   Construction of the valence orbitals of a C_{2v} ML$_4$ fragment from octahedral ML$_6$.

19.2   19.3

Most of this is due to changing the ligand field (the coordinate system has also been changed from that normally given for an octahedron so that the ML$_4$ fragment orbitals are simpler combinations). The components of $1a_1$ and $3a_1$ intermix so that $3a_1$ retains maximal antibonding to the ligands and $1a_1$ remains nonbonding.

A d^6 ML$_4$ fragment is set up to interact with either two external ligands or a single four-electron donor [say a cyclobutadiene or butadiene ligand; both cyclobutadiene–Cr(CO)$_4$ and butadiene–Cr(CO)$_4$ are well-known organometallic compounds] via its empty $2a_1$ and b_2 fragment orbitals. An octahedral-like splitting pattern will be restored. A more interesting situation arises with the addition of two electrons. A low spin d^8 species would have the b_2 level filled. There is not so great of an energy difference between b_2 and the empty $2a_1$ orbital. We are back again to the singlet–triplet energy difference we first encountered with methylene (Sec-

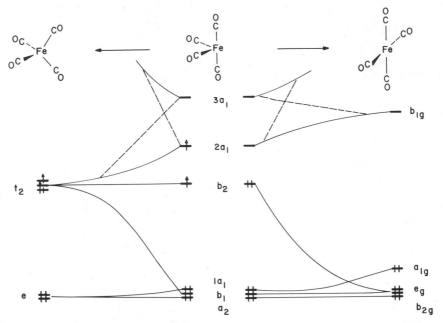

FIGURE 19.2.   A Walsh diagram for bending triplet Fe(CO)$_4$ toward a tetrahedron (left) and singlet Fe(CO)$_4$ to a square plane (right).

tion 8.8). Actually there are a number of similarities between Fe(CO)$_4$ and CH$_2$ besides the spin multiplicity questions. Both species have orbitals of a_1 and b_2 (or b_1) symmetry with a small energy difference between them. The singlet–triplet energy difference in each case is sensitive to the geometry. Figure 19.2 shows two distortion pathways that are of interest to us. The left side pertains to the triplet Fe(CO)$_4$ molecule. The b_1 orbital as shown in **19.4** is destabilized, becoming one member of the t_2 set on distortion to a tetrahedron. Another member is b_2. Finally, $2a_1$ undergoes an avoided crossing with $3a_1$. There is also considerable interorbital mixing with $1a_1$ so that ultimately $2a_1$ becomes the third member of t_2, **19.5**. The $3a_1$ level is converted into a metal s orbital of a_1 symmetry. a_2 and $1a_1$ basically stay at constant energy yielding the e set. On the surface it seems that there should be no tendency for triplet Fe(CO)$_4$ to adopt a tetrahedral geometry. The one electron in $2a_1$ which is stabilized certainly cannot overrule the two electrons in b_1 that rise in energy. However, recall from related distortions (see, for example, Section 15.4) that there is a bonding equivalent to **19.4** which *is* stabilized. This partially offsets the upward slope of b_1. The optimum geometry of

19.4   **19.5**   **19.6**

triplet Fe(CO)$_4$ then lies somewhere between these two extremes; it is given by **19.6**.[2] The addition of two more electrons fully populates $2a_1$ and b_2. The structure of a complex such as Na$_2$ Fe(CO)$_4$ then lies[3] very close to the tetrahedron. Returning to Fe(CO)$_4$ now with a singlet electronic configuration, we can see on the right side of Figure 19.2 that distortion to a square planar structure should be stabilizing. The hybridized b_2 orbital becomes one member of e_g, **19.7**. Antibond-

19.7

ing between metal d and the ligands is reduced which stabilizes the b_2 level. It also must lose its hybridization. There is again a complicated intermixing between $1a_1$, $2a_1$, and $3a_1$ so that ultimately $1a_1$ becomes a_{1g} (z^2 using the normal coordinate system of a square plane), $2a_1$ becomes a_{2u} (z), and $3a_1$ becomes b_{1g} ($x^2 - y^2$). Unfortunately nothing is experimentally known about singlet Fe(CO)$_4$, although it apparently is less stable than the triplet form. Certainly on the basis of electron counting one would expect a square planar complex for this 16-electron system.

19.2.   OLEFIN–ML$_4$ COMPLEXES

The d^8 C_{2v} ML$_4$ fragment presents an anisotropic electronic environment to a ligand. The three lower orbitals in Figure 19.1, $a_2 + b_1 + 1a_1$, are filled along with the higher lying b_2 orbital. The asymmetry comes about from the $b_2 - b_1$ difference. The b_2 orbital lies at a higher energy and is hybridized toward the missing ligand(s). Therefore, it will be a better donor orbital than b_1. Our probe ligand where this effect can nicely be illustrated is ethylene. In Section 17.3 we developed a rationale why olefins lie in the equatorial plane for d^8 trigonal bipyramidal complexes rather than orienting themselves along the axial direction. We return to the problem here from a fragment orbital perspective. Figure 19.3 shows orbital interaction diagrams for ethylene–Fe(CO)$_4$ in the two possible orientations, **19.8** and **19.9**. In both conformations $1a_1$ and $2a_1$ interact with the ethylene π level. That produces three molecular orbitals; the lower two shown in Figure 19.3 are filled. The middle level can be identified with the e' set in a trigonal bipyramidal splitting pattern and the lowest level is mainly ethylene π with some $1a_1$ and $2a_1$ character mixed into it in a bonding fashion. What is important is that the overlap between these three fragment orbitals is essentially invariant to rotation. In other words, all are cylindrically symmetrical with respect to the Fe–olefin axis. Therefore, energies of the three molecular orbitals must also be constant as a function of rotation. The a_2 Fe(CO)$_4$ fragment orbital is nonbonding at conformations **19.8** and **19.9**. This leaves us with the b_1, b_2 pair of Fe(CO)$_4$ and ethylene π^*. In the

19.8   **19.9**

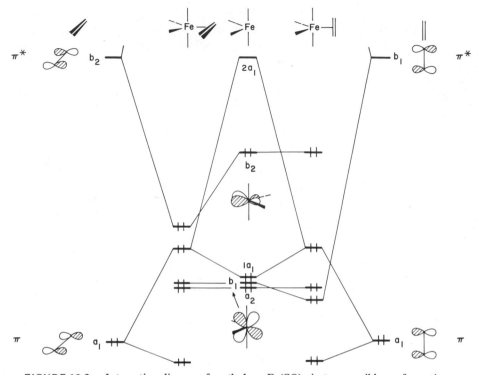

FIGURE 19.3.   Interaction diagrams for ethylene–Fe(CO)$_4$ in two possible conformations.

conformation **19.8** as shown on the left side of Figure 19.3 the π^* orbital is of b_2 symmetry. Its overlap with the Fe(CO)$_4$ b_2 fragment orbital is large and the small energy gap between them creates a stable two-electron–two-orbital bonding situation. The filled bonding orbital can be identified with the other e' component of a trigonal bipyramidal complex, strongly backbonding to ethylene π^*. When the ethylene ligand is rotated by 90° to **19.9** the π^* orbital becomes of b_1 symmetry. It then forms a bonding combination with the b_1 fragment on Fe(CO)$_4$. The b_2 Fe(CO)$_4$ level is left nonbonding. Which conformation is more stable? The real question is which of the two bonding combinations, $b_2 + \pi^*$ in **19.8** or $b_1 + \pi^*$ in **19.9**, is more stabilizing? The b_2 level is closer in energy to π^* than b_1 by virtue of the antibonding between metal d and the two equatorial lone pairs. Furthermore b_2 is hybridized (see **19.2**) out toward the ethylene, whereas b_1 is not. The hybridization creates a larger overlap with π^*. Therefore, both energy gap and overlap factors make the $b_2 + \pi^*$ interaction stronger than that for $b_1 + \pi^*$, and so conformation **19.8** with the olefin lying in the equatorial plane is of lower energy than **19.9** where ethylene is in the axial plane. Steric factors also reinforce this preference. The energy difference is actually calculated to be quite large—about 30 kcal/mol at the extended Hückel[4] and *ab initio*[5] levels. Indeed, all d^8 olefin-ML$_4$ complexes adopt this geometry.[6] However, NMR measurements[7] have put the rotational barrier for olefin–Fe(CO)$_4$ complexes to be in the 10–15 kcal/mol range.

Why is there so large a discrepancy? The energy and hybridization differences of b_1 and b_2 are at the heart of the barrier problem but we have looked at a rigid rotation. There is a geometrical way to equivalence b_1 and b_2 by carrying out a pseudo-rotation motion, **19.10**. The arrows indicate the direction the carbonyl ligands take going from the C_{2v} to a C_{4v} structure where both *trans* C—Fe—C angles are identical. As the two equatorial carbonyls move apart the lone pairs move toward the node of the metal d component in b_2. Consequently the b_2 fragment orbital drops in energy, **19.11**. On the other hand, as the axial carbonyls move closer together the lone pair hybrids overlap with the b_1 fragment orbital in an antibonding way so the energy rises and hybridization is turned on. The b_2 and b_1 levels meet to form an e set at the C_{4v} structure, **19.12**. Now the ethylene π^* acceptor orbital will not dis-

19.10   19.11   19.12

criminate between the two fragment orbitals because they are members of an e set and there should be free rotation at this geometry. The reaction path these complexes take is complicated. A coupled olefin-rotation, pseudorotation path is followed from **19.13** to the square pyramidal structure, **19.14**, which serves as a transition state. It will cost energy to attain **19.14**; notice in **19.11** and **19.12** that the energy of b_1 rises and b_2 becomes a poorer π donor (it is stabilized less by π^*). Continuing the rotation–pseudorotation motion in the same direction leads to the trigonal bipyramidal structure **19.15**. The equatorial and axial carbonyl groups have

19.13   19.14   19.15

exchanged sites. In order for this to be the correct mechanism, equatorial–axial carbonyl exchange must be (and is from the NMR studies) present. Notice that a rigid rotation mechanism would not exchange equatorial and axial groups. Another sequence of rotation–pseudorotation steps brings **19.15** to the original configuration, **19.13**.

Allene-Fe(CO)$_4$ complexes undergo another kind of intramolecular rearrangement wherein all four π-faces of the allene become equivalenced, **19.16**.[8] Drawing

19.16

from our past experience with ethylene–Fe(CO)$_4$ we can make some easy predictions about the mechanism for this rearrangement. The dominant interaction in ethylene–Fe(CO)$_4$ was between b_2 and π^*. So too will it be for this case. This is illustrated in **19.17** along with the π^* orbital for the uncoordinated C—C double bond. If we assume that the mechanism for the rearrangement in **19.16** proceeds via a structure where the Fe(CO)$_4$ group is most strongly coordinated to the central allene carbon and more weakly, but equally, bound to the outer carbons, then the allene must rotate about its axis by 45°. That is only reasonable since the π faces of allene are orthogonal. But in order for bonding to be maintained with the Fe(CO)$_4$ b_2 orbital there must also be a 90° rotation about the allene–Fe axis. The requisite geometry is given in **19.18**. Both π^* orbitals (in reality a linear combina-

19.17   **19.18**

tion of the two) overlap with b_2 at the transition state **19.18**. So the reaction path for **19.16** is technically complicated with the Fe(CO)$_4$ slipping toward the center of the allene ligand and two separate rotational motions.

19.3.   THE C_{2v} ML$_2$ FRAGMENT

Figure 19.4 shows the derivation of the orbitals of a C_{2v} ML$_2$ fragment by removal of two *cis* ligands from a square planar ML$_4$ complex. The e_g and a_{2u} levels of the square plane are totally unaffected by the perturbation. The $2a_{1g}$ and b_{2g} molecular orbitals intermix a little since the symmetry of each fragment orbital becomes a_1. But the major change comes from $2b_{1g}$ and $3a_{1g}$. Both orbitals lose one-half of their antibonding character to ligand lone pairs and so they are stabilized considerably. Metal p character is also mixed into each in a way which is bonding to the remaining ligand lone pairs. Just like in the C_{4v} ML$_5$ and C_{2v} ML$_3$ fragments there is a correspondence between the orbitals of C_{2v} ML$_2$ and ML$_4$ (compare Figure 19.1 with 19.4). Both have a set of three d-based orbitals at low energy of $a_1 + b_1 + a_2$ symmetry. There are two hybrid orbitals at higher energy which point away from the remaining ligands. They are of a_1 and b_2 symmetry and were derived by removal of two *cis* ligands. The ML$_2$ unit has one additional orbital at low

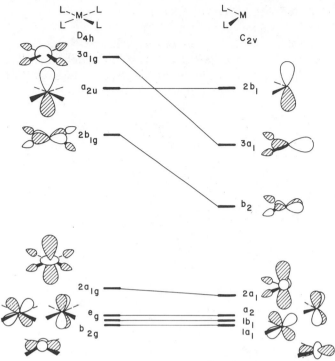

FIGURE 19.4.   Derivation of the orbitals of a C_{2v} ML_2 fragment from the molecular orbitals of square planar ML_4.

energy, $2a_1$, which corresponds to the high lying, antibonding $3a_1$ level in ML_4. Finally, ML_2 has a high energy metal p orbital, $2b_1$. That orbital along with $2a_1$ is destabilized when two axial ligands are added to form a C_{2v} ML_4 fragment. This analogy then pairs a C_{2v} d^6 ML_4 fragment with a d^8 ML_2 unit and d^8 ML_4 with d^{10} ML_2. A singlet ground state for d^{10} $Ni(CO)_2$ is guaranteed (the species has been observed by matrix isolation[2]) because the molecule is greatly stabilized by distortion to a linear, $D_{\infty h}$ geometry. The b_2 HOMO goes down in energy, meeting the block of the other four d orbitals. This is then just a linear 14-electron system (see the discussion around **16.6**). One could envision a chelating bisphosphine (an example is provided by **19.19**), where the P—Ni—P angle is forced to be less than $180°$. The b_2 level then is energetically close to $3a_1$ and a triplet state with one electron in b_2 and the other in $3a_1$ should be possible.

$$R_2$$
$$P$$
$$Ni \qquad 19.19$$
$$P$$
$$R_2$$

19.4. POLYENE–ML$_2$ COMPLEXES

The bonding and conformational preference of ethylene–Ni(PR$_3$)$_2$ is very similar to the case study of ethylene–Fe(CO)$_4$. The orbital interaction diagram for an "in-plane" conformation is shown in **19.20**. Ethylene π interacts with $1a_1$ and $3a_1$ to produce three molecular orbitals; the two filled lower ones are explicitly shown

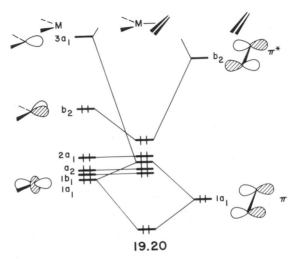

19.20

in **19.20**. The $1b_1$, a_2, and $2a_1$ orbitals are nonbonding. This leaves us with b_2 and ethylene π^*. They combine to form a strong bonding interaction. Rotation of the olefin to an "out-of-plane" geometry, **19.21**, will be energetically costly. At **19.21** ethylene π^* has b_1 symmetry and will stabilize the ML$_2$ b_1 fragment orbital, **19.22**. We are again at a point where the $b_2 + \pi^*$ combination in **19.23** for the in-plane

19.21   19.22   19.23

conformer needs to be compared with the $b_1 + \pi^*$ one, **19.22**, for the out-of-plane geometry. For the same energy gap and overlap reasons as in ethylene–Fe(CO)$_4$, the interaction in **19.23** is much stronger than that in **19.22**. The a_1 orbitals are cylindrically symmetric, so these molecular orbitals remain at constant energy during rotation. The actual barrier is quite high, ~20–25 kcal/mol,[4] because unlike ethylene–Fe(CO)$_4$ there is no good geometric way here to make the b_1 orbital equivalent to b_2. In a sense this result should not surprise us too much. Consider the ethylene to be a *bidentate* ligand. The in-plane conformation then corresponds to a 16-electron square planar system (one could either think of the olefin as a two-electron neutral donor or as a *dianionic* four-electron donor). The out-of-plane geometry would correspond to an unstable 16-electron tetrahedral complex (see Section 16.4). An extension of this analysis[9] also shows that the sterically much more encumbered structure, **19.24**, for trisethylene nickel is more stable than **19.25**. This has been determined to be experimentally true by X-ray diffraction.[10]

19.24   **19.25**

In the past two chapters we have analyzed metal-olefin bonding in four different systems. Let us present a general argument which will tie this discussion together and pursue some of its ramifications. Any ML$_n$ fragment has an empty orbital of a_1 symmetry, **19.26**. It interacts with the filled π orbital of the olefin. There is also a filled metal orbital of b_2 (or b_1) symmetry, **19.27**, available for backbonding to ethylene π^*. We have taken an olefin-metal respresentation, **19.28**, for these complexes. But perhaps they are viewed better as metallacyclopropane complexes, **19.29**. Putting this question in another way, is there any difference be-

19.26   **19.27**   **19.28**   **19.29**

tween the formulations of **19.28** and **19.29**? The metallacyclopropane model can be represented by the two localized metal-carbon σ bonds in **19.30a**. These are made symmetry correct by taking in-phase and out-of-phase combinations. The resultant orbitals, **19.30c** and **19.30b**, have a_1 and b_2 symmetry, respectively. Clearly, **19.30c** corresponds to **19.26** and **19.30b** to **19.27**. So there appears to be

19.30a

19.30c

19.30b

no difference between the metal-olefin and metallacyclopropane structures. But a further consequence of the metallacyclopropane formation is that one immediately expects the substituents on the olefinic carbons to be bent back from a plane containing the olefinic carbons and away from the metal. This is illustrated in **19.29**. Does the metal-olefin formulation lead to this distortion? When the hydrogens in ethylene are pyramidalized (see Section 10.3) a higher σ^* level, **19.31a**, mixes into π^* in a way so that the hydrogen s components are bonding with respect to π^*. This hybridizes π^* in the sense shown by **19.31b**. The addition of carbon s character and tilting of the two orbitals as shown leads to better overlap with the filled metal b_2 orbital. This mixing also lowers the energy of π^* so the energy gap be-

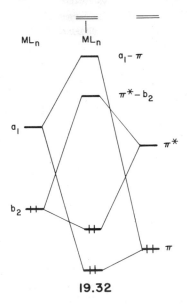

tween it and metal b_2 is smaller. Therefore, bending back the hydrogens makes the metal b_2-ethylene π^* interaction, **19.27**, stronger. Again there is no difference between the olefin–metal and metallacyclopropane representations.

What does change is the relative amount of $a_1 + \pi$ versus $b_2 + \pi^*$ interaction. A weaker $b_2 + \pi^*$ interaction will diminish the tendency for the olefinic substituents to bend back. Notice that an increased $a_1 + \pi$ interaction, **19.26**, will cause the C—C bond to lengthen as will increased $b_2 + \pi^*$ bonding, **19.27**. In the former case electron density is shifted from the carbon–carbon (π) bonding region toward the metal. For the latter case electron density is transferred from the metal to a fragment orbital which is carbon–carbon (π) antibonding. The relative amounts of forward and back donation can also play a role in reactivity questions. Take, for example, nucleophilic attack on a olefin–metal complex.[11] Our generalized bonding model for olefin–metal complexes is presented again in **19.32**. The filled, lone-pair

orbital of a nucleophile will seek maximal bonding with the LUMO of olefin-ML$_n$. This is shown in **19.32** as being the antibonding combination of π^* with metal b_2, $\pi^* - b_2$. That orbital is concentrated on the olefinic portion of the molecule and, thus, a large overlap with an incoming nucleophile lone-pair orbital is expected. The antibonding $a_1 - \pi$ level is normally at a higher energy and it is concentrated at the metal. Nevertheless, one could utilize this orbital as an acceptor for the nucleophile, respectively. The stabilization of the HOMO, **19.41**, causes the pyramidal geometry for CpMn(CO)$_2$ to be strongly favored over a planar one. The LUMO, **19.42**, is beautifully hybridized to interact with the lone pair of a third carbonyl giving CpMn(CO)$_3$ –a well-known, 18 electron complex (see Section 20.1). There is an amusing antithetic relationship here between these organometallic CpML$_2$ complexes and main group analogs. Consider the Cp unit to be one "ligand." The 16-electron L'—ML$_2$ (L' = Cp) complexes are pyramidal whereas 6-electron compounds like CH$_3^+$ and BR$_3$ are planar (see Section 9.2). Both classes, of course, have a low lying LUMO and are strong Lewis acids. The addition of two electrons causes the L'—ML$_2$ series to become planar (**19.42** is stabilized); however, 8 electron main group compounds like CH$_3^-$ and NH$_3$ are pyramidal. Again both classes have a high lying HOMO and are Lewis bases.

19.5. REDUCTIVE ELIMINATION

In a reductive elimination reaction a dialkyl transition metal complex, symbolized by **19.43**, decomposes into an alkane and a coordinatively unsaturated complex. The reaction is synthetically useful under catalytic and stoichiometric conditions[16]

$$\begin{array}{c} \text{ML}_{n-2} \\ R \quad R \end{array} \longrightarrow \text{ML}_{n-2} + R\text{—}R$$

19.43

and the reverse reaction, oxidative addition of an unsaturated metal complex into a C—C or C—H bond, offers a way to functionalize alkanes. Consequently, there has been much research on the mechanistic aspects of the reaction in the academic and industrial communities. The electronic aspects of this reaction and its reverse have also been studied extensively.[17] The wide variety of metals, number of ligands, and the electronic properties of the ligand set in **19.43** create a tremendous diversity in terms of reaction rates and even mechanistic details.

The discussion here starts in a very generalized sense. There are four electrons in the two M—R bonds of **19.43**. We would formally assign them to the alkyl groups. Two of the electrons are used to form the C—C bond in the alkane product of the reaction; the remaining two electrons become localized on the coordinatively unsaturated metal complex. Linear combination of the two M—R bonds produces **19.44** and **19.45**. The splitting between **19.44** and **19.45** is expected to be small. Both orbitals will be concentrated on the alkyl groups with some metal d and p character (**19.44** will also contain metal s character). As the reductive elimination reaction proceeds **19.44** smoothly correlates to a σ C—C bond while **19.45** evolves into a nonbonding metal d orbital, **19.47**. The metal plays two roles in this reaction: it serves as a geometrical template holding the two alkyl groups in close proximity, and it is a repository for the other two electrons. There is a crucial difference here in the latter role compared to organic and main group compounds.

as well. It is clear that the lower $\pi^* - b_2$ is in energy (the less π^* interacts and is destabilized by b_2), the greater will be its interaction with the attacking nucleophile. There are obvious ways to accomplish this by perturbations within the ML$_n$ unit and substitutional factors at the olefin; however, it would seem that nucleophilic attack on olefin–ML$_n$ complexes should never proceed at a rate that is faster than on the uncoordinated olefin. The ML$_n$ b_2 orbital will always destabilize π^* to some extent. Yet, a number of ML$_n$ groups accelerate nucleophilic attack.[11] The point we have missed is that the ML$_n$ group slips from an η^2 position to η^1 in the product, **19.33** to **19.34**, as the nucleophile attacks. It is this slipping motion that

19.33   **19.34**

activates the olefin to nucleophilic attack. Let us examine the form of the crucial HOMO–LUMO interaction between the nucleophile, Nuc$^-$, and olefin–ML$_n$ complex at some point intermediate to **19.33** and **19.34**. The HOMO lone pair of the nucleophile will still interact mainly with the $\pi^* - b_2$ LUMO of the complex as shown in **19.35**. That stabilizes the lone pair. Slipping the ML$_n$ group toward η^1 lowers the local symmetry in the olefin–metal region and so the unoccupied $a_1 - \pi$ orbital can also mix into **19.35**. It will do so in a way given by **19.36** which is bonding to the in-coming nucleophile. The resultant orbital is shown in **19.37**. There are

(HOMO)

(LUMO)

19.35   **19.36**   **19.37**

two factors at work here. First, $\pi^* - b_2$ is lowered by the slipping motion since the overlap between π^* and b_2 is maximized at the η^2 geometry. It is lowered further in energy by the first-order mixing of $a_1 - \pi$. Secondly, the mixing of **19.36** into **19.35** induces a polarization on the olefinic carbons (compare this to nucleophilic attack of olefin vs. carbonyl compounds in Section 10.5). The atomic p coefficient at the carbon atom being attacked increases—see **19.37**. That results in a larger overlap between the LUMO and the lone-pair HOMO. Therefore, slipping from η^2 to η^1 activates attack by the nucleophile both by energy gap and overlap factors. This reaction type has been described here in a very general fashion. The number and kinds of ligands, the charge on the transition metal, and so on, will set the relative energies of $\pi^* - b_2$ and $a_1 - \pi$ and their composition. This, in turn, varies

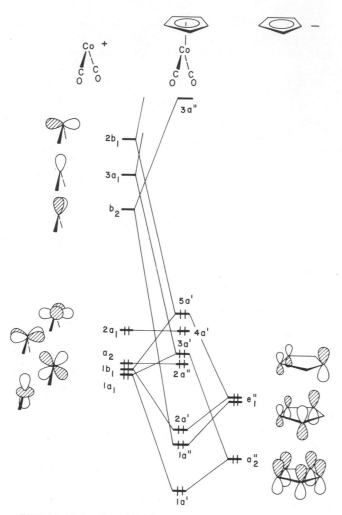

FIGURE 19.5. An orbital interaction diagram for CpCo(CO)₂.

the extent of **19.35–19.36** intermixing and consequently the propensity toward nucleophilic attack.[11]

Let us move on to a more complicated polyene–ML₂ complex. An orbital interaction diagram for CpCo(CO)₂[12] is presented in Figure 19.5. The Cp ligand has been treated as an anionic, six-electron donor. The three filled π orbitals that serve this function are displayed on the right side of the figure. This requires the metal to be formally Co(+1), d^8, and with the two CO ligands an 18-electron count at the metal is achieved. The form of the resultant MOs is fairly easy to derive. The a_2'' orbital of Cp⁻ interacts primarily with $1a_1$ and $3a_1$ on Co(CO)₂⁺. Two of the three composite MOs, labeled $1a'$ and $3a'$ in the figure, stay at low energy and are filled. The $2a_1$ and a_2 fragment orbitals of Co(CO)₂⁺ have δ symmetry with respect to

Cp⁻. This means that they overlap with the two highest π^* orbitals of Cp (not shown in the figure, see Figure 12.5). However, the energy gap is quite large and the overlap is small since it is of the δ type, so the interaction is quite weak. Therefore, $2a_1$ and a_2 are essentially nonbonding. One component of the e_1'' set on Cp⁻ interacts strongly with the *empty* b_2 fragment orbital of Co(CO)₂⁺. This generates the $1a''$ and $3a''$ MOs in the molecule. The former MO is filled and the latter is empty, see Figure 19.5. The other component of e_1'' interacts primarily with $1b_1$. Notice that $1b_1$ and e_1'' are *both* filled, therefore the bonding, $2a'$, and antibonding, $5a'$, combinations are occupied. The composition of $5a'$ is primarily given by $1b_1$ with e_1'' mixed into it in an antibonding fashion, **19.38**. Additionally some $2b_1$ character from the Co(CO)₂⁺ fragment mixes in second order with the phase relationship shown in **19.39** (bonding to e_1''). The resultant MO, **19.40**, is heavily

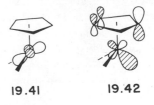

19.38   **19.39**   **19.40**

weighted on the metal and hybridized away from the Cp ligand. The reader may be puzzled why the $b_1 - e_1''$ orbital, $5a'$, stays at such a low energy compared to the $b_2 - e_1''$ analog, $3a''$. The b_2 level is initially at a higher energy than $1b_1$. Furthermore, it is hybridized whereas $1b_1$ is not, which makes the overlap between b_2 and e_1'' larger than that between $1b_1$ and e_1''. Another reason is the second-order mixing that $2b_1$ offers. This keeps the energy of $5a'$ low and b_2 has no low lying counterpart which accomplishes an analogous task. Nevertheless, $5a'$ does lie at moderately high energies because of metal d–Cp π antibonding. CpCo(CO)₂ is consequently a strong base.[13] Puckering the Cp ligand to diminish this antibonding is energetically favorable for CpCo(CO)₂[14] as well as for other 18-electron polyene–ML₂ complexes where an analogous situation occurs.[15] Alternatively the ML₂ unit can slip to a lower coordination number to relieve this antibonding interaction. But suppose the two electrons in $5a'$ were removed, yielding a 16-electron complex, for example CpMn(CO)₂. Reference back to Figure 19.5 shows that the now empty $5a'$ orbital lies fairly close in energy to the HOMO, $4a'$. Pyramidalization causes the two orbitals to mix[12] which lowers the energy of $4a'$ and raises the energy of the empty $5a'$. The form of $4a'$ and $5a'$ at a pyramidal geometry is given by **19.41** and **19.42**

19.41   **19.42**

as well. It is clear that the lower π^* – b_2 is in energy (the less π^* interacts and is destabilized by b_2), the greater will be its interaction with the attacking nucleophile. There are obvious ways to accomplish this by perturbations within the ML$_n$ unit and substitutional factors at the olefin; however, it would seem that nucleophilic attack on olefin–ML$_n$ complexes should never proceed at a rate that is faster than on the uncoordinated olefin. The ML$_n$ b_2 orbital will always destabilize π^* to some extent. Yet, a number of ML$_n$ groups accelerate nucleophilic attack.[11] The point we have missed is that the ML$_n$ group slips from an η^2 position to η^1 in the product, **19.33** to **19.34**, as the nucleophile attacks. It is this slipping motion that

19.33   **19.34**

activates the olefin to nucleophilic attack. Let us examine the form of the crucial HOMO–LUMO interaction between the nucleophile, Nuc⁻, and olefin–ML$_n$ complex at some point intermediate to **19.33** and **19.34**. The HOMO lone pair of the nucleophile will still interact mainly with the π^* – b_2 LUMO of the complex as shown in **19.35**. That stabilizes the lone pair. Slipping the ML$_n$ group toward η^1 lowers the local symmetry in the olefin–metal region and so the unoccupied a_1 – π orbital can also mix into **19.35**. It will do so in a way given by **19.36** which is bonding to the in-coming nucleophile. The resultant orbital is shown in **19.37**. There are

(HOMO)

(LUMO)

19.35   **19.36**   **19.37**

two factors at work here. First, π^* – b_2 is lowered by the slipping motion since the overlap between π^* and b_2 is maximized at the η^2 geometry. It is lowered further in energy by the first-order mixing of a_1 – π. Secondly, the mixing of **19.36** into **19.35** induces a polarization on the olefinic carbons (compare this to nucleophilic attack of olefin vs. carbonyl compounds in Section 10.5). The atomic p coefficient at the carbon atom being attacked increases—see **19.37**. That results in a larger overlap between the LUMO and the lone-pair HOMO. Therefore, slipping from η^2 to η^1 activates attack by the nucleophile both by energy gap and overlap factors. This reaction type has been described here in a very general fashion. The number and kinds of ligands, the charge on the transition metal, and so on, will set the relative energies of π^* – b_2 and a_1 – π and their composition. This, in turn, varies

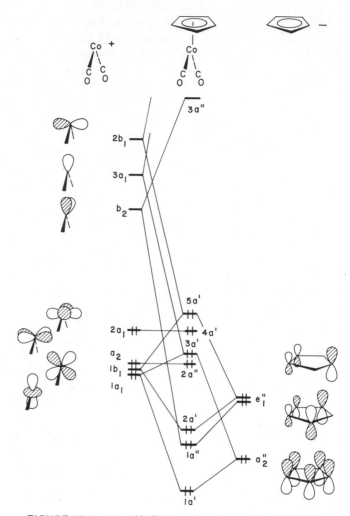

FIGURE 19.5.   An orbital interaction diagram for CpCo(CO)₂.

the extent of **19.35–19.36** intermixing and consequently the propensity toward nucleophilic attack.[11]

Let us move on to a more complicated polyene–ML₂ complex. An orbital interaction diagram for CpCo(CO)₂[12] is presented in Figure 19.5. The Cp ligand has been treated as an anionic, six-electron donor. The three filled π orbitals that serve this function are displayed on the right side of the figure. This requires the metal to be formally Co(+1), d^8, and with the two CO ligands an 18-electron count at the metal is achieved. The form of the resultant MOs is fairly easy to derive. The a_2'' orbital of Cp⁻ interacts primarily with $1a_1$ and $3a_1$ on Co(CO)₂⁺. Two of the three composite MOs, labeled $1a'$ and $3a'$ in the figure, stay at low energy and are filled. The $2a_1$ and a_2 fragment orbitals of Co(CO)₂⁺ have δ symmetry with respect to

Cp$^-$. This means that they overlap with the two highest π^* orbitals of Cp (not shown in the figure, see Figure 12.5). However, the energy gap is quite large and the overlap is small since it is of the δ type, so the interaction is quite weak. Therefore, $2a_1$ and a_2 are essentially nonbonding. One component of the e_1'' set on Cp$^-$ interacts strongly with the *empty* b_2 fragment orbital of Co(CO)$_2^+$. This generates the $1a''$ and $3a''$ MOs in the molecule. The former MO is filled and the latter is empty, see Figure 19.5. The other component of e_1'' interacts primarily with $1b_1$. Notice that $1b_1$ and e_1'' are *both* filled, therefore the bonding, $2a'$, and antibonding, $5a'$, combinations are occupied. The composition of $5a'$ is primarily given by $1b_1$ with e_1'' mixed into it in an antibonding fashion, **19.38**. Additionally some $2b_1$ character from the Co(CO)$_2^+$ fragment mixes in second order with the phase relationship shown in **19.39** (bonding to e_1''). The resultant MO, **19.40**, is heavily

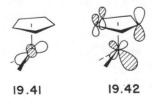

19.38   **19.39**   **19.40**

weighted on the metal and hybridized away from the Cp ligand. The reader may be puzzled why the $b_1 - e_1''$ orbital, $5a'$, stays at such a low energy compared to the $b_2 - e_1''$ analog, $3a''$. The b_2 level is initially at a higher energy than $1b_1$. Furthermore, it is hybridized whereas $1b_1$ is not, which makes the overlap between b_2 and e_1'' larger than that between $1b_1$ and e_1''. Another reason is the second-order mixing that $2b_1$ offers. This keeps the energy of $5a'$ low and b_2 has no low lying counterpart which accomplishes an analogous task. Nevertheless, $5a'$ does lie at moderately high energies because of metal d–Cp π antibonding. CpCo(CO)$_2$ is consequently a strong base.[13] Puckering the Cp ligand to diminish this antibonding is energetically favorable for CpCo(CO)$_2$[14] as well as for other 18-electron polyene-ML$_2$ complexes where an analogous situation occurs.[15] Alternatively the ML$_2$ unit can slip to a lower coordination number to relieve this antibonding interaction. But suppose the two electrons in $5a'$ were removed, yielding a 16-electron complex, for example, CpMn(CO)$_2$. Reference back to Figure 19.5 shows that the now empty $5a'$ orbital lies fairly close in energy to the HOMO, $4a'$. Pyramidalization causes the two orbitals to mix[12] which lowers the energy of $4a'$ and raises the energy of the empty $5a'$. The form of $4a'$ and $5a'$ at a pyramidal geometry is given by **19.41** and **19.42**,

19.41   **19.42**

respectively. The stabilization of the HOMO, **19.41**, causes the pyramidal geometry for CpMn(CO)$_2$ to be strongly favored over a planar one. The LUMO, **19.42**, is beautifully hybridized to interact with the lone pair of a third carbonyl giving CpMn(CO)$_3$ – a well-known, 18 electron complex (see Section 20.1). There is an amusing antithetic relationship here between these organometallic CpML$_2$ complexes and main group analogs. Consider the Cp unit to be one "ligand." The 16-electron L$'$—ML$_2$ (L$'$ = Cp) complexes are pyramidal whereas 6-electron compounds like CH$_3^+$ and BR$_3$ are planar (see Section 9.2). Both classes, of course, have a low lying LUMO and are strong Lewis acids. The addition of two electrons causes the L$'$—ML$_2$ series to become planar (**19.42** is stabilized); however, 8 electron main group compounds like CH$_3^-$ and NH$_3$ are pyramidal. Again both classes have a high lying HOMO and are Lewis bases.

19.5.   REDUCTIVE ELIMINATION

In a reductive elimination reaction a dialkyl transition metal complex, symbolized by **19.43**, decomposes into an alkane and a coordinatively unsaturated complex. The reaction is synthetically useful under catalytic and stoichiometric conditions[16]

$$\underset{R}{\overset{ML_{n-2}}{\diagup}}\underset{R}{\diagdown} \quad \longrightarrow \quad ML_{n-2} \;+\; R\text{–}R$$

19.43

and the reverse reaction, oxidative addition of an unsaturated metal complex into a C—C or C—H bond, offers a way to functionalize alkanes. Consequently, there has been much research on the mechanistic aspects of the reaction in the academic and industrial communities. The electronic aspects of this reaction and its reverse have also been studied extensively.[17] The wide variety of metals, number of ligands, and the electronic properties of the ligand set in **19.43** create a tremendous diversity in terms of reaction rates and even mechanistic details.

The discussion here starts in a very generalized sense. There are four electrons in the two M—R bonds of **19.43**. We would formally assign them to the alkyl groups. Two of the electrons are used to form the C—C bond in the alkane product of the reaction; the remaining two electrons become localized on the coordinately unsaturated metal complex. Linear combination of the two M—R bonds produces **19.44** and **19.45**. The splitting between **19.44** and **19.45** is expected to be small. Both orbitals will be concentrated on the alkyl groups with some metal d and p character (**19.44** will also contain metal s character). As the reductive elimination reaction proceeds **19.44** smoothly correlates to a σ C—C bond while **19.45** evolves into a nonbonding metal d orbital, **19.47**. The metal plays two roles in this reaction: it serves as a geometrical template holding the two alkyl groups in close proximity, and it is a repository for the other two electrons. There is a crucial difference here in the latter role compared to organic and main group compounds.

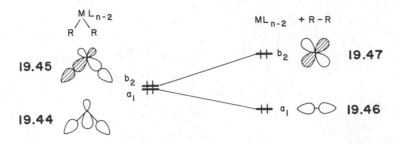

Consider an analogous least-motion reaction for the reductive elimination process in CH_2R_2 which yields CH_2 and R—R. The MOs displayed in **19.48** and **19.49** are analogs of **19.44** and **19.45**, respectively. They have been labeled according to the C_{2v} symmetry of the molecule. A least-motion path then conserves C_{2v} symmetry. The a_1 level, **19.48**, again becomes the C—C σ level, **19.50**, of the resulting alkane. However, now the b_2 orbital, **19.49**, correlates with a p orbital of CH_2, **19.51**. We know from Section 7.2 that the a_1 level, **19.52**, of CH_2 lies below b_2 in energy. **19.52** evolved from an empty σ^* orbital, and therefore, the least-motion path for this reaction or the reverse—insertion of CH_2 into a C—C bond—is symmetry forbidden. The CH_2 group must undergo a sideways rocking motion as the two alkyl groups couple. The symmetry of the activated complex is lowered to C_s. An analogous pathway, the reaction of CH_2 with ethylene to form cyclopropane, was covered in Section 11.2.

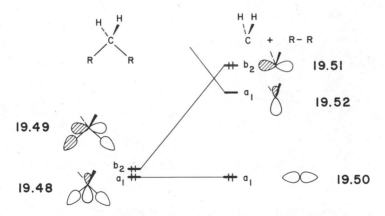

Returning back to the reductive elimination process for transition metal complexes, the evolution of **19.45** to **19.47** seems to be a bit mysterious. Originally **19.45** is concentrated on the alkyl groups. One might think that this orbital should ultimately become the σ^* level of R—R. At some point along the reaction path electron density must be shifted from the alkyl groups, toward the metal. We have conveniently left out a number of orbitals in this analysis, one of which serves to redistribute the electron density in **19.45**. **19.45** is crucial, because its upward slope will figure heavily in setting the activation energy for the reaction.[17a, b] Let us look

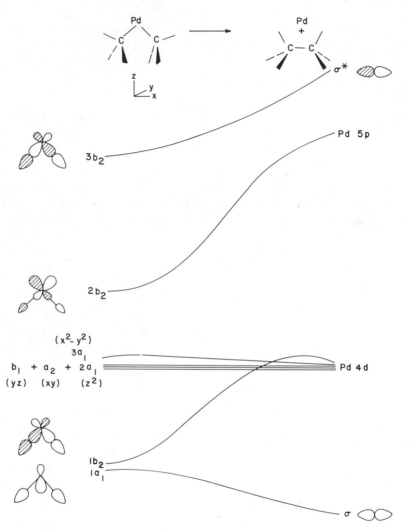

FIGURE 19.6.   A Walsh diagram for reductive elimination in Pd(CH$_3$)$_2$.

at the decomposition of MR$_2$ into a naked metal atom and R—R. Figure 19.6[17a] shows this for R=CH$_3$ and M=Pd. This is not a reaction which is likely to occur on thermodynamic grounds; however, it contains all of the elements present in a more realistic case. On the left side of this figure are the orbitals of Pd(CH$_3$)$_2$. They are identical with those derived for a ML$_2$ fragment in Figure 19.4 except that they have been redrawn so that the MR$_2$ "molecule" lies in the plane of the paper. The lower two M–R σ levels which correspond to **19.44** and **19.45** (1a_1 and 1b_2, respectively) have also been included in the Walsh diagram. The 1a_1 orbital has been drawn very stylistically. It actually consists of metal s, z, $x^2 - y^2$, and z^2 bonding to the in-phase combination of alkyl lone-pair orbitals. There are now three orbitals of b_2 symmetry: $1b_2$ is concentrated on the alkyl groups bonding to metal x and xz; $2b_2$ corresponds to the b_2 valence orbital in the ML_2 fragment— see Figure 19.4; finally, $3b_2$ is the fully antibonding analog of $1b_2$. There is also a block of four nonbonding metal d orbitals, $2a_1 + b_1 + a_2 + 3a_1$, at moderate energy. A d^8 complex like $Pd(CH_3)_2$ will have $3a_1$ as the HOMO and $2b_2$ as the LUMO.

When the C—Pd—C angle decreases and the two methyl groups pivot toward each other the energy of $1a_1$ goes down. It smoothly correlates to the σ C—C bond of ethane. The $1b_2$ level rises in energy; metal–carbon bonding is lost and some antibonding between the methyl groups is introduced until finally $1b_2$ evolves into the Pd xz atomic orbital. Nothing much happens to the block of four metal d orbitals. They essentially stay nonbonding with respect to the methyl groups along the reaction path. The $2b_2$ and $3b_2$ levels behave differently. Initially $2b_2$ is primarily metal xz; as the methyl groups move toward each other antibonding between them and xz is diminished, therefore, $2b_2$ stays at relatively constant energy, or in a more realistic system it may even be stabilized. Ultimately it becomes a Pd x atomic orbital; consequently it rises in energy. The $3b_2$ MO behaves in a similar manner; it becomes the C—C σ^* orbital of ethane. The three b_2 levels undergo avoided crossings. Let us concentrate only on $1b_2$ and $2b_2$. There is a natural correlation between $1b_2$ and σ^* along with $2b_2$ descending to the metal d block. This is indicated by the dashed line in **19.53**. However, remember that two molecular orbitals of the same symmetry may never cross (Section 4.7). An avoided crossing occurs so that $1b_2$ becomes metal xz. Another avoided crossing takes place between $2b_2$ and $3b_2$ so that $2b_2$ actually correlates to the Pd x AO, and $3b_2$ becomes σ^*. For simplicity we will disregard the latter avoided crossing, and as shown in **19.53** $2b_2$ will yield σ^*. The intermixing of $1b_2$ and $2b_2$ can be treated in a typical perturbation mode. The amount of mixing depends directly on the amount of overlap between the two unperturbed MOs that is introduced along the reaction path and inversely on the energy difference between the molecular orbitals. The quantity ΔE in **19.53** is related to the origin of the activation barrier for reductive elimination considering only the changes in orbital energies. The situation shown in

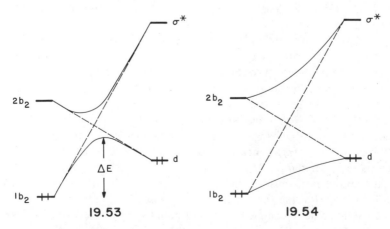

σ^* $2b_2$ ΔE d $1b_2$ 19.53 σ^* $2b_2$ d $1b_2$ 19.54

19.53 is a weakly avoided crossing. There is not much intermixing between $1b_2$ and $2b_2$ until just before the crossing would have taken place. The pattern illustrated by **19.54** is indicative of a strongly avoided crossing. Here there is substantial inter-mixing between $1b_2$ and $2b_2$ all along the reaction path. The avoided crossing for Pd$(CH_3)_2$ in Figure 19.6 lies somewhere between these two extremes and is complicated by the intermixing with $3b_2$. Nonetheless, there is significant $1b_2$, $2b_2$ mixing and the phase relationship is given by the major components in both molecular orbitals—the methyl hybrids in $1b_2$ and metal xz in $2b_2$. The $2b_2$ MO lies above $1b_2$, and therefore it mixes into $1b_2$ in a net bonding manner, **19.55**. The resultant MO, **19.56**, contains increased metal xz and decreased methyl character. The intermixing continues until $1b_2$ becomes a pure metal xz orbital. The $1b_2$ level will also mix into $2b_2$ with the opposite phase relationship to that shown in **19.55**. Cancellation of metal x and xz character and reinforcement of methyl character occurs so that $2b_2$ becomes σ^* (neglecting the influence of $3b_2$).

Let us now turn our attention to a couple of more realistic models, **19.57** and **19.58**, where in both cases the metal is d^8 in the starting complex and L is an arbitrary two-electron donor ligand (e.g., PPh$_3$). Walsh diagrams for these examples are displayed in Figure 19.7. On the left side is the case for the trigonal L—MR$_2$ species. It is essentially identical to the Pd$(CH_3)_2$ case in Figure 19.6. The extra ligand L with its donor orbital of a_1 symmetry cannot mix into any of the crucial b_2 orbitals (notice that $2a_1$ and $3a_1$ are destabilized slighly by the donor orbital of L). The calculated total energy for the reaction is plotted at the bottom left of Figure 19.7. The moderate activation energy is clearly due to the rise in energy of $1b_2$, counterbalanced by the stabilization in $1a_1$. The L$_2$MR$_2$ system is plotted on the right side of this figure. First of all, the calculated total energy is much greater than that for L—MR$_2$. Notice that $1b_2$ rises to a much higher energy in L$_2$Pd$(CH_3)_2$. The reason behind this change is that one combination of the ligand lone-pair hybrids has b_2 symmetry. It destabilizes $2b_2$ greatly and restores a square planar splitting pattern. Therefore, $2b_2$ corresponds now to $2b_{1g}$ in Figure 19.4. **19.59** shows the essential details of the avoided crossing in the L—Pd$(CH_3)_2$ (or Pd$(CH_3)_2$) model. Raising the energy of $2b_2$ causes the intersection of the dashed lines in **19.59** to go up in energy. This is indicated by the arrows. The avoided cross-

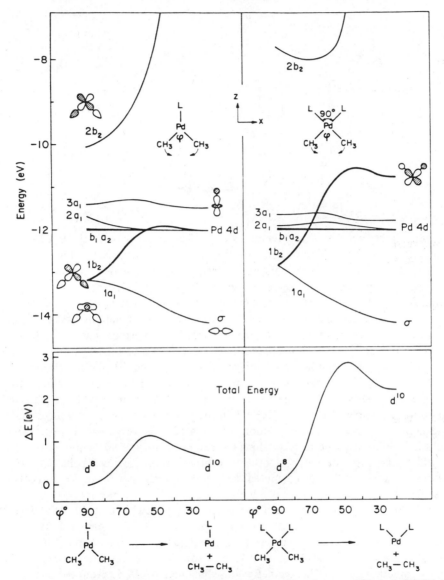

FIGURE 19.7. Walsh diagrams for reductive elimination in a three coordinate L–Pd(CH$_3$)$_2$ and L$_2$Pd(CH$_3$)$_2$ complex. For simplicity the L groups were taken to be H$^-$ with their valence state ionization potentials adjusted to match the lone pair orbital of PH$_3$. Here ϕ is defined as the C–Pd–C angle; this was varied in concert with stretching the Pd–C bonds and rocking the methyl groups off from the Pd–C axis, toward each other. Reprinted with permission from R. Hoffmann, in *Frontiers of Chemistry*, K. J. Laidler, editor, Pergamon Press, Oxford (1982), pp. 247–263.

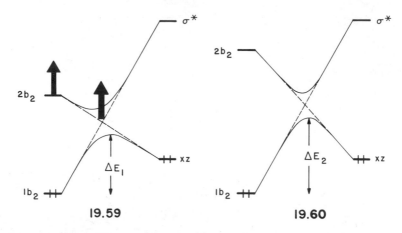

19.59   19.60

ing now corresponds to that given in **19.60** and is appropriate for the $L_2Pd(CH_3)_2$ reaction on the right side of Figure 19.7. The $1b_2$ level must rise to a higher energy before the avoided crossing occurs, and consequently the reaction requires a greater activation energy. However, there are two important qualifications that should be given for this analysis. First of all, there must be approximately the same intermixing of $1b_2$ and $2b_2$ along the reaction path in the two reactions. Secondly, the avoided crossing cannot be of the strongly avoided type since that would predict equal activation energies. These conditions are fulfilled in this example. One can also think of other complications. For example, the L—Pd—L angle in Figure 19.7 was kept at 90°. This is unreasonable since the product of the reductive elimination for the $L_2Pd(CH_3)_2$ case is the 14-electron L—Pd—L complex which is expected to be linear. Notice that the $1b_2$ orbital on the product side lies at very high energy and this is the reason why the reaction is calculated to be very endothermic. Allowing the L—Pd—L angle to relax along the reaction path to 180° will stabilize $1b_2$; it will merge with the block of other d orbitals. This also lowers the activation energy; however, it is still computed to be larger than that in the L—Pd(CH_3)$_2$ system.[17a, b]

The theoretical prediction we have made here is somewhat unusual. What we are saying is that reductive elimination from a 16-electron L_2MR_2 complex, **19.61**, to the 14-electron L_2M species is less facile than reductive elimination from a 14-electron LMR_2 complex, **19.62** to a 12-electron LM intermediate. Moreover, we know something about the geometry and dynamics of **19.62** from Section 18.2. Inspection of Figure 18.2 and the discussion around it tells us that a 14-electron LMR_2 complex will be stable at a "T" or "Y" structure. A trigonal geometry where the L—M—R and R—M—R angles are approximately equal is a high energy point on the potential surface. (Although the symmetry of LMR_2 is likely to be lower than D_{3h}, the surface will still strongly resemble that in Figure 7.5; references 17a,b give some practical examples.) The Y structure, **19.63**, where the R—M—R angle is acute, will serve as the exit geometry for reductive elimination. The path from **19.61** to **19.63** has been experimentally demonstrated for Pd(2+) and Au(3+) systems.[16] The dissociation of a ligand from **19.61** to **19.62** must be an endothermic

19.61

$$\begin{array}{c} L \diagdown M \diagup L \\ R \diagup \diagdown R \\ 16\ e^- \end{array} \xrightarrow{-L} \begin{array}{c} L \\ | \\ M \\ R \diagup \diagdown R \\ 14\ e^- \end{array} \longrightarrow \begin{array}{c} L \\ | \\ M \\ R \diagup \diagdown R \end{array}$$ 19.63

19.62

$$L-M-L + R-R \qquad\qquad L-M + R-R$$
$$14\ e^- \qquad\qquad\qquad 12\ e^-$$

process. Therefore, a delicate balance can exist for reductive elimination from **19.61** or **19.63**, as well. If the avoided crossing between $1b_2$ and $2b_2$ is made to be more strongly avoided (or if ligand dissociation becomes too endothermic), then direct reductive elimination from the four-coordinate **19.61** will be more favorable. Examples exist primarily for Ni(2+) complexes.[16] One can also see that the electronic nature of the auxiliary ligands determines the eventual choice of reaction paths. As L becomes a stronger donor, the $2b_2$ orbital is increasingly destabilized and reductive elimination is rendered less facile. On the other hand, making the alkyl groups better σ donors will actually lower the activation energy. This due to the fact that the energy of $1b_2$ (concentrated on the alkyl lone-pair functions) rises when R is a better σ donor. Finally, the model that we have presented can easily be extended to other systems. In the past sections of this chapter we have highlighted the close resemblance between the orbitals of C_{2v} ML_2 and C_{2v} ML_4 fragments. Our discussion in this chapter could have started from the orbitals of a d^6 L_2MR_2 species, **19.64**. The orbital pattern and occupation is equivalent to that for the d^8 MR_2 system in Figure 19.6. One or two extra ligands could be added and we would have come to an equivalent prediction: reductive elimination from the d^6, 18-electron L_4MR_2 complex, **19.65**, requires a larger activation energy than that for the 16-electron L_3MR_2 complex, **19.66**.[17b] Reductive elimination for a metallacy-

19.64   **19.65**   **19.66**

clopentane complex forming cyclobutane as opposed to a fragmentation reaction which yields a bis-olefin complex[17c] (the reverse of the reaction we studied in Section 18.2) or oxidative addition[17d,h,k] present special problems that we will not cover. But it should be clear to the reader that a conceptually simple model with wide versatility can be constructed for this reaction type.

REFERENCES

1. M. Elian and R. Hoffmann, *Inorg. Chem.*, **14**, 1058 (1975); J. K. Burdett, ibid., **14**, 375 (1975); *J. Chem. Soc., Faraday Trans. 2*, **70**, 1599 (1974).

2. B. Davies, A. McNeish, M. Poliakoff, and J. J. Turner, *J. Am. Chem. Soc.*, 99, 7573 (1977); J. K. Burdett, *Coord. Chem. Rev.*, 27, 1 (1978); M. Poliakoff and J. J. Turner, *J. Chem. Soc., Dalton Trans*, 2276 (1974).

3. R. G. Teller, R. G. Finke, J. P. Collman, H. B. Chin, and R. Bau, *J. Am. Chem. Soc.*, 99, 1104 (1977).

4. T. A. Albright, R. Hoffmann, J. C. Thibeault, and D. L. Thorn, *J. Am. Chem. Soc.*, 101, 3801 (1979).

5. J. Demuynck, A. Strich, and A. Veillard, *Nouv. J. Chim.*, 1, 217 (1977).

6. For a review see, S. D. Ittel and J. A. Ibers, *Adv. Organomet. Chem.*, 14, 33 (1976).

7. See, for example, L. Kruczynski, L. K. K. Li ShingMan, and J. Takats, *J. Am. Chem. Soc.*, 96, 4006 (1974); S. T. Wilson, N. J. Coville, J. R. Shapley, and J. A. Osborn, ibid., 96, 4038 (1974); J. A. Segal, and B. F. G. Johnson, *J. Chem. Soc., Dalton Trans.*, 677, 1990 (1975).

8. R. Ben-Shoshan and R. Pettit, *J. Am. Chem. Soc.*, 89, 2231 (1967).

9. N. Rösch and R. Hoffmann, *Inorg. Chem.*, 13, 2656 (1974).

10. K. Jonas, P. Misbach, R. Stabba, and G. Wilke, *Angew. Chem.*, 85, 1002 (1973); E. G. Hoffman, P. W. Jolly, A. Kusters, R. Mynott, and G. Wilke, *Z. Naturforsch.*, 31b, 1712 (1976); see also J. A. K. Howard, J. L. Spencer, and S. A. Mason, *Proc. Roy. Soc. Lond A*, 386, 145 (1983).

11. O. Eisenstein and R. Hoffmann, *J. Am. Chem. Soc.*, 103, 4308 (1981); H. Fujimoto and N. Koga, *Tet. Letts.*, 4357 (1982); see also the related case of nucleophilic substitution on π-allyl complexes, S. Sakaki, M. Nishikawa, and A. Ohyoshi, *J. Am. Chem. Soc.*, 102, 4062 (1980); and coordinated carbon monoxide, S. Nakamura and A. Dedieu, *Theoret. Chim. Acta*, 61, 587 (1982).

12. P. Hofmann, *Angew. Chem.*, 89, 551 (1977); Habilitationschrift, Universität Erlangen-Nurnberg (1978).

13. H. Werner and W. Hofmann, *Angew. Chem.*, 89, 835 (1977); W. Hofmann, W. Buchner, and H. Werner, ibid., 89, 836 (1977); N. Dudeney, J. C. Green, P. Grebenik, and O. N. Kirehner, *J. Organomet. Chem.*, 252, 221 (1983).

14. L. R. Byers and L. F. Dahl, *Inorg. Chem.*, 19, 1760 (1979).

15. T. A. Albright and R. Hoffmann, *Chem. Ber.*, 111, 1578 (1978); T. A. Albright, *J. Organomet. Chem.*, 198, 159 (1980); T. A. Albright, R. Hoffmann, Y.-C. Tse, and T. D'Ottavio, *J. Am. Chem. Soc.*, 101, 3812 (1979); L. J. Radonovich, F. J. Koch, and T. A. Albright, *Inorg. Chem.*, 19, 3373 (1980); D. M. P. Mingos, *J. Chem. Soc., Dalton Trans.*, 602 (1977); D. M. P. Mingos, M. I. Forsyth, and A. J. Welch, ibid., 1363 (1978); D. M. P. Mingos and A. J. Welch, ibid., 1674 (1980); C. Mealli, S. Midollini, S. Moneti, L. Sacconi, J. Silvestre, and T. A. Albright, *J. Am. Chem. Soc.*, 104, 95 (1982).

16. J. P. Collman and L. S. Hegedus, *Principles and Applications of Organotransition Metal Chemistry*, University Science Books, Mill Valley, CA (1980), pp. 177–258.

17. (a) R. Hoffmann in, *Frontiers of Chemistry*, K. J. Laidler, editor, Pergamon Press, New York (1982), pp. 247–263.

    (b) K. Tatsumi, R. Hoffmann, A. Yamamoto, and J. K. Stille, *Bull. Chem. Soc. Japan*, 54, 1857 (1981); S. Kominya, T. A. Albright, R. Hoffmann, and J. K. Kochi, *J. Am. Chem. Soc.*, 98, 7255 (1976); 99, 8440 (1977).

    (c) R. J. McKinney, D. L. Thorn, R. Hoffmann, and A. Stokis, ibid., 103, 2595 (1981).

    (d) J. O. Noell and P. J. Hay, ibid., 104, 4578 (1982).

    (e) B. Åkermark and A. Ljungqvist, *J. Organomet. Chem.*, 182, 47, 59 (1979).

    (f) A. Flores-Riveros and O. Novaro, ibid., 235, 383 (1982).

    (g) B. Åkermark, H. Johanson, B. Roos, and U. Wahlgen, *J. Am. Chem. Soc.*, 101, 5876 (1979).

    (h) K. Kitaura, S. Obara, and K. Morokuma, ibid., 103, 2891 (1981).

    (i) A. C. Balazs, K. H. Johnson, and G. M. Whitesides, *Inorg. Chem.*, 21, 2162 (1982).

    (j) P. S. Braterman and R. J. Cross, *Chem. Soc. Rev.*, 2, 271 (1973).

    (k) A. Sevin, *Nouv. J. Chim.*, 5, 233 (1981); A. Sevin and P. Chaquin, ibid., 7, 353 (1983).

CHAPTER TWENTY

Complexes of ML₃, MCp, and Cp₂M

20.1. DERIVATION OF ORBITALS FOR A C_{3v} ML₃ FRAGMENT

Following the pattern established in the previous chapters we can construct the valence orbitals of a C_{3v} ML₃ fragment by the removal of three *fac* ligands in an octahedral complex, **20.1–20.2**. Three *empty* hybrid orbitals will then be formed, labeled ϕ_1, ϕ_2, and ϕ_3 in **20.2**. These point toward the missing ligands. If the original octahedron was a d^6, 18-electron complex, then **20.2** will also possess three

| 20.I | 20.2 |

filled valence orbitals which closely correspond to the t_{2g} set (Section 15.2) of the octahedron. Recall that L is taken to be an arbitrary ligand with only σ-donating capability, therefore, the t_{2g} set in **20.1** is rigorously nonbonding. Those three orbitals will consequently remain unperturbed when the three *fac* ligands are removed. Alternatively one can easily establish that if the three ligands have π acceptor functions, then the three members of t_{2g} will rise in energy on going to **20.2** since π backbonding is lost. The three localized bond orbitals, illustrated in **20.2**, are highly directional. They are a convenient set to be used for conformational problems. Linear combinations of them form a set of symmetry-correct fragment orbitals. This is shown in **20.3**. One orbital of a_1 symmetry and an e set are formed by the

$$\psi_1 \propto \phi_1 + \phi_2 + \phi_3 \quad = \qquad \equiv$$

$$\psi_2 \propto 2\phi_1 - \phi_2 - \phi_3 \quad = \qquad \equiv$$

$$\psi_3 \propto \phi_2 - \phi_3 \quad =$$

20.3

linear combinations (see Section 4.3). For $Cr(CO)_3$ the $a_1 + e$ triad in **20.3** is empty and a set of three orbitals of $a_1 + e$ symmetry which are analogous to the t_{2g} set in $Cr(CO)_6$ are filled. In the $Fe(CO)_3$ fragment there are two more electrons. It becomes problematic whether they are associated with the a_1 or e orbitals. In other words, does the a_1 level lie energetically above or below the e set? The situation is most clearly seen by looking carefully at what happens to the MOs of an octahedron when three *fac* ligands are removed. This is done in Figure 20.1 for the generalized ML_6 to ML_3 conversion. The t_{2g} and e_g sets, displayed on the left side of this figure, have a different composition from what we have previously used. This is due to a change of coordinate system, shown at the top center of Figure 20.1. The z axis coincides with a threefold rotational axis of the octahedron. The orbitals are exactly the same as those derived in Section 15.2; however, their atomic composition has changed. The members of t_{2g} become:[1]

$$z^2$$
$$\sqrt{\tfrac{2}{3}}\,(x^2 - y^2) - \sqrt{\tfrac{1}{3}}\,yz$$
$$\sqrt{\tfrac{2}{3}}\,xy - \sqrt{\tfrac{1}{3}}\,xz$$

and the metal component of e_g is given by:

$$\sqrt{\tfrac{1}{3}}\,(x^2 - y^2) + \sqrt{\tfrac{2}{3}}\,yz$$
$$\sqrt{\tfrac{1}{3}}\,xy + \sqrt{\tfrac{2}{3}}\,xz$$

The members of the t_{2g} set that are predominately $x^2 - y^2$ and xy have some yz and xz character, respectively, mixed into them so that they are reoriented to lie between the M—L bonds. The e_g set is mainly xz and yz. The xy and $x^2 - y^2$ character mixed into them provides maximal antibonding to the ligand lone-pair functions. The intermixing of atomic functions is due to nothing more than our choice of a coordinate system. It is certainly not the normal one for an octahedron; however, it is the natural choice for a C_{3v} ML_3 unit.

When the three *fac* ligands are removed from ML_6 the t_{2g} set is unperturbed. The z^2 component is labeled $1a_1$ and the other two are listed as the $1e$ set at the right of Figure 20.1. The e_g set is stabilized considerably on going to ML_3 since

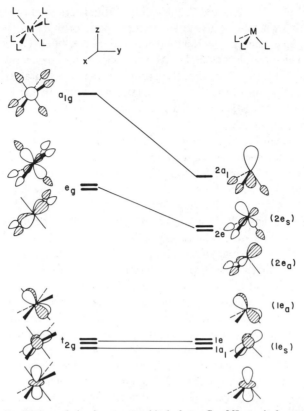

FIGURE 20.1. Derivation of the fragment orbitals for a C_{3v} ML$_3$ unit from an octahedron.

one-half of the antibonding from the ligand lone-pair hybrids to metal is lost. Because the symmetry of the molecule is lowered from O_h to C_{3v}, some hybridization also comes into play. Metal x and y character mix into what was e_g in a way that is bonding to the remaining lone-pair functions; see **20.4**. Another way to think

20.4

about this hybridization is that reduction of the symmetry allows the higher lying $2t_{1u}$ set (see Figure 15.1) to mix into e_g in a bonding way. The resultant orbitals, labeled $2e$ in Figure 20.1, are hybridized away from the remaining ligands. The a_{1g} orbital in ML$_6$ also loses one-half of its antibonding when the ligands are removed. So that orbital, $2a_1$ in Figure 20.1, is also lowered in energy and considerable metal z character mixes into it in a way which is bonding to the lone-pair hybrids on the remaining ligands. The $2a_1$ and $2e$ molecular orbitals are readily identified as the $a_1 + e$ set in **20.3** which were generated from the bond orbital approach. Notice that $2a_1$ lies at a higher energy than the $2e$ combination. A d^8 Fe(CO)$_3$ molecule would then have two electrons housed in $2e$. Another way to rationalize this level ordering is to note that the $2a_1$ and $2e$ orbitals are primarily metal in character. Furthermore, the major contribution to the $2e$ set is metal d while that in $2a_1$ is mainly metal s and p. The level ordering is then a natural consequence of the energetics of the constituent AOs on the metal.

The intermixing of $x^2 - y^2$ with yz and xy with xz in the $1e$ and $2e$ sets can be derived along other lines.[2] It is also sensitive to the pyramidality of the ML$_3$ fragment. That is most apparent by comparing the ML$_3$ orbitals at a L—M—L angle of 90° in Figure 20.1 with those for a planar ML$_3$ unit where L—M—L = 120° at the middle of Figure 18.2. Starting from the planar, D_{3h} species the a_2'' orbital evolves into $2a_1$ (Figure 20.1) upon pyramidalization. The perturbation is exactly analogous to the AH$_3$ pyramidalization in Section 9.2B. The a_2'' level is stabilized and hybridized by the second-order mixing of a higher lying metal s orbital. The a_1', z^2, level is also stabilized on pyramidalization since the ligand lone-pair functions move off from the torus, toward the nodal plane of z^2 (see **1.3**). At the planar geometry the e' set consists of $x^2 - y^2$ and xy (see **18.6** and **18.8**). The lower energy e'' set is metal xz and yz. At the pseudo-octahedral geometry the $2e$ set is mainly xz and yz, whereas the lower lying $1e$ set consists of primarily $x^2 - y^2$ and xy character. In other words, pyramidalization induces an intermixing of the character in the two e sets and an avoided crossing occurs. So the exact composition of $1e$ and $2e$ is given only at the planar, D_{3h}, and pyramidal geometry when L—M—L = 90°. For the ML$_3$ complexes that we shall study, the L—M—L angles are close to 90°, so the level ordering and orbital shape shown on the right side of Figure 20.1 is appropriate.

An example where the ML$_3$ valence orbitals are utilized is given by cymantrene, cyclopentadienyl-Mn(CO)$_3$. An orbital interaction diagram is shown in Figure 20.2. The complex has been divided into cyclopentadienyl (Cp) anion and Mn(CO)$_3$ cation fragments. The a_2'' orbital of Cp$^-$ and $1a_1$ and $2a_1$ from Mn(CO)$_3^+$ enter into a three orbital pattern. The lowest molecular level is primarily a_2'' stabilized by $1a_1$ and $2a_1$. The middle member is mainly $1a_1$. Some a_2'' character mixes into the molecular orbital in an antibonding manner. Furthermore, $2a_1$ mixes into it in second order (bonding with respect to a_2''). It is the second-order mixing that keeps $1a_1$ at moderate energy. The e_1'' set of Cp$^-$ is stabilized greatly by $2e$ on Mn(CO)$_3^+$. Finally there is a weak interaction between the $1e$ and e_2'' levels. That stabilizes the $1e$ set, but not by much. First of all, the overlap between $1e$ and e_2'' is primarily of the δ type and consequently is much smaller than the predominately π type between e_1'' and $2e$ or the σ type in the $1a_1 + 2a_1 + a_2''$ combinations. Secondly, there are relatively high lying σ orbitals on the Cp fragment which overlap with and destabilize the $1e$ set. Therefore, the $1e$ and $1a_1$ based molecular levels are not expected to be split apart significantly in energy. The photoelectron spectrum of CpMn(CO)$_3$ shows this to be true.[3] Notice that the symmetry of this complex is only C_s. However, the "apparent" symmetry is higher. This is because the symmetry of Cp$^-$

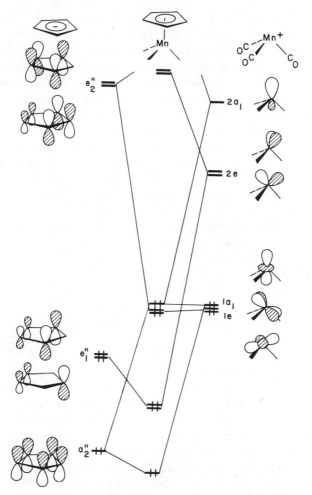

FIGURE 20.2.   An orbital interaction diagram for $CpMn(CO)_3$.

is D_{5h} and $Mn(CO)_3^+$ is C_{3v}. The e sets of Cp^- find good overlap matches with the e sets on $Mn(CO)_3^+$. In actual fact the $1e$ set of $Mn(CO)_3^+$ has an overlap with both e_1'' and e_2'' of Cp^-. The same is true for the $2e$ set on $Mn(CO)_3^+$. But these are technical details that we will not pursue here.

The astute reader will have noticed something very familiar about the orbital interaction diagram for $CpMn(CO)_3$ in Figure 20.2. There are three closely spaced, filled orbitals at moderate energies. The two lowest unoccupied orbitals are primarily metal $2e$ antibonding to the e_1'' set on Cp^-. Furthermore, $2e$ is comprised of metal d antibonding to the carbonyl donor orbitals (see Figure 20.1). The octahedral splitting pattern has been restored! In other words, there is not much difference between the interaction diagram in Figure 20.2 and the one for $Cr(CO)_3$ (also a d^6 ML_3 fragment) interacting with three carbonyls. The three carbonyl σ donor orbitals form symmetry-adapted linear combinations that are topologically analogous to the $a_2'' + e_1''$ set of Cp$^-$. Replacement of a Cp$^-$ fragment for three carbonyl ligands is a useful concept and is one that we shall extensively develop; however, it is important to realize that there are differences between the two fragments.[4] Perhaps the most important difference lies in the fact that the three carbonyl ligands are excellent π acceptors. To be sure, the Cp$^-$ ligand does possess the e_2'' set for metal backbonding, but its spatial extent does not allow for maximal overlap with $1a_1$ and $1e$ as is present for the π^* combinations on a *fac* carbonyl set. This brings up a final point about the molecular orbital description of CpMn(CO)$_3$. It is obvious that there are six bonds to Cr in Cr(CO)$_6$ or any d^6 ML$_6$ complex. The delocalized molecular orbital picture in Figure 15.1 also shows six filled M—L bonding orbitals. In CpMn(CO)$_3$ there will be three orbitals that are bonding between Mn and CO donor functions, not shown in Figure 20.2. There are also three filled molecular orbitals that are bonding between the Cp$^-$ π set and the Mn(CO)$_3^+$ fragment. Therefore, it is conceptually useful to imagine that there are three bonds between Mn and Cp unit. The single line between Cp and Mn at the top center of Figure 20.2 does *not* imply a bond order of one between the two units. Rather, it indicates delocalized bonding between Mn and five carbons.

Actually the basic orbital pattern for any 18-electron polyene-ML$_3$ complex will be very similar to that found in CpMn(CO)$_3$. Figure 20.3 illustrates the situation for cyclobutadiene–Fe(CO)$_3$. The e_g set on cyclobutadiene is stabilized by $2e$ on Fe(CO)$_3$. Likewise, the a_{2u} orbital is stabilized by the $1a_1$ and $2a_1$ levels on Fe(CO)$_3$. Three metal-centered orbitals are left "nonbonding." Notice that the two fragments have been partitioned to both be neutral. The e_g set on cyclobutadiene and $2e$ set on Fe(CO)$_3$ are each half-filled. It is reasonable to assume that e_g lies at a lower energy than $2e$ (recall that the $2e$ set is carbonyl σ-metal d antibonding). Therefore, the electron density in the molecular orbitals that result from the union of these two fragment orbitals is more concentrated on the cyclobutadiene ligand. To take an extreme view one could say that the two electrons in the $2e$ levels of Fe(CO)$_3$ are transferred to the e_g set located on the cyclobutadiene fragment, making it a six-π-electron aromatic system. That is certainly an overstatement but it does point to the fact that electrophilic substitution reactions on the cyclobutadiene ligand are very common.[5] Electrophilic substitution on the Cp ligand for CpMn(CO)$_3$ is also facile and this reactivity has been used to support the concept of "metalloaromaticity"[6] for CpMn(CO)$_3$ and cyclobutadiene–Fe(CO)$_3$. However, what is not clear at the present time is whether an electrophile directly attacks the polyene ring or whether attack occurs at the metal with subsequent migration of the electrophile to the polyene ring. From the interaction diagrams of Figure 20.2 and 20.3 it is clear that the molecular orbitals involving polyene π-metal d interactions are concentrated on the polyene portion of these molecules. That would favor direct attack by an electrophile on the polyene. However, there are also three occupied molecular orbitals in each complex which are the remnants of the octahedral t_{2g} set. Since these lie at relatively high energies and are concentrated at the metal, they too will overlap effectively and find a good energy match with the LUMO of an electrophile. This interaction, of course, leads to attack at the metal. Which re-

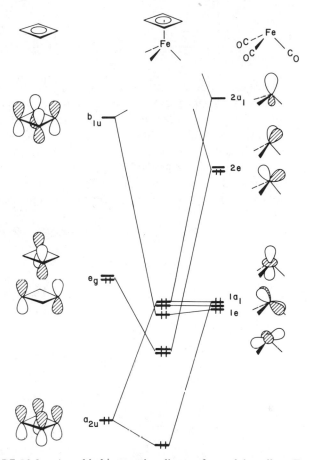

FIGURE 20.3.   An orbital interaction diagram for cyclobutadiene–Fe(CO)$_3$.

action path occurs is not clear; furthermore, it will be sensitive to the metal, the electronic features of the ligands, and the steric bulk of the electrophile. One can also envision reaction paths wherein the electrophile LUMO interacts with a polyene π-based MO and metal "t_{2g}" MO simultaneously (an analogous situation was found for protonation of H_2O in Section 7.5B). In other words, the electrophile directly attacks the polyene ring from the same side as the metal is coordinated.

20.2.   THE CpM FRAGMENT ORBITALS

Suppose the three carbonyl ligands are removed from CpMn(CO)$_3$; this generates the CpMn fragment, **20.5**. Three empty hybrid orbitals are produced which point toward the missing carbonyls. This is exactly the same pattern produced by the removal of three *fac* carbonyls from Cr(CO)$_6$, see **20.2**. In other words, the CpMn fragment is expected to be very similar to Cr(CO)$_3$. That result should not be too

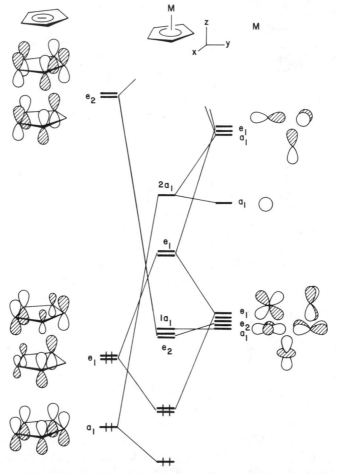

20.5

surprising. In the previous section it was shown how a Cp$^-$ ligand is topologically equivalent to three *fac* carbonyls. Therefore, replacing the carbonyls with Cp$^-$ in Cr(CO)$_3$ generates CpCr$^-$ which is isoelectronic to CpMn. The orbitals of an arbitrary CpM fragment are constructed in Figure 20.4 by interacting a Cp$^-$ ligand with an M atom. The symmetry labels used for the orbitals of Cp$^-$ and M correspond to the C_{5v} symmetry of CpM. The lowest π level of Cp$^-$, a_1, is stabilized by metal s and z (see coordinate system at the top of this figure). The e_1 set on Cp$^-$ is stabilized primarily by metal xz and yz and to a lesser extent (because of the larger energy

FIGURE 20.4.   An orbital interaction diagram for the MCp fragment, which shows the orbital occupancy for the d^0 case.

gap) by metal x and y. Therefore, a total of the three Cp–metal bonding orbitals are occupied for any CpM fragment. There are also three metal-centered orbitals at moderate energy. The $x^2 - y^2$ and xy levels are stabilized to a small extent by the e_2 set on Cp$^-$. The δ overlap and large energy gap make this interaction relatively weak. Although the a_1 π level of Cp$^-$ and z^2 have the same symmetry, they overlap with each other to a minor extent. The a_1 π level lies approximately on the nodal plane of z^2 (see **1.3** for the nodal properties of z^2). Consequently z^2 is left nonbonding. Metal xz and yz are significantly destabilized by the Cp$^-$ e_1 set. However, metal x and y mix into the *molecular orbital* labeled e_1 in Figure 20.4 bonding to the Cp e_1 set in second order. This shown in **20.6**. This second-order mixing keeps

20.6

molecular e_1 at moderate energy and hybridizes the metal-centered orbitals away from the Cp ligand. Inspection of Figure 20.4 shows another molecular orbital, labeled $2a_1$, at moderate energy. It is the middle member of the metal s, z, and Cp a_1 union. It arises via a first-order orbital mixing between metal s and the Cp a_1 orbital (antibonding) with a second-order mixing of metal z (bonding between metal and Cp) as in **20.7**. Molecular $2a_1$ is again hybridized out, away from the Cp

20.7

ligand. Not shown in Figure 20.4 are three levels primarily of metal x, y, and z character antibonding to Cp, $a_1 + e_1$, and the Cp-based e_2 set which are destabilized by $x^2 - y^2$ and xy.

The valence levels of a CpM fragment are then $e_2 + 1a_1 + e_1 + 2a_1$ in Figure 20.4. A d^6 CpM fragment (like CpMn) will have $e_2 + 1a_1$ filled. The $e_1 + 2a_1$ set of three levels are empty and because of their hybridization (see **20.6** and **20.7**) will form the strongest interactions with extra ligands.

The reader should carefully compare these valence orbitals with those of a ML$_3$ fragment in Figure 20.1. The ML$_3$ fragment has the same three below, three above level pattern and almost identical atomic composition at the metal. In the CpM fragment it is clear that the e_2 set is of δ symmetry and e_1 is of π symmetry. However, $1e$ in ML$_3$ is primarily δ with some π character and $2e$ is mainly π with some δ character. The e sets in ML$_3$ are tilted off from the xz plane whereas the e sets in MCp are not. The tilting in ML$_3$ comes about because of its octahedral parentage. The three σ donor orbitals of the L groups are highly localized and are situated at three corners of an octahedron. On the other hand, the three donor orbitals of Cp$^-$ are delocalized, of course, over the entire Cp ring and have a cylindrical symmetry. Therefore, metal δ and π functions remain distinct. In most cases it makes no difference whether the e sets in ML$_3$ and MCp are tilted or not, thus for the same electron count the fragments can be interchanged. This is a critical factor, however, in polyene–ML$_3$ and polyene–MCp rotational barriers.[7] When a polyene possesses a threefold localization of π donor orbitals, then the interaction with the $2e$ acceptor set in ML$_3$ will be maximized in one conformation. Examples of polyenes where this is found are given by hexa-alkylborazines, **20.8** (see Section 12.4) and the trimethylenemethane dianion, **20.9**. Therefore, a Cr(CO)$_3$ complex of **20.8** and Fe(CO)$_3^{2+}$ complex of **20.9** have substantial rotational barriers.[7]

20.8   **20.9**

The advantage in knowing the form of the valence orbitals in ML$_3$ and MCp fragments can be illustrated for triple-decker sandwiches, **20.10**. Based on our pre-

20.10

vious experience we expect that a stable metal configuration will be one where six electrons occupy the "t_{2g}-like" set. The ligands in **20.10** will present a total of 18 electrons to the metals, thus an electron count of 30 is anticipated to be a stable one. This is true, but complexes with up to four more electrons also exist. It is easy to see how this comes about.[8] An orbital interaction diagram for the Cp$_3$M$_2$ example is given in Figure 20.5. In-phase and out-of-phase combinations of the valence orbitals for a CpM dimer are indicated on the left side of the figure. They are not split apart much in energy because of the large distance between the metal atoms. So there is a nest of six levels at low energy which correspond to the $1a_1 + e_2$ set in MCp. At higher energy are the combinations derived from the $2a_1 + e_1$ set. On the right side of this figure are shown the three donor orbitals of the middle Cp$^-$ unit. They find good overlap matches with the a_2'' and e_1'' fragment orbitals that are drawn for the CpM dimer. The six metal-centered orbitals of the Cp$_2$M$_2$ fragment, $a_1' + e_2' + e_2'' + a_2''$, are left nonbonding along with the in-phase combina-

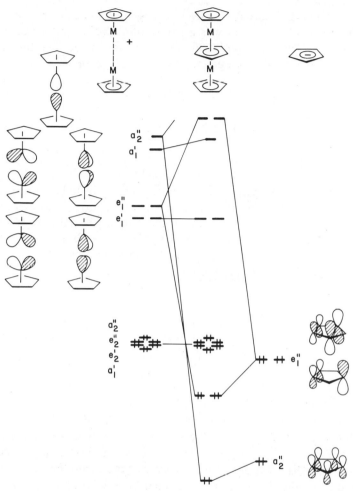

FIGURE 20.5. An orbital interaction diagram for a triple decker sandwich complex in D_{5h} symmetry.

tion of the CpM e_1 set which has e'_1 symmetry. Figure 20.5 shows the occupancy for a 30-electron case (remember that there are six occupied levels not shown in this figure which are Cp—M bonding for the end Cp units). Notice that the Cp_2M_2 e'_1 set lies at moderate energy and is well separated from the antibonding combination of Cp_2M_2 e''_1 and the e''_1 set from the central Cp^- ligand. Threfore, complexes with four more electrons are also stable; $Cp_3Ni_2^+$ is one such example.[9] One might naively think that since there are formally 17 electrons associated with each metal atom in the 34-electron systems, then there ought to be a direct metal–metal single bond. However, the distance between the nickel atoms in $Cp_3Ni_2^+$ is 3.58 Å,[10] which is far too long to permit any substantial metal–metal interaction. Likewise, in the 30-electron case, it is clear that there can be no metal–metal triple bond (see Figure

20.5). The middle Cp^- ligand effectively couples the electrons between the two CpM units. In other words, there is a strong through-bond rather than through-space interaction. The 18-electron rule obviously cannot be used for these situations. An MO-based, delocalized description like that presented in Figure 20.5 must be utilized.

20.3. Cp_2M AND METALLOCENES

We have previously discussed how a *fac* L_3 set in an octahedrally based complex is equivalent to a Cp^- ligand. Thus, the level splitting pattern for $Cr(CO)_6$, **20.11**, (see Figure 15.1) is very similar to that in $CpMn(CO)_3$, **20.12** (see Figure 20.2). One can quibble about minor (although chemically important) differences, for example, the *fac* carbonyl set is a better π acceptor which stabilizes the metal t_{2g} levels more than the e_2 acceptor set does in Cp^-. However, the basic three below two level pattern for the valence, metal-centered orbitals and their nodal structure occurs in both compounds. Replacing the tnree carbonyls in **20.12** with Cp^- yields Cp_2Mn^- which is isoelectronic with ferrocence, **20.13**. The six (localized) Cr—CO

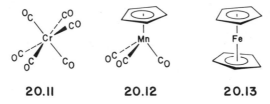

20.11   **20.12**   **20.13**

bonds in $Cr(CO)_6$ are obvious. There should also be six Fe—C bonds in ferrocene. That is difficult to see in a localized sense. Figure 20.6 shows an orbital interaction diagram for ferrocene at a staggered (the Cp rings are staggered with respect to each other), D_{5d}, geometry. The molecule has been partitioned into Fe^{2+} and Cp_2^{2-} units. A short-hand notation has been used for the orbitals of the Cp^- dimer on the left side of this figure which emphasizes the nodal characteristics and phase relationships to the metal atom, as illustrated for a couple of examples in **20.14**. The Cp_2^{2-}

20.14

levels of a_{1g} and a_{2u} symmetry are stabilized by metal s and z, respectively. Likewise the Cp π sets, e_{1u} and e_{1g} are stabilized by metal x, y, and xz, yz, respectively. We have just described six occupied MOs which are bonding between Fe and the Cp rings. These are the six bonds that are analogous to the Cr—CO bonds in $Cr(CO)_6$ of $1a_{1g}$, $1t_{1u}$, and $1e_g$ symmetry (see Figure 15.1). At moderate energy are the molecular orbitals labeled e_{2g} and a_{1g}. They are basically the nonbonding $x^2 - y^2$,

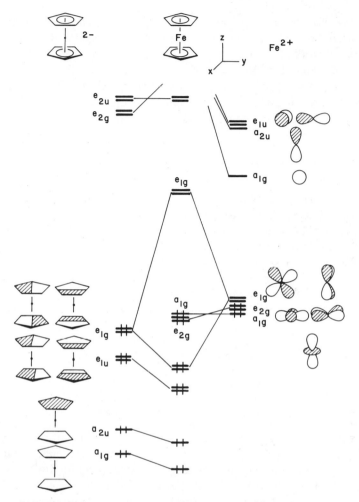

FIGURE 20.6.   Construction of the molecular orbitals of ferrocene.

xy, and z^2 set. Finally at higher energy is the molecular e_{1g} level. This is the anti-bonding combination of metal xz and yz with the e_{1g} set of Cp_2^{2-}. Notice that the octahedral splitting pattern of three below two has again been established.

Ferrocene, an 18-electron complex, has the bottom nine MOs filled through a_{1g}. Actually Cp_2V with 15 and Cp_2Cr with 16 electrons also exist. They have three and four electrons, respectively in the $a_{1g} + e_{2g}$ manifold. Cp_2Co with 19 electrons and Cp_2Ni with 20 electrons have also been prepared. For Cp_2Ni the extra two electrons must be placed in the molecular e_{1g} levels. Therefore, the molecule has triplet ground state. The molecular e_{1g} set (see Figure 20.6) is antibonding between the Cp rings and Fe. Therefore, occupation of e_{1g} will cause the metal–carbon bond distance to increase; for example, the Fe—C distance in Cp_2Fe is 2.05 Å[11] while that in Cp_2Ni is 2.19 Å.[12] In Cp_2Co and Cp_2V—Cp_2Mn there is also the possibility of Jahn–Teller distortions. The distortion here is one where the Cp rings buckle and the metal may be shifted away from a perfect η^5 coordination. Experimental and theoretical work on this topic has been carried out by Ammeter and coworkers.[13] Interaction diagrams for bis(benzene)Cr or any other metallocene could also be constructed. They show very similar splitting patterns. The photoelectron spectra[14] for a wide variety of metallocenes including the Cp_2V—Cp_2Ni series have been analyzed with some care and confirm this expectation.

20.4. Cp_2ML_n COMPLEXES

The last mononuclear transition metal fragment that we shall study in some depth is bent Cp_2M. There exists a vast body of chemical information on Cp_2ML_n complexes where $n = 1$–3 and M is an early transition metal atom (e.g., Ti, V. Zr, Hf, Mo, etc). There have been many theoretical treatments of these molecules;[15] the one we shall follow was given by Lauher and Hoffmann.[16] These Cp_2ML_n complexes are unique for several reasons; the L_n groups are forced to lie in a common plane, see 20.15 for one example. There are obvious steric requirements; in 20.15 the H—Nb—H angle is only 61°.[17] Cp_2ZrCl_2, 20.16, is illustrative of another com-

20.15   20.16

mon problem it is a 16-electron complex and yet is even air stable! A 16-electron count at the metal is perhaps more common for these compounds than the 18-electron one. Finally, allowing for the fact that each Cp^- ligand forms three bonds to the metal, 20.15 and 20.16 are then 9 and 8 coordinate, respectively. These are high coordination numbers for a nonlanthanide metal; coordination numbers of 4-6 are far more common.

In 20.15 and 20.16 the Cp rings are bent back from the parallel geometry treated for metallocenes in the previous section. This is a pervasive feature for Cp_2ML_n complexes. It is easiest to derive the fragment orbitals of a bent Cp_2M unit from the parallel geometry. The molecular orbitals of the metallocene that we shall study are the e_{2g} and a_{1g} orbitals of Figure 20.6. The other orbitals are either Cp centered or lie at too high an energy (for example the molecular e_{1g} set in Figure 20.6). The three initial levels of a parallel Cp$_2$M system are shown on the left side of Figure 20.7. It does not matter whether the geometry of the Cp rings is staggered or eclipsed. For convenience the orbitals are labeled in C_{2v} symmetry when the Cp rings are bent back as if we had utilized an eclipsed geometry. Bending back the Cp rings splits the e_{2g} set into orbitals of a_1 and b_2 symmetry. The b_2 orbital rises in energy as the Cp rings are bent back. It loses some of the bonding to the Cp π^* orbitals and repulsive interactions with Cp σ levels are turned on. The same would

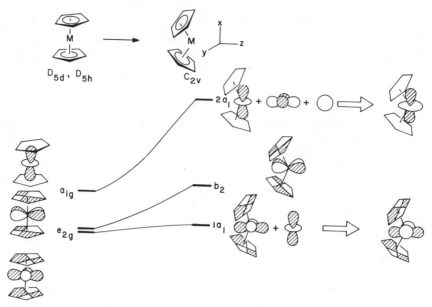

FIGURE 20.7. A Walsh diagram for bending back the Cp rings in the Cp$_2$M fragment.

happen to the other component of e_{2g} which is labeled $1a_1$ in Figure 20.7. How-ever, it mixes with what was a_{1g}, now labeled $2a_1$. This mixing between $1a_1$ and $2a_1$ is shown on the right side of the figure and it serves to keep $1a_1$ at constant energy. The $2a_1$ orbital, however, rises rapidly in energy. This is partly due to the avoided crossing with $1a_1$ and the fact that the a_{1g} combination of Cp π levels (see Figure 20.6) has a greater overlap with metal z^2 at a bent geometry. Some metal s character and $x^2 - y^2$ from $1a_1$ mix into $2a_1$ so that the torus of z^2 becomes hybridized away from the Cp ligands. As mentioned previously the L groups in Cp$_2$ML$_n$ com-plexes lie in the yz plane (for the coordinate system see Figure 20.7). A convenient view of these orbitals in the yz plane that we shall utilize for the remaining discus-sion is given in **20.17**. The reader should recall that the levels in **20.17** are derived from the t_{2g}-like set of a metallocene. They now have been split apart in energy and are somewhat hybridized in the direction of the L$_n$ set.

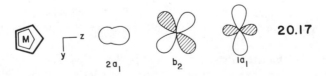

20.17

In Cp$_2$NbH$_3$, **20.15** the metal is formally d^0. Therefore, $1a_1$, b_2, and $2a_1$ are empty. The three hydride ligands form symmetry-adapted combinations shown in **20.18–20.20**. They nicely match the nodal properties of **20.17**: $1a_1$ with **20.20**, b_2 with **20.19**, and $2a_1$ with **20.18**. So all three donor functions are stabilized, yielding a stable d^0 complex with an 18-electron count.

20.18   **20.19**   **20.20**

Construction of the molecular orbitals for a Cp_2ML_2 complex yields an analogous pattern with one important difference. The ligand set brings with it donor functions of a_1 and b_2 symmetry, as shown at the right side of **20.21**. The b_2 ligand orbital

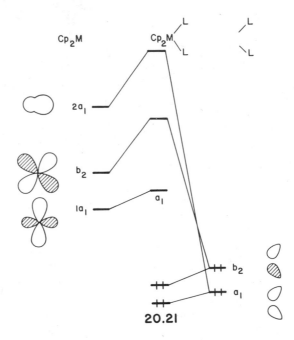

20.21

finds an excellent match with b_2 on Cp_2M. The a_1 donor orbital has maximum overlap with the $2a_1$ fragment orbital. The lone-pair functions lie almost on the nodal planes of $1a_1$ and so the resultant overlap is small. The Cp_2M $1a_1$ orbital remains nonbonding at moderate energy. It is well separated from the destabilized b_2 and $2a_1$ levels and, therefore, an 18-electron complex like Cp_2MoCl_2 is stable where $1a_1$ is filled. Notice also in **20.21** that $1a_1$ is well separated from the two M—L bonding orbitals. A 16-electron complex with $1a_1$ empty is then also a stable complex, an example being Cp_2ZrCl_2.

There are important reactivity differences between Cp_2ML_2 systems, depending upon whether the $1a_1$ orbital is filled or not. Consider reductive elimination in a dialkyl substituted complex. In **20.22** the orbitals of a d^0 Cp_2MR_2 complex are listed. A correlation is drawn for reductive elimination to a d^2 Cp_2M complex along with the σ and σ^* levels of an alkane in **20.23** where a least-motion pathway that conserves C_{2v} symmetry is followed. The reaction is symmetry forbidden, a non-least-motion pathway must be followed, not unlike reductive elimination from an

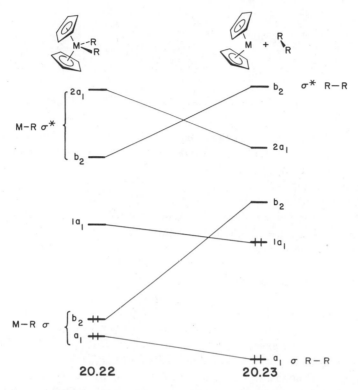

20.22   **20.23**

alkane to give a carbene and H$_2$, for example. The symmetry prohibition would be removed starting from an 18-electron, d^2 Cp$_2$MR$_2$ complex where $1a_1$ is filled. However, notice that reductive elimination may still be energetically unfavorable since one M—R bonding orbital (b_2) must rise to high energy ultimately becoming the b_2 orbital of Cp$_2$M. There is a further complication, namely that there is no reason for the Cp$_2$M product to remain bent. A state correlation diagram which incorporates this feature has been developed[18] and the important conclusion is that reductive elimination from 18-electron complexes should actually be photochemically initiated. The reverse path, insertion of d^4 Cp$_2$M into a C—C (or C—H) bond, is expected to be more favorable and there is some evidence for this reaction.[19] The reader should refer back to Section 19.5 and compare the situation for reductive elimination there with the correlation in **20.22-20.23**. There are many differences because the ligand set has changed.

There are interesting π-bonding effects in Cp$_2$ML$_n$ complexes. Consider a Cp$_2$M(CH$_2$)L complex. The carbene may be oriented as given by **20.24** or **20.25**. There are greater steric problems with **20.24**; the hydrogens on the carbenes are quite close to the Cp rings. Yet **20.24** is more stable than **20.25** and rotational barriers about the metal–carbene bond in the range of 15 to greater than 20 kcal/mol have been observed.[20] The electronic structure for the σ component in this molecule will correspond to that found for Cp$_2$ML$_2$ in **20.21**.[16] In structure **20.24** the carbene p orbital may also interact with $1a_1$ as shown by **20.26**.[21] In **20.25** no such in-

20.24   20.25   20.26

teraction exists with the valence orbitals. Consequently a d^2 complex where $1a_1$ is filled (and the carbene p orbital is formally empty) will provide a maximum stabilizing interaction at the geometry given by **20.24**. This is indeed the electron count where carbene complexes of this type have been prepared.

We saw in Section 15.3 that the orbitals of CH_2 were topologically like those in ethylene. Each has a donor orbital of a_1 symmetry and a low-lying acceptor orbital. The electronic structure of an (alkyl) (olefin)MCp₂ complex is then similar to the Cp₂M (carbene) ones which we have just treated. **20.27** shows the $1a_1 - \pi^*$ interaction which is analogous to the π interaction in **20.26** for the carbene case. A key intermediate for olefin polymerization using Ziegler–Natta catalysts are these (alkyl) (olefin)MCp₂ species,[22] **20.28**. The alkyl group inserts into the metal olefin bond yielding **20.29** which picks up another olefin, and so on. The olefin π and alkyl lone-pair orbitals present two donor functions to the metal. A pattern like that in **20.21** is again found except for the stabilization of $1a_1$ by olefin π^* (**20.27**). On going from **20.28** to **20.29** the two donor-based orbitals smoothly transform[16] into M—C and R—C bonds. However, in **20.29**, the $1a_1$ orbital is no longer stabilized by the olefin π^* level and it along with the $2a_1$ orbital of Cp₂M overlap considerably with the lone-pair orbital of the newly formed alkyl group. The $1a_1$ level will be destabilized and lie at a much higher energy than it did in **20.28**. Consequently an efficient catalyst will be one with $1a_1$ empty—a d^0, 16-electron system.

20.27   20.28   20.29

The oxidative coupling reaction, **20.30–20.31**, offers an interesting contrast to the (olefin)₂Fe(CO)₃ rearrangement in Section 18.3. Both reactions involve 18-electron bis-olefin complexes rearranging to 16-electron metallacyclopentanes. The Fe(CO)₃ case was shown to be symmetry forbidden. The orbitals of **20.31** are going to be very similar to those for Cp₂ML₂ in **20.21**. This is a d^0 complex, so $1a_1$ is empty. The orbitals of **20.30** are constructed in Figure 20.8.[16] In-phase and out-of-phase combinations of the ethylene π and π^* orbitals are of a_1 and b_2 symmetry. The a_1 and b_2 combinations of the ethylene π set are stabilized by the Cp₂M $2a_1$ and b_2 fragment orbitals, respectively. The resultant molecular orbitals are shown in **20.32** and **20.33**. The $1a_1$ Cp₂M level is especially stabilized by the a_1 combina-

20.30   20.31

a_1 b_2 $1a_1 + \pi^*$
20.32   20.33   20.34

(20.34) $1a_1 + \pi^*$

(20.33)
(20.32) b_2
         a_1 b_2 Ti—C σ
                                            a_1

20.35 a_1 C—C σ

tion of ethylene π^*, as shown in **20.34**. As the two olefins are coupled a new C—C bond is formed. In other words, **20.32**, which is concentrated on the two olefins, is stabilized and will form the C—C σ bond in the titanacyclopentane. A partial correlation for a least-motion pathway that conserves C_{2v} symmetry is shown in **20.35**. **20.33** and **20.34** smoothly correlate to the Ti—C σ orbitals of b_2 and a_1 symmetry, respectively. Therefore, the reaction is symmetry allowed. For the same electron count in the reactant and product we saw in Section 18.3 that a C_{2v} Fe(CO)$_3$ complex gave a symmetry-forbidden reaction. The electronic details of a reaction are strongly influenced by the number and geometrical disposition of the auxiliary ligands. The transition metal cannot be viewed simply as an electronic black-box that supplies and accepts electrons from the organic portion of the molecule. Unfortunately this attitude has been taken too many times in the past. One must not forget that the remaining ligands, be they multidentate like Cp or of the σ donor type like CO or phosphines, tailor the metal orbitals to specific shapes and energy

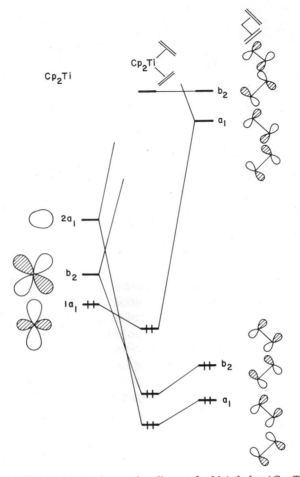

FIGURE 20.8. An interaction diagram for bis(ethylene)Cp_2Ti.

patterns. Even with this provision there are a tremendous number of analogies that can be used to simplify reactivity problems. A very important one that links organic, organometallic, inorganic, and main group chemistry is the subject of the next chapter.

REFERENCES

1. L. E. Orgel, *An Introduction to Transition-Metal Chemistry*, Wiley, New York (1960), p. 174.
2. M. Elian and R. Hoffmann, *Inorg. Chem.*, **14**, 1058 (1975); J. K. Burdett, ibid., **14**, 375 (1975); *J. Chem. Soc., Faraday Trans. II*, **70**, 1599 (1974).
3. D. L. Lichtenberger and R. Fenske, *J. Am. Chem. Soc.*, **98**, 50 (1976); D. C. Calabro and D. L. Lichtenberger, ibid., **103**, 6846 (1981).

4. M. Elian, M. M. L. Chen, D. M. P. Mingos, and R. Hoffmann, *Inorg. Chem.*, **15,** 1148 (1976).

5. See, for example, J. P. Collman and L. S. Hegedus, *Principles and Applications of Organo-transition Metal Chemistry*, University Science Books, Mill Valley, CA; R. Pettit, *J. Organomet. Chem.*, **100,** 205 (1975).

6. B. E. Bursten and R. F. Fenske, *Inorg. Chem.*, **18,** 1760 (1979).

7. T. A. Albright, P. Hofmann, and R. Hoffmann, *J. Am. Chem. Soc.*, **99,** 7546 (1977); T. A. Albright, *Accts. Chem. Res.*, **15,** 149 (1982).

8. J. W. Lauher, M. Elian, R. H. Summerville, and R. Hoffmann, *J. Am. Chem. Soc.*, **98,** 3219 (1976).

9. A. Salzer and H. Werner, *Angew. Chem.*, **84,** 949 (1972); see also W. Siebert, *Adv. Organomet. Chem.*, **18,** 301 (1980) for several other systems of the 30-and 34-electron type.

10. E. Dubler, M. Textor, H. -R. Oswald, and A. Salzer, *Angew. Chem.*, **86,** 125 (1974).

11. P. Seiler and J. D. Dunitz, *Acta Crystallogr.*, **B35,** 2020 (1979) and references therein.

12. P. Seiler and J. D. Dunitz, ibid., **B36,** 2255 (1980).

13. J. H. Ammeter, N. Oswald, and R. Bucher, *Helv. Chim. Acta*, **58,** 671 (1975); J. H. Ammeter, *Nouv. J. Chim.*, **4,** 631 (1980) and references therein.

14. J. C. Green, *Structure and Bonding*, **43,** 37 (1981).

15. (a) J. L. Petersen, D. L. Lichtenberger, R. F. Fenske, and M. B. Hall, *J. Am. Chem. Soc.*, **97,** 6433 (1975); J. L. Petersen and L. F. Dahl, ibid., **96,** 2248 (1974); **97,** 6416, 6422 (1975).

    (b) H. H. Brintzinger, L. L. Lohr, Jr., and K. T. Wong, ibid., **97,** 5146 (1975); H. H. Brintzinger, *J. Organomet. Chem.*, **171,** 337 (1979).

    (c) N. W. Alcock, *J. Chem. Soc. A*, 2001 (1967).

    (d) For photoelectron spectroscopy studies on Cp_2ML_n complexes, see C. Cauletti, J. P. Clark, J. C. Green, S. E. Jackson, I. L. Fragala, E. Gilberto, and A. W. Coleman, *J. Electron Spectr.*, **18,** 61 (1980); J. C. Green, S. E. Jackson, and B. Higginson, *J. Chem. Soc. Dalton Trans.*, 403 (1975); H. van Dam, A. Terpstra, A. Oskam, and J. H. Teuben, *Z. Naturforsch.*, **36b,** 420 (1981).

16. J. W. Lauher and R. Hoffmann, *J. Am. Chem. Soc.*, **98,** 1729 (1976).

17. R. D. Wilson, T. F. Koetzle, D. W. Hart, A. Kvick, D. L. Tipton, and R. Bau, ibid., **99,** 1775 (1977); also see S. B. Jones and J. L. Petersen, *Inorg. Chem.*, **20,** 2889 (1981) for a more exotic example.

18. A. Veillard, *Nouv. J. Chim.*, **5,** 599 (1981).

19. G. P. Pez, *J. Am. Chem. Soc.*, **98,** 8072 (1976) and references therein.

20. R. R. Schrock, *Science*, **219,** 13 (1983) and references therein.

21. The electronic structure of related metal–carbene complexes have also been studied by M. M. Francl, W. J. Pietro, R. F. Hout, Jr., and W. J. Hehre, *Organometallics*, **2,** 281 (1983); M. M. Francl and W. J. Hehre, *ibid*, **2,** 457 (1983).

22. J. Boor, *Ziegler–Natta Catalysts and Polymerizations* Academic Press, New York (1979).

CHAPTER TWENTY-ONE

The Isolobal Analogy

21.1. INTRODUCTION

In the preceding four chapters we have developed a catalog of orbitals for various ML_n fragments. A recurring theme among what seems to be a myriad of fragment orbitals is the existence of relationships between them. These have paired the valence orbitals of a C_{4v} ML_5 fragment with those of C_{2v} ML_3, the C_{2v} ML_4 orbitals with those of C_{2v} ML_2, and the levels of C_{3v} ML_3 with those of MCp. The first two relationships come from the correspondence between the octahedral and square planar splitting patterns (Section 16.2). The third is a consequence of the fact that Cp^- presents three donor orbitals to the metal that are topologically equivalent to a *fac* L_3 donor set. There exist many more which we will develop in a generalized way in this chapter. This relationship is called the isolobal analogy.[1] Two fragments (or molecules) are said to be isolobal if the number, symmetry properties, occupation by electrons, and approximate energy of their *frontier* orbitals are similar. What makes this relationship so useful is that it relates the orbitals of inorganic/ organometallic fragments to those in the organic/main group areas, namely the AH_n fragment orbitals that were developed in Chapters 7 and 9 to those of ML_n. The isolobal relationship is symbolized by a double-headed arrow with a tear-drop, as shown in **21.1**.

$$ML_n \quad \longleftrightarrow \quad AH_n$$

21.1

The utility of the isolobal analogy is that if a ML_n fragment is isolobal with a particular AH_n arrangement, then one should be able to replace the ML_n fragment in a molecule with a AH_n unit to produce a new compound (on paper at least). The two molecules should have a very similar electronic description and in a very qualitative sense have similar reactivity. The isolobal analogy is, therefore, a useful codifier of electronic, structural, and reactivity data, as well as a tool for predicting new compounds and reactions. The idea had its origins in Halpern's perceptive comments[2] about the similarities that exist between several organic and transition metal intermediates. The structural and stability patterns in boranes and transitional metal clusters developed by Wade[3] and Mingos[4] also played a role. But it was Hoffmann[1] who showed the full utility of the isolobal analogy and extended it greatly.

21.2.   GENERATION OF ISOLOBAL FRAGMENTS

The easiest way to see how particular fragments are isolobal is illustrated in Scheme 21.1. The archtypical molecule of organic chemistry is methane, **21.2**. Breaking one C—H bond in a homolytic sense generates the methyl radical, **21.3**. It has one frontier orbital pointed in the direction of the missing hydrogen atom with one electron in it (Section 9.2). Removing another hydrogen atom creates methylene, **21.4**. There are now two hybrid orbitals which in a localized sense point toward the two missing hydrogens (Section 7.5A). Removing still another hydrogen generates a methine fragment, **21.5**. It has three frontier orbitals (Section 9.1) with three electrons partitioned between them in some way. **21.3–21.5** are representative of the three fragments (AH_3, AH_2, AH where A is a main group element) used through all of organic or main group chemistry with the exception of electron-rich, hypervalent molecules. Another starting point is CrL_6, **21.6**. The use of Cr here is totally arbitrary. What is important is that we have started with a d^6 octahedral complex (consequently it is a saturated 18-electron complex just as **21.2** is an 8-electron compound). We now break one Cr—L bond in a *homolytic* fashion. One electron leaves with L and the other remains at Cr. Therefore, a L_5Cr^- fragment is formed and L^+ is lost (recall that L is a neutral two-electron donor). In order to make the number of charges manageable, let us move one element to the right in the periodic table; neutral MnL_5 is isoelectronic to CrL_5^-. What is again important here is that **21.7** is a d^7, C_{4v} ML_5 fragment. It has one hybrid orbital directed toward the missing ligand (Section 17.2) with one electron in it. The hybrid orbital is composed of mainly metal d with some s and p character mixed into it and is of a_1 symmetry. The frontier orbital of the methyl radical is also of a_1 symmetry and is hybridized away from the remaining hydrogens. It is close to the traditional sp^3 hybrid orbital. But what is important is that the single frontier orbital (with one electron in it) for both MnL_5 and CH_3 is directed out away from the remaining "ligands" and is of a_1 symmetry. This means that MnL_5 and CH_3 are isolobal. So too is $Co(CN)_5^{3-}$, another d^7 system. We shall shortly explore the consequences of this, but for now we note that both CH_3 and several d^7 ML_5 molecules have been studied by ESR spectroscopy and have been trapped in low-temperature matrices.[5] They display very similar chemistry, namely free radical abstraction reactions and dimerization.[2] Removing a second ligand from MnL_5 and moving one element to the right in the periodic table generates the C_{2v} FeL_4 fragment, **21.8**. It is isolobal to CH_2. Removing a third ligand generates the C_{3v} CoL_3 fragment which is isolobal to CH.

SCHEME 21.1

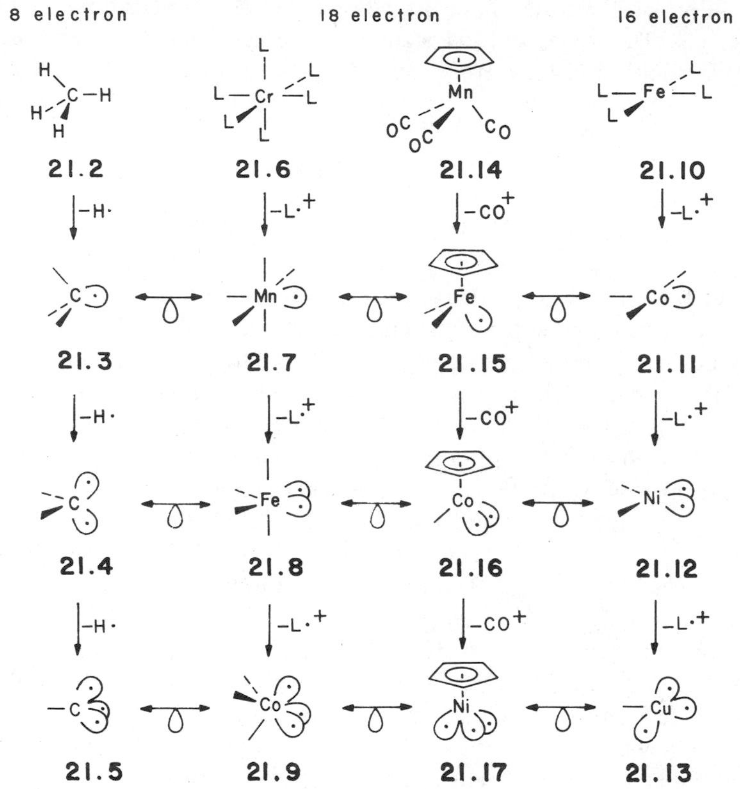

8 electron   18 electron   16 electron

21.2   21.6   21.14   21.10

21.3   21.7   21.15   21.11

21.4   21.8   21.16   21.12

21.5   21.9   21.17   21.13

Our other starting point was the 16-electron square planar complex, **21.10**. The fragments that are generated by the removal of one to three L^+ units are given by **21.11–21.13**. They will also find isolobal partners in the CH_3 to CH series. There are many other starting points that could be utilized. For example, $CpMn(CO)_3$, **21.14**, is an 18-electron complex with an octahedral splitting pattern (Section 20.1). Sequential removal of the three CO^+ ligands generates **21.15–21.17**. The reader can easily verify that replacement of a *fac* L_3 donor set by Cp^- in **21.7–21.9** generates the same fragments. One could just as easily have started from the 18-electron benzene–$Cr(CO)_3$ or cyclobutadiene–$Fe(CO)_3$ complexes. For now, *any* octahedrally based 18-electron or square planar 16-electron complex can be used as a starting point to derive fragments. One can also go down the column in the periodic table to generate alternative isolobal fragments. Thus, CH_2 is isolobal to C_{2v} $Fe(CO)_4$, $Ru(CO)_4$, and $Os(CO)_4$. Likewise $(CH_3)_3Sn$ is isolobal to $Mn(CO)_5$, $Fe(CO)_5^+$, or $Mo(CO)_5^-$. Adjustments in the electron count of the frontier orbitals can also be made. For example, CH_3^+ or BH_3 will have one empty frontier orbital, so too will $Mn(CO)_5^+$, $Cr(CO)_5$, $CpMn(CO)_2$, $(Ph_3P)_3Rh^+$, and $Mn(CO)_4Cl$. Likewise, CH_3^-, $(CH_3)_3Si^-$, or NH_3 are isolobal to $Mn(CO)_5^-$, $Fe(CO)_5$, or $Rh(CO)_5^+$

where all of the metal complexes are C_{4v} square pyramidal (not trigonal bipyrami-dal) fragments. It is important that when one considers the molecules in Scheme 21.1 as compounds in their own right, then there may well be some adjustments in the geometry. For example, CH_3^+ by itself is trigonal while $CpMn(CO)_2$ is pyramidal and CH_3^- is pyramidal while $CpCo(CO)_2$ is trigonal. This was covered in some detail in Section 19.4. The ground state for a d^{10} ML_2 compound is linear[5] (see Section 16.3), not bent as shown in **21.12**. While a C_{4v} $Fe(CO)_5$ fragment is isolobal to CH_3^-, its ground state is certainly the D_{3h} trigonal bipyramidal geometry. But given the geometries of the molecules shown by **21.4**, **21.8**, **12.12**, and **21.16**, all have singlet and triplet electronic states. The actual geometry of both states and the singlet–triplet energy difference in CH_2 has been the subject of a tremendous amount of research (Section 8.8); very little is known (experimentally or theoreti-cally) about the other members of this series. Moreover one can readily envision many other examples which are isolobal to CH_2. So the isolobal analogy anticipates the behavior of those very reactive intermediates. However, our main concern is to use the isolobal analogy to generate alternative fragments in molecules. So the geometry of the fragment *in a molecule*, not as an isolated species, is important.

The usefulness of the isolobal analogy is highlighted when a fragment is replaced with its isolobal analog in a molecule. Since CH_3 is isolobal with $Mn(CO)_5$, one can replace one or both CH_3 groups in ethane to give $CH_3Mn(CO)_5$ and $Mn_2(CO)_{10}$. These are well-known molecules. The electronic description of the three is similar in that there is a σ bond present between C—C, C—Mn, or Mn—Mn. There are cer-tainly other molecular orbitals present, particularly for the transition metal com-plexes where the t_{2g} set at the metals has been ignored. Also notice that an octa-hedral splitting pattern is established at each metal center.

Let us take a closer look at the composition of the frontier orbitals that are dis-played in Scheme 21.1. Since CH_2, $Fe(CO)_4$, $Ni(PPh_3)_2$, and $CpCo(CO)$ are isolobal, then each should have two orbitals with the same symmetry properties. The localized orbitals in Scheme 21.1 are not necessarily the proper symmetry-adapted orbitals that will be most useful for us. Linear combinations of the local-ized orbitals, however, can easily be generated. These are shown in **21.18**. Each

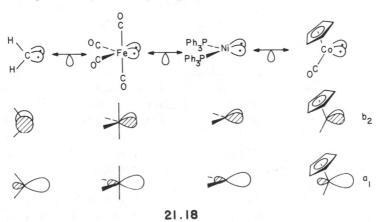

21.18

fragment has one orbital of a_1 and of b_2 symmetry. For Fe(CO)$_4$ and Ni(PPh$_3$)$_2$ we saw that b_2 was lower in energy than the a_1 orbital (Sections 19.1, 19.3). This is a natural consequence of the fact that the b_2 orbital consists primarily of metal d character and a_1 is mainly metal s and p. In CH$_2$ it is a_1 that lies lower than b_2 (Section 7.3). So for a CH$_2$ fragment we would assign the two electrons to the a_1 orbital. For the three ML$_n$ examples in **21.18**, the two electrons would go into the b_2 level. That does not make much of a difference in a real molecule. The occupation of the fragment orbitals that make up a molecule is arbitrary. It is the electron distribution in the final molecule which is important, not the details in electron occupancy of these conceptual fragments. For example, consider cyclopropane, **21.19**, as being derived from the interaction of ethylene with a CH$_2$ fragment. The overall bonding picture that we get should be comparable to that derived for any of the molecules in **21.20** where one CH$_2$ fragment has been replaced by an isolobal partner from **21.18**. That this is true can be established from **21.21**. In each instance the b_2 orbital interacts with ethylene π^* and an a_1

21.19   21.20

21.21

orbital with ethylene π. Two bonding and two antibonding molecular orbitals are formed. Two electrons come from ethylene and two from the CH$_2$ or ML$_n$ fragment which just fill the two bonding MOs. We recognize that the structures in **21.20** are just metallacyclopropane formulations of metal–olefin complexes and **21.21** is the Dewar–Chatt–Duncanson model of bonding as described more fully in Section 19.2 and 19.4.

We will cover several more complexes that can be traced back to cyclopropane in the next section. For now, we note that the isolobal analogy nicely solves conformational issues. Let us rotate the ethylene unit in **21.19** and **21.20** by 90°. The ethylene π^* orbital is now of b_1 symmetry and the b_2 orbital in CH$_2$ or ML$_n$ is left occupied and nonbonding. Thus, **21.22** and **21.23** are high-energy structures. **21.22**

21.22   21.23

contains one square planar carbon and the nonbonding b_2 orbital is readily identified with the nonbonding p orbital in square planar CH_4 (Section 9.4). Of course
the actual energy difference between 21.19 and 21.22 in cyclopropane will not be
the same as the rotational barriers in the olefin complexes. The rotational barrier in
each case depends on the strength of the $b_2 + \pi^*$ interaction. As the interaction becomes stronger the barrier becomes larger.

There are two important caveats that one must keep in mind when using the
isolobal analogy. A normal operational order consists of replacing a fragment in a
molecule with an isolobal one. We must be careful to establish the coordination
mode (or number) at the transition metal because there is not an exact mapping of
fragments in the isolobal analogy. If we use $Fe(CO)_5$, 21.24, as an 18-electron
starting point, then one can readily derive the sequence of fragments given by
21.25 and 21.26. $Co(CO)_4$ and $Ni(CO)_3$ are isolobal with CH_3 and CH_2, respectively. Suppose however, we started with $Ni(CO)_4$, 21.27, another 18-electron
complex which is tetrahedral. Removal of CO (in a heterolytic sense) generates
$Ni(CO)_3$, 21.28, which is predicted to have one empty frontier orbital and is therefore isolobal to CH_3^+. Now we have a problem—is $Ni(CO)_3$ isolobal to CH_2 or CH_3^+?

21.24   21.27

21.25   21.28

21.26

In fact it is isolobal to both. If we use $Ni(CO)_3$ to reconstruct a trigonal bipyramid,
then it is isolobal to CH_2; however, if it is used to form a tetrahedral complex then
it is isolobal to CH_3^+. Suppose we remove an equatorial CO ligand from $Fe(CO)_5$,
21.24. That creates a C_{2v} $Fe(CO)_4$ ligand with one empty hybrid orbital and the
$Fe(CO)_4$ group is isolobal to CH_3^+. Ethylene-$Fe(CO)_4$, 21.29, is then equivalent to
protonated cyclopropane, 21.30. But we had said that ethylene-$Fe(CO)_4$, drawn
in the metallacyclopropane form in 21.20, was equivalent to cyclopropane! The
metallacyclopropane drawing emphasizes octahedral coordination at iron. The
OC—Fe—CO angle in the equatorial plane should be $\sim90°$. In molecular orbital

terms this means that $a_1 + \pi$ and $b_2 + \pi^*$ interactions are both strong and of comparable magnitudes (i.e., there is substantial delocalization between the fragment orbitals in each bonding molecular orbital). Using the olefin–metal formulation in **21.29** emphasizes trigonal bipyramidal coordination at iron. The equatorial OC—Fe—CO angle then should be $\sim 120°$. The important interaction here in molecular orbital terms is that between the empty a_1 orbital on $Fe(CO)_4$ and the filled π orbital of ethylene. The $b_2 + \pi^*$ interaction is neglected (or treated to be of minor importance). In this context it is useful to rewrite protonated cyclopropane as in **21.31** which emphasizes the principal mode of bonding, donation of electron density from the filled π orbital of ethylene to the empty hybrid orbital of CH_3^+. This is the essence of the bonding model we developed for the bridged geometry in the ethyl cation (Section 10.5) which came about in a totally different way! In actual fact the OC—Fe—CO angles in olefin-$Fe(CO)_4$ complexes[6] are normally $\sim 110°$ which is intermediate between two extremes that have been presented here. So the inexact mapping of isolobal fragments, while it is complicating, nonetheless gives us different perspectives about the bonding in molecules that we may well have overlooked.

21.29   **21.30**   **21.31**

The most common ligand in transition metal chemistry is CO. It has a high lying, nicely directional σ donor orbital and a pair of low lying acceptor levels. These are ideal components for a strong metal–carbon bond. CO also brings a real problem to the organometallic chemist, namely, it is one of the few ligands which can either bridge two or three metal centers or bond terminally to one transition metal. More importantly pairwise bridging-terminal CO exchange is a very common feature in transition metal dimers and clusters and there is often little energy difference between the two structural possibilities. This creates a second caveat when using the isolobal analogy. A pertinent example is offered by ethane. We have previously established that CH_3 is isolobal to $Mn(CO)_5$ and, therefore, ethane, **21.32**, is equivalent to $Mn_2(CO)_{10}$, **21.33**. But $CpFe(CO)_2$ (**21.15** in Scheme 21.1) is also isolobal to CH_3. There should also be a $Cp_2Fe_2(CO)_4$ complex with a structure like **21.34**. Such a compound does exist but it is not the ground-state structure. The bridged form, **21.35**, is more stable than **21.34** and there is a facile equilibrium between the two.[7] That certainly does not happen in ethane (Section 10.2). Bridging hydrogens

21.32   **21.33**   **21.34**   **21.35**

and alkyl groups are found in carbonium ion and borane chemistry where the molecules are electron deficient, but not for saturated compounds. In a very qualitative sense the valence orbital structure for **21.34** and **21.35** are similar. It is the metal t_{2g} and carbonyl π^* orbitals that are most perturbed as the carbonyls bridge. The organic/main group analogs, of course, do not have this set of "lone-pair" orbitals at the central atom. Remember that the metal t_{2g} set is neglected (for now) when using the isolobal analogy, therefore, **21.35** is regarded as being electronically equivalent to **21.34**. With these two qualifications in mind we turn to examples where the isolobal analogy can be used.

21.3.   ILLUSTRATIONS OF THE ISOLOBAL ANALOGY

We replaced one CH_2 in cyclopropane, **21.19**, with an isolobal $Fe(CO)_4$, $Ni(PPh_3)_2$, and $CpCo(CO)$ group to generate the metal–olefin complexes given by **21.20**. Replacing two or all three CH_2 groups by the isolobal $Fe(CO)_4$ or $CpRh(CO)$ fragments generates the set of compounds shown in **21.36**, all of which are known.

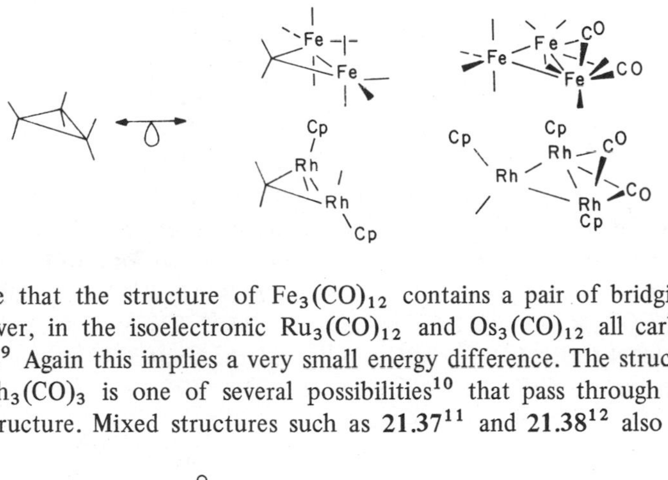

21.36

Notice that the structure of $Fe_3(CO)_{12}$ contains a pair of bridging carbonyls.[8] However, in the isoelectronic $Ru_3(CO)_{12}$ and $Os_3(CO)_{12}$ all carbonyls are terminal.[9] Again this implies a very small energy difference. The structure shown for $Cp_3Rh_3(CO)_3$ is one of several possibilities[10] that pass through an all-terminal CO structure. Mixed structures such as **21.37**[11] and **21.38**[12] also have been pre-

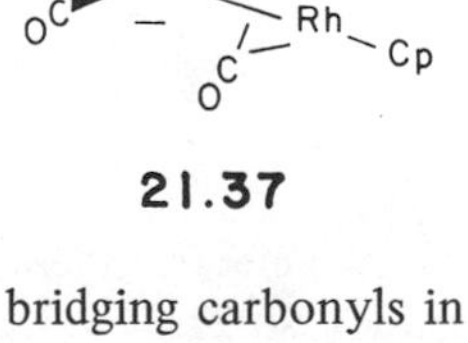

21.37 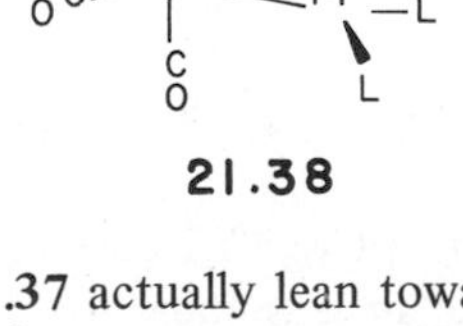   **21.38** $L = P(OPh)_3$

pared. The bridging carbonyls in **21.37** actually lean toward the $Rh(CO)_2$ unit for reasons that are covered elsewhere,[10] but the important point is that electronically the compounds bear a marked resemblance to cyclopropane.[10,13,14] We need not stop with cyclopropane, for **21.39** has recently been prepared and structurally categorized.[15] $Os(CO)_4$ is isolobal to CH_2 so **21.39** is immediately recognized as

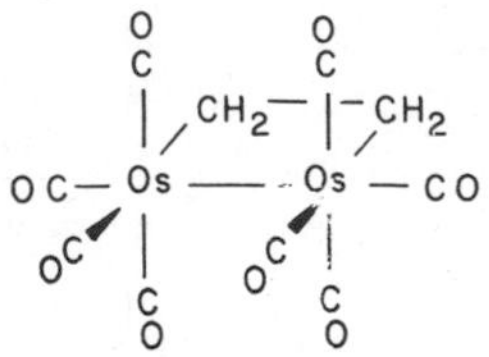

21.39

being equivalent to cyclobutane. Undoubtedly higher homologs will eventually be prepared.

The most simple, stable organic compound containing the CH_2 group is ethylene. It is reasonable to expect that the compounds given by **21.40–21.43** can be

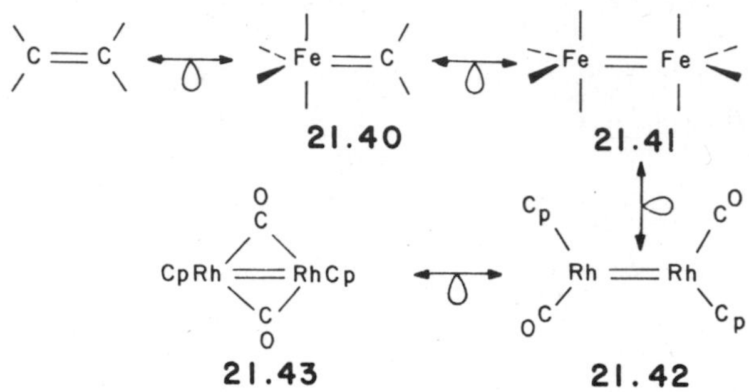

21.40   **21.41**

21.43   **21.42**

made. Derivatives of **21.40** have been prepared[16] and **21.41** has been isolated in a low temperature matrix.[17] Their *kinetic* stability is certainly limited. A high lying π orbital and low lying π^* especially in **21.41–21.43** make these molecules extraordinarily reactive. It is easy to recognize that the all-terminal form of $Fe_3(CO)_{12}$ (see **21.36**) is just the $Fe(CO)_4$ complex of **21.41**. **21.37** is a d^{10} ML_2 complex of **21.43**. The $Os_2(CO)_8$ analog of **21.41** appears in **21.44**.[18] Replacing the $Os(CO)_4$ groups with isolobal $CpRh(CO)$ and CH_2 units and the central Sn atom with an isoelectronic carbon generates **21.45**.[19] Both complexes are equivalent to spiropentane, **21.46**.

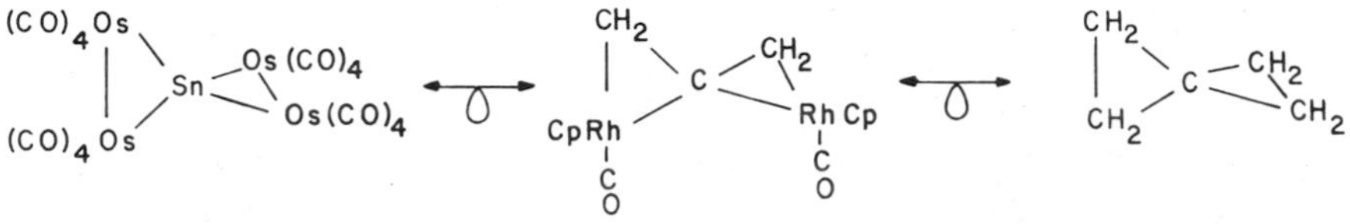

21.44   **21.45**   **21.46**

A compound containing a C_{2v} ML_4 complex with two electrons less is shown in **21.47**.[20] A $Cr(CO)_4^-$ fragment is isolobal to CH_2^+ or BH_2, so **21.47** is just an analog of diborane, **21.48**. The orientation of the $Cr(CO)_4$ group in **21.47** just matches that of BH_2 in **21.48** as can be seen by a comparison of the two sets of frontier orbitals in **21.18**. It is tempting to speculate on the existence of **21.49**.

21.47   21.48   21.49

So far we have concentrated on isolobal analogs of CH_2. Let us now turn our attention to other fragments. In Section 17.4 we showed that for the hydrido bridged $Cr(CO)_5$ dimer, 21.50, a "closed" structure with a Cr—H—Cr angle of less than $180°$ should be favored with a very low bending potential. All of the elements of an electron deficient, two-electron-three-center bonding pattern are present in 21.50. Replacing each $Cr(CO)_5$ group by the isolobal CH_3^+ yields 21.51. A cyclic derivative of 21.51 has recently been prepared.[21] Its actual structure is uncertain; however, it has been established that the hydrogen atom must be bridged symmetrically between the two carbons. Calculations on 21.51 at various levels[22] have found that the most stable structure is highly bent and it perhaps can be regarded as a protonated ethane. An isoelectronic compound, 21.52, has been structurally categorized in the form of a $(Ph_3P)_2N^+$ salt.[23] The B—H—B angle was found to be $136°$. There must be a very low bending potential (as is found for the many structures of 21.50 and isoelectronic compounds—see Section 17.4) since 21.53 is lin-

21.50   21.51   21.52   21.53

ear.[24] We suspect that a variety of structures will be found for 21.52 by varying the counter cation which in turn can modify intermolecular packing forces. It would be interesting to find conditions to stabilize 21.54. It may well exist in a low temperature matrix.[5] Isoelectronic BH_4^- complexes are known.[25] There are several more complicated examples[26] of two-electron-three-center M—H—C bonding which can be traced by the isolobal analogy back to 21.54. Here again the M—H—C angle in 21.55 was found to be $111°$.[27] A $(Ph_3P)Au^+$ fragment can be regarded as being

21.54   21.55

derived from a linear, 14-electron, d^{10} ML_2 species. It is then isolobal to CH_3^+ or $Cr(CO)_5$ so 21.55 can be related to 21.50. CH_3^+ and $(Ph_3P)Au^+$ each have one vacant frontier orbital which also makes them isolobal to H^+. In fact, there is a beautiful mapping between the structures of transition metal clusters which contain the $Au(PR_3)$ unit and those with an H atom.[28,29]

Scheme 21.1 shows that a CH fragment is isolobal to $Co(CO)_3$; both have three frontier orbitals with three electrons in them (21.5 and 21.9). The entire set of compounds given by 21.56–21.60 where CR groups have been sequentially replaced

21.56   21.57   21.58

21.59   21.60

by $Co(CO)_3$ units has been prepared. Normally 21.57 is written as a cyclopropenium complex, and 21.58 as an acetylene-bridged $Co_2(CO)_6$ dimer. It is a straightforward matter to take linear combinations of the Co—C σ bonds to generate a set of orbitals that are equivalent to those derived from a cyclopropenium or acetylene–metal formulation. The structures of 21.60 and $Rh_4(CO)_{12}$ actually contain *three* dibridging carbonyls[30] and both compounds show very facile CO scrambling. $Ir_4(CO)_{12}$ does indeed have the all-terminal structure[31] indicated in 21.60. Notice that the acetylene ligand in 21.58 lies perpendicular to the Co—Co vector as it should for a metallatetrahedrane derivative. Rotation of the acetylene by 90° gives 21.61 which is now isolobal to cyclobutadiene 21.62. The orbital structure of 21.61 contains

21.61   21.62

four levels of π symmetry; the two middle ones are nearly degenerate and constitute the HOMO and LUMO for the molecule.[32] That is exactly the picture one gets for 21.62 (see Figure 12.5). Rotation of one "acetylene" unit by 90° in tetrahedrane to give cyclobutadiene is a symmetry-forbidden reaction. The same restriction applies to the 21.58 to 21.61 conversion.[32]

A fascinating series of organic compounds are produced when the ML_3 unit is replaced in polyene-ML_3 complexes with an isolobal analog. Consider butadiene-

Fe(CO)$_3$, **21.63**, as being constructed by the interaction of the four π electrons in the butadiene ligand with an Fe(CO)$_3$ which is isolobal to CH$^+$. The CH$^+$ fragment has three frontier orbitals with two electrons, shown in a localized manner by **21.64**. Pairing the electrons to form two C—C σ bonds and one C—C π bond generates the nonclassical form of the 3-cyclopentenium cation, **21.65**. That sort

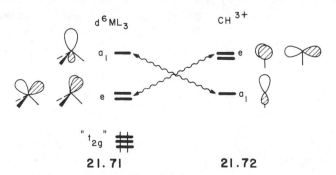

21.63   **21.64**   **21.65**   **21.66**

21.67   **21.68**   **21.69**   **21.70**

of representation has actually been used for butadiene–Fe(CO)$_3$ complexes,[33] see **21.66**. One can argue how much the donation from the π orbital in **21.65** to the empty hybrid orbital is worth, but suppose we start with a relative of **21.63**, namely cyclobutadiene–Fe(CO)$_3$, **21.67**. Its isolobal analog is the C$_5$H$_5^+$ cation, **21.68**. A Mn(CO)$_3$ group is isolobal to CH^{2+}, thus, **21.69** is isolobal to **21.70**, the Hogeveen–Kwant dication.[34] The electrons in both sets of complexes can be partitioned into a d^6 ML$_3$ (i.e., Fe(CO)$_3^{2+}$ or Mn(CO)$_3^+$) or CH^{3+} units which interact with the cyclobutadiene^{2-} or Cp$^-$ ligand. The polyene ligands then each have three filled π orbitals. The frontier orbitals of a d^6 ML$_3$ and CH^{3+} fragment are shown in **21.71** and **21.72**, respectively. Both fragments contain three empty frontier orbitals

d^6ML$_3$ CH^{3+}

a_1 e

e a_1

"t_{2g}"

21.71   **21.72**

of $a_1 + e$ symmetry. They will just match the nodal properties of the three filled π levels of the two polyenes. The reader should carefully compare the orbital interaction diagrams in Figure 20.3 for **21.67** with that in Figure 11.6 for **21.68** and Figure 20.2 for **21.69** with Figure 11.7 for **21.70**. The basic ordering of the bonding

molecular orbitals is indeed identical going from the organometallic to the all organic complex. Species such as the $C_5H_5^+$ cation or $C_6H_6^{2+}$ dication are certainly
unusual and uncomfortable structures for organic chemists. Yet all available experimental[34] and theoretical[35] evidence suggests that the apical carbon in each is
symmetrically bonded to the four and five membered ring. Many other all-organic,
π-bonded compounds have been proposed.[35] In fact, there is a recent structure of
a derivative of $C_6H_6^{2+}$, namely **21.73**.[36] Deprotonating the apical carbon in **21.70**

and replacing it with an isoelectronic Sn atom gives **21.73**. The Sn atom lies in the
center of the pentamethyl–cyclopentadienyl ring and the five central carbon atoms
in the ring lie in a common plane. The apical CR^{2+} unit in **21.70** can also be substituted with the isoelectronic $P—Me^+$ and $Be—Cl$ groups to form stable compounds
with a similar structure.

One needs to be adaptable in viewing the structures of molecules. Normally
21.74 is considered to be an acetylene coordinated to a trinuclear cluster.[37] There

are some advantages in adopting this viewpoint but we have drawn the structure in
a different way. A CpNi fragment is isolobal to CH, so **21.74** is analogous to cyclo-
butadiene-$Fe(CO)_3$ or $C_5H_5^+$. Replacing the "basal" $Fe(CO)_3$ group for CH^+ in the
ferrole dimer, **21.75**, yields $CpFe(CO)_3^+$, **21.76**, which in turn is isolobal to ferrocene, **21.77**, when the three carbonyl ligands are replaced by Cp^-. All three complexes are stable and well known. But suppose we substitute the "apical" $Fe(CO)_3$
fragment in **21.75** by CH^+? That generates **21.78** and further substitution of a Cp^-
ligand gives **21.79**, a nonclassical isomer of ferrocene! No complex of this structure

has yet been prepared; MO calculations[38] indicate a substantial HOMO–LUMO gap which implies some kinetic stability. The CH interactions with the ferrole ring are very similar in magnitude to those between the FeCp and Cp fragments in ferrocene. Since Fe is quite electropositive in comparison to C, there are two high lying Fe—C σ levels in the nonclassical isomer, **21.79**, which are not matched in ferrocene. Consequently **21.79** is computed to be much less stable than the classical structure of ferrocene. Provided that there is a symmetry-allowed pathway from **21.79** to **21.77** (and there is one), there is probably not much chance of ever experimentally observing the nonclassical structure. However, given that there is nothing "wrong" with the basic orbital pattern in **21.79**, it might be possible to devise strategies that will overcome the thermodynamic problem and enable the synthesis of a derivative of **21.79**.[38] The important point to recognize here is that the isolobal analogy has led us to consider a very unusual structure for a well-known compound. Several other examples where a $Fe(CO)_3$ group in transition metal complexes has been replaced by CH^+ or BH can be cited[39] which lead to unusual structural possibilities. The isolobal analogy does not tell us whether or not these structures are stable ones, however. Rearrangement to a lower energy structure should always be considered. The isolobal analogy does give us compounds that have a similar electronic structure in terms of the number, symmetry type, and occupation of frontier molecular orbitals.

In the preceding discussion we have used the isolobal analogy to generate new compounds and to provide conceptual links between classes of known molecules. It can also be used to provide clues about reaction mechanisms. Two problems are briefly outlined here. Circumambulation or the walk rearrangement is a reaction in organic chemistry wherein a CH_2 (or CR_2) group migrates around the periphery of a polyene ring by a sequence of sigmatropic rearrangements. Three examples are shown in the top half of Scheme 21.2. The reactions given by **21.80** and **21.81** have been studied extensively by theoreticians[40] and experimentalists.[41] The **21.82** to **21.83** circumambulation is experimentally complicated by an electrocyclic ring-opening reaction to **21.84**. There are analogous reactions, called ring-whizzing, in organometallic chemistry. Replacing the migrating CH_2 unit in **21.80** by $Pt(PR_3)_2$ yields an η^2-benzene complex, drawn in **21.85** with a metallacyclopropane formulation. A hexakis(trifluoromethyl) derivative has been prepared and the rearrangement shown in **21.85** is quite facile.[42] Several cyclobutadiene–NiL_2 complexes have been prepared.[43] Are their ground-state structures η^4 (as has been formulated[43]) or are they η^2 with a very facile rearrangement which is analogous to **21.81**? Calculations[44] favor the latter proposal although an η^4 intermediate is also found on the potential energy surface with an energy not much higher than that found for the η^2 isomer. The rearrangement of **21.87** to **21.88** has been experimentally observed[45] where M is Pt, Pd, Ni. The X-ray structures[45] of four members in this class of compounds show a progression of geometries from the η^2 species, **21.87**, to the transition state for the rearrangement. It is therefore possible to chart the reaction path experimentally *via* these structures. There are also good indications[46] that **21.87** can undergo ring opening to **21.89**. Thus, the reaction manifold of d^{10}-cyclopropenium–ML_2^+ complexes is matched by the isolobal bi-

SCHEME 21.2

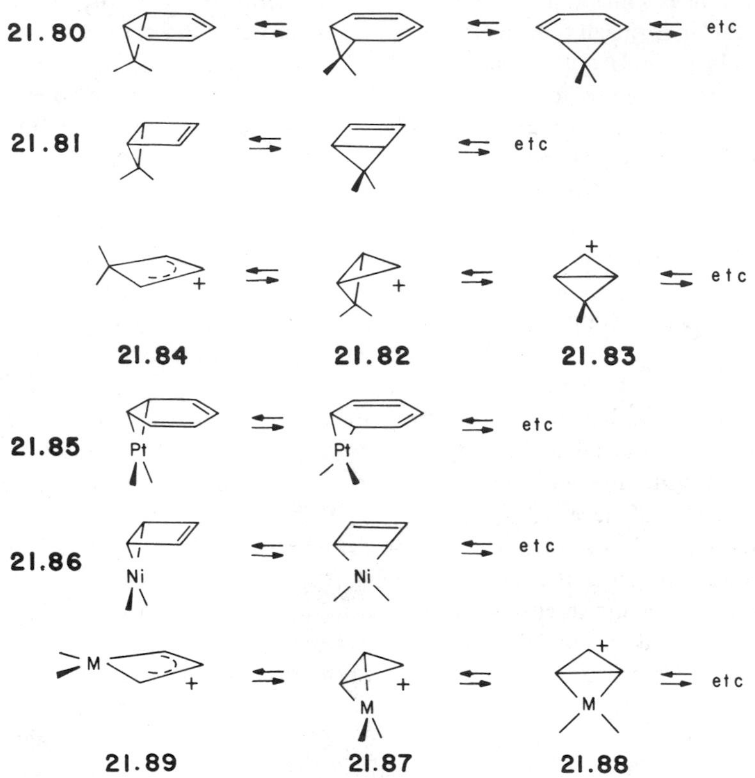
21.80
etc
21.81
etc
+
etc
21.84
21.82
21.83
21.85
Pt
Pt
etc
21.86
Ni
Ni
etc
M
+
M
M
etc
21.89
21.87
21.88

cyclobutyl cation. There are two stereochemically distinct ways that the walk rearrangement and ring-whizzing process can occur. The ML_2 or CH_2 group can rotate with respect to the polyene as it migrates or it can simply slide from one "coordination site" to another. In each reaction of Scheme 21.2 there is a direct correspondence between the stereochemistry of reaction for the organic systems and its isolobal organometallic analog.[44]

21.4. EXTENSIONS

The compounds that we have discussed in this chapter were either saturated or, at most, two electrons short of being saturated. The next chapter treats borane and transition metal clusters which are very electron deficient. A tool for comparing these compounds and predicting their structures will be described which is related to these isolobal ideas. We have also only used the isolobal analogy for transition metal complexes with coordination numbers up to six. For coordination numbers of 7–9 the members of the t_{2g} set must be utilized. Reference back to the generalized interaction diagram of Figure 16.2 shows that when n ligand donor levels interact with a transition metal, n M—L bonding and n M—L antibonding molecular orbitals are formed. Left behind are $9 - n$ nonbonding orbitals that are localized on the transition metal. Suppose $n = 8$, then a d^2 complex is of the saturated, 18-electron type and one member of what corresponded to the octahedral t_{2g} set is left nonbonding. The other two members of t_{2g} are used to form interactions with the surrounding ligands. Removal of three ligands from this 8-coordinate complex generates an ML_5 fragment with three *empty* frontier orbitals. Therefore, a d^2 ML_5 fragment is isolobal to CH^{3+}. This is shown in **21.90** where for convenience we have

21.90

used an ML_5 fragment of C_{4v} symmetry (Section 17.2). Compare the relationship in **21.90** with that in **21.71** and **21.72**. A d^6 ML_3 fragment is isolobal to d^2 ML_5! Again, the isolobal mapping of fragments is not an exact one; many permutations are possible and the coordination number at the metal in a compound must be established before making isolobal replacements. The reader may object to the fact that the metal t_{2g} set in **21.71** lies at much lower energy than $a_1 + e$ for CH^{3+}. In **21.90** they are at comparable energies. We are making very qualitative arguments here, but this is not so bad of an approximation. A d^2 metal will be one of the far left of the periodic table; therefore, it is more electropositive than the d^6 metal in a ML_3 fragment. Three carbonyls can be replaced by a Cp^- ligand, and so $V(CO)_5$ where three electrons are distributed among the $a_1 + e$ set in **21.90** is isolobal to $CpW(CO)_2$ and CH. **21.91**[47] can easily be related to the cyclopropene **21.92** by using this relationship. The cyclopropenium complex **21.93**[48] is nothing more than another tetrahedrane, like **21.57–21.60**. There are many more patterns that can be established when the restriction of octahedral and square planar coordination is removed.[1b]

21.91

21.92

L⌒L = bipyr
R = Ph

21.93

A second extension can be made for boranes and transition metal clusters. Bridging hydrogen atoms are often found in both of these species. In general it is not very helpful to rearrange them to a terminal geometry, as we did for the bridging carbonyl groups. However, it is useful to remove bridging hydrogens *as protons*. That may seem to be unreasonable, but actually bridging hydrogens do not change the underlying orbital structure in terms of symmetry. For example, the frontier molecular orbitals of diborane, **21.94**, were discussed in Section 10.2B. Removal of two bridging protons gives $B_2H_4^{2-}$ which is isoelectronic to ethylene, **21.95**. The reader should compare the orbital structure of diborane in **10.9** with

$$H \overset{\displaystyle H}{\underset{\displaystyle H}{\searrow}} B \overset{H}{\underset{H}{\diagup}} B \overset{-H}{\underset{H}{\diagdown}} \quad \xrightarrow{-2H^+} \quad H \overset{\displaystyle }{\underset{H}{\searrow}} B \!-\! B \overset{2-}{\overset{-H}{\underset{H}{\diagdown}}} \equiv \quad H \overset{\displaystyle }{\underset{H}{\searrow}} C \!=\! C \overset{-H}{\underset{H}{\diagdown}}$$

21.94   **21.95**

ethylene in **10.17**. Both have a LUMO of b_{2g} symmetry and occupied valence orbitals of b_{1g}, a_g, b_{1u}, and b_{2u} symmetry. To be sure, there is not an exact match in the level ordering of occupied orbitals. For example, the b_{1u} (π) orbital of ethylene is stabilized by the bridging hydrogens in diborane, but we need not worry too much about these quantitative details. The π orbitals of $B_3H_7^{2-}$, **21.96**, resemble those for the allyl anion, **21.97**.[49] Thus, the basic orbital interactions in **21.98**[50] are similar to those in **21.99**. There are some differences, but many similarities in the orbital structure of $H_3M_3(CO)_9(CR)$, **21.100**, and $Co_3(CO)_9(CR)$, **21.59**.[51]

$$\underset{\textstyle \text{21.96}}{\overset{\textstyle 2-}{H \overset{\displaystyle }{\underset{\displaystyle }{\searrow}} B - \\ -B \overset{H}{\diagdown} \\ B -}} \quad \equiv \quad \underset{\textstyle \text{21.97}}{\overset{\textstyle -}{- C \overset{C-}{\diagdown} \\ C -}}$$

$$\underset{\textstyle \text{21.98}}{\overset{\displaystyle H \searrow B - \\ -B \diagdown H \\ B - \\ | \\ Pt \\ \diagup | \diagdown \\ R_3P \;\; PR_3}} \quad \underset{\textstyle \text{21.99}}{\overset{\displaystyle - C \diagdown \\ -C \diagdown \\ | C - \\ | \\ Pt^+ \\ \diagup | \diagdown \\ R_3P \;\; PR_3}}$$

$$M = Fe, Ru \qquad (CO)_3M \overset{\overset{\textstyle R}{\overset{\textstyle |}{C}}}{\underset{\underset{\textstyle (CO)_3}{M}}{\diagup H \diagdown}} M(CO)_3 \qquad \text{21.100}$$

Still further connections can be made. Removing one cap from an icosahedral molecule gives a cage structure with a pentagonal open face, **21.101**. This is the structure for $C_2B_9H_{11}^{2-}$. Hawthorne[52] first suggested that the $C_2B_9H_{11}^{2-}$ cage was

21.101

equivalent to Cp⁻. In orbital terms, **21.101** has five frontier orbitals emanating from the open face which have the same symmetry and occupation by electrons as the π orbitals of Cp⁻. So compounds like $(C_2B_9H_{11})_2Fe^{2-}$, $C_2B_9H_{11}$—FeCp⁻, and $C_2B_9H_{11}$—Mn(CO)$_3$ have the basic ferrocene splitting pattern and orbital shapes.[53] How this comes about or, more precisely, the derivation of the orbital structures of large cluster compounds like **21.101** is the topic of the next chapter.

REFERENCES

1. (a) M. Elian, M. M. L. Chen, D. M. P. Mingos, and R. Hoffmann, *Inorg. Chem.*, **15**, 1148 (1976).
    (b) R. Hoffmann, *Angew. Chem.*, **94**, 725 (1982); *Angew. Chem. Int. Ed.*, **21**, 711 (1982).
2. J. Halpern, in *Advances in Chemistry; Homogeneous Catalysis*, No. 70, American Chemical Society, Washington, D.C., 1968, pp. 1–24.
3. K. Wade, *Chem. Commun.*, 792 (1971); *Inorg. Nucl. Chem. Letts.*, **8**, 559, 563 (1972); *Adv. Inorg. Radiochem.*, **18**, 1 (1976).
4. D. M. P. Mingos, *Nature (London), Phys. Sci.*, **236**, 99 (1972).
5. J. K. Burdett, *Coord. Chem. Rev.*, **27**, 1 (1978).
6. S. D. Ittel and J. A. Ibers, *Adv. Organomet. Chem.*, **14**, 33 (1976).
7. A. Mitschler, B. Rees, and M. S. Lehmann, *J. Am. Chem. Soc.*, **100**, 3390 (1978); For a general review of bridging terminal exchange, see R. D. Adams and F. A. Cotton, *Dynamic Nuclear Magnetic Resonance Spectroscopy*, L. M. Jackman and F. A. Cotton, editors, Academic Press, New York (1975), pp. 489–522.
8. F. A. Cotton and J. M. Troup, *J. Am. Chem. Soc.*, **96**, 4155 (1974).
9. M. R. Churchhill, F. J. Hollander, and J. P. Hutchinson, *Inorg. Chem.*, **16**, 2655 (1977); M. R. Churchill and B. G. DeBoer, ibid., **16**, 878 (1977).
10. A. R. Pinhas, T. A. Albright, P. Hofmann, and R. Hoffmann, *Helv. Chim. Acta*, **63**, 1 (1980) and references therein.
11. W. D. Jones, M. A. White, and R. G. Bergman, *J. Am. Chem. Soc.*, **100**, 6772 (1978).
12. V. G. Albano, G. Ciani, M. I. Bruce, G. Shaw, and F. G. A. Stone, *J. Organomet. Chem.*, **42**, C99 (1972).
13. P. Hofmann, *Angew. Chem.*, **91**, 591 (1979); *Angew. Chem. Int. Ed.*, **18**, 554 (1979); private communications.
14. M. Benard, *Inorg. Chem.*, **18**, 2782 (1979).
15. M. R. Burke, J. Takats, F.-W. Grevels, and J. G. A. Reuvers, *J. Am. Chem. Soc.*, **105**, 4092 (1983); K. M. Motyl, J. R. Norton, C. K. Schauer, and O. P. Anderson, ibid., **104**, 7325 (1982).
16. E. O. Fischer, H.-J. Beck, C. G. Kreiter, J. Lynch, J. Muller, and E. Winkler, *Chem. Ber.*, **105**, 162 (1972); H. LeBozec, A. Gorgues and P. H. Dixneuf, *J. Am. Chem. Soc.*, **100**, 3946 (1978).
17. M. Poliakoff and J. J. Turner, *J. Chem. Soc.* (A), 2403 (1971).
18. J. D. Cotton, S. A. R. Knox, I. Paul, and F. G. A. Stone, *J. Chem. Soc.* (A), 264 (1967).
19. The structure of an η^5 indenyl derivative may be found in Y. N. Al-Obaidi, P. K. Baker, M. Green, N. D. White, and G. E. Taylor, *J. Chem. Soc., Dalton Trans.*, 2321 (1981).
20. M. Y. Darensbourg, R. Bau, M. W. Marks, R. R. Burch, Jr., J. C. Deaton, and S. Slater, *J. Am. Chem. Soc.*, **104**, 6961 (1982).
21. R. P. Kirchen, K. Ranganayakulu, A. Rauk, B. P. Singh, and T. S. Sorensen, ibid., **103**, 588 (1981); R. P. Kirchen, N. Okazawa, K. Ranganayakulu, A. Rauk, and T. S. Sorensen, ibid., **103**, 597 (1981).
22. R. M. Minyaev and V. I. Pavlov, *J. Mol. Structure, Theochem.*, **29**, 205 (1983); *Zh. Organ.*

Khim., **18**, 1595 (1982); W. A. Lathan, W. J. Hehre, and J. A. Pople, *J. Am. Chem. Soc.*, **93**, 808 (1971); K. Raghavachari, R. A. Whiteside, J. A. Pople, and P. V. R. Schleyer, ibid., **103**, 5649 (1981).

23. S. G. Shore, S. H. Lawrence, M. I. Watkins, and R. Bau, ibid., **104**, 7669 (1982).

24. J. L. Atwood, D. C. Hrncir, R. D. Rogers, and J. A. K. Howard, ibid., **103**, 6787 (1981); for calculations on this system see R. A. Chiles and C. E. Dykstra, *Chem. Phys. Letts.*, **92**, 471 (1982).

25. T. J. Marks and J. R. Kolb, *Chem. Rev.*, **77**, 263 (1977).

26. A. J. Schultz, R. G. Teller, M. A. Beno, J. M. Williams, M. Brookhart, W. Lamanna, and M. B. Humphrey, *Science*, **220**, 197 (1983); R. K. Brown, J. M. Williams, A. J. Schultz, G. D. Stucky, S. D. Ittel, and R. L. Harlow, *J. Am. Chem. Soc.*, **102**, 981 (1980).

27. M. Green, A. G. Orpen, I. D. Slater, and F. G. A. Stone, *J. Chem. Soc., Chem. Commun.*, 813 (1982).

28. D. G. Evans and D. M. P. Mingos, *J. Organomet. Chem.*, **232**, 171 (1982).

29. J. W. Lauher and K. Wald, *J. Am. Chem. Soc.*, **103**, 7648 (1981); F. E. Simon and J. W. Lauher, *Inorg. Chem.*, **19**, 2338 (1980).

30. F. H. Carve, F. A. Cotton, and B. A. Frenz, *Inorg. Chem.*, **15**, 380 (1976); C. H. Wei, ibid., **8**, 2384 (1969).

31. M. R. Churchill and J. P. Hutchinson, ibid., **17**, 3528 (1978).

32. D. M. Hoffman, R. Hoffmann, and C. R. Fisel, *J. Am. Chem. Soc.*, **104**, 3858 (1982); D. M. Hoffman and R. Hoffmann, *J. Chem. Soc., Dalton Trans.*, 1471 (1982).

33. See, for example, R. Mason and G. Wilkinson, *Experientia, Suppl.*, **9**, 233 (1964).

34. H. Hogeveen and P. W. Kwant, *Acc. Chem. Res.*, **8**, 413 (1975); H. Carnadi, C. Giordano, R. F. Heldeweg, H. Hogeveen, and E. M. G. A. van Kruchten, *Isr. J. Chem.*, **21**, 229 (1981), and references therein.

35. E. D. Jemmis, *J. Am. Chem. Soc.*, **104**, 7017 (1982); E. D. Jemmis and P. V. R. Schleyer, ibid., **104**, 4781 (1982); V. I. Minkin and R. M. Minyaev, *Russ. Chem. Rev.*, **51**, 332 (1982); *Usp. Khim.*, **51**, 586 (1982); H. Schwarz, *Angew. Chem. Int. Ed*, **20**, 991 (1981); *Angew. Chem.*, **93**, 1046 (1981).

36. P. Jutzi, F. Kohl, P. Hofmann, C. Kruger, and Y.-H. Ysay, *Chem. Ber.*, **113**, 757 (1980).

37. G. Jaouen, A. Marinetti, B. Mentzen, R. Mutin, J.-Y. Saillard, B. G. Sayer, and M. J. McGlinchey, *Organometallics*, **1**, 753 (1982).

38. J. Silvestre, T. A. Albright, and R. Hoffmann, to be submitted for publication.

39. A. Sevin and A. Devaquet, *Nouv. J. Chim.*, **1**, 357 (1977); T. Clark and P. V. R. Schleyer, ibid., **2**, 665 (1978); K. Krogh-Jespersen, D. Cremer, D. Poppinger, J. A. Pople, P. v. R. Schleyer, and J. Chandrasekhar, *J. Am. Chem. Soc.*, **101**, 4843 (1979); J. Chandreskhar , P. v. R. Schleyer, and H. B. Schlegel, *Tet. Letts.*, 3393 (1978); J. A. Ulman, E. L. Andersen, and T. P. Fehlner, *J. Am. Chem. Soc.*, **100**, 456 (1978); R. L. DeKock and T. P. Fehlner, *Polyhedron*, **1**, 521 (1982); P. Brint, K. Pelin, and R. T. Spalding, *Inorg. Nucl. Chem. Letts.*, **16**, 391 (1980); *J. Chem. Soc., Dalton Trans.*, 546 (1981).

40. W. J. Hehre and A. J. P. Devaquet, *J. Am. Chem. Soc.*, **98**, 4370 (1976); W. J. Hehre, ibid., **96**, 5207 (1974); W. W. Schoeller, ibid., **97**, 1978 (1975); W. L. Jorgensen, ibid., **98**, 6784 (1976) and references therein.

41. F.-G. Klarner and B. Brassel, ibid., **102**, 2469 (1980); F.-G Klarner and F. Adamsky, *Chem. Ber.*, **116**, 299 (1983) and references therein; a general review may be found in T. H. Lowry and K. S. Richardson, *Mechanism and Theory in Organic Chemistry*, second edition, Harper & Row, New York (1981), pp. 867–893.

42. J. Browning, M. Green, J. L. Spencer, and F. G. A. Stone, *J. Chem. Soc., Dalton Trans.*, 97 (1974); J. Browning and B. R. Penfold, *J. Cryst. and Mol. Structure*, **4**, 335 (1974).

43. U. Griebsch and H. Hoberg, *Angew. Chem.*, **90**, 1014 (1978); H. Hoberg and C. Fröhlich, ibid., **92**, 131 (1980); H. Hoberg and W. Richter, *J. Organomet. Chem.*, **195**, 347, 355 (1980); H. Hoberg and C. Fröhlich, ibid., **209**, C69 (1981); H. Hoberg, W. Richter, and C. Fröhlich, ibid., **213**, C49 (1981).

44. J. Silvestre and T. A. Albright, to be published.
45. C. Mealli, S. Midollini, S. Moneti, L. Sacconi, J. Silvestre, and T. A. Albright, *J. Am. Soc.*, **104**, 95 (1982).
46. E. D. Jemmis and R. Hoffmann, ibid., **102**, 2570 (1980).
47. G. A. Carriedo, D. Hodgson, J. A. K. Howard, K. Marsden, F. G. A. Stone, M. J. Went, and P. Woodward, *J. Chem. Soc., Chem. Commun.*, 1006 (1982).
48. M. G. B. Drew, B. J. Brisdon, and A. Day, *J. Chem. Soc., Dalton Trans.*, 1310 (1981).
49. C. E. Housecroft and T. P. Fehlner, *Inorg. Chem.*, **21**, 1739 (1982); R. L. DeKock, P. Deshmukh, T. P. Fehlner, C. E. Housecroft, J. S. Plotkin, and S. G. Shore, *J. Am. Chem. Soc.*, **105**, 815 (1983).
50. L. J. Guggenberger, A. R. Kane, and E. L. Muetterties, ibid., **94**, 5665 (1972).
51. K. S. Wong, K. J. Haller, T. K. Dutta, D. M. Chipman, and T. P. Fehlner, *Inorg. Chem.*, **21**, 3197 (1982); R. L. DeKock, K. S. Wong, and T. P. Fehlner, ibid., **21**, 3203 (1982); D. E. Sherwood, Jr. and M. B. Hall, *Organometallics*, **1**, 1519 (1982).
52. M. F. Hawthorne and P. L. Pilling, *J. Am. Chem. Soc.*, 87, 3987 (1965); M. F. Hawthorne, D. C. Young, T. D. Andrews, D. V. Howe, P. L. Pilling, A. D. Pitts, M. Reintjes, L. F. Warrent, Jr., and P. A. Wegner, ibid., **90**, 879 (1968)
53. D. A. Brown, M. O. Fanning, and N. J. Fitzpatrick, *Inorg. Chem.*, 17, 1620 (1978); **19**, 1822 (1980).

CHAPTER TWENTY-TWO

Cluster Compounds

22.1. TYPES OF CLUSTER COMPOUNDS

An important class of compounds is one where the atoms are arranged at the vertices of a polyhedron to give a cage or cluster compound.[1-8] In previous chapters we have mentioned several species of this type without drawing special attention to their cluster nature. The $(CH)_5^+$ and $(CH)_6^{2+}$ species of Section 11.4 are cluster compounds as is, for example, the molecule of **21.74.** In this chapter we will describe in a simplistic way the skeletal orbitals of a large class of cage molecules, those based on the deltahedra of Figure 22.1. As a result we will be able to derive a set of electron counting rules which enable predictions to be made concerning the geometries of such species.[5]

Such cluster compounds divide naturally into four major types. (a) Main group compounds (**22.1-22.3**). These include the boranes,[4] carboranes, and polyhedral

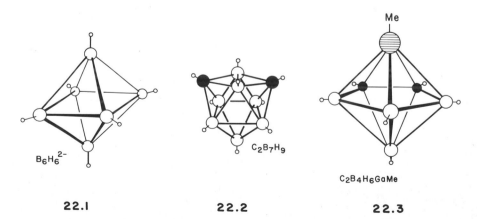

$B_6H_6^{2-}$

$C_2B_7H_9$

Me

$C_2B_4H_6GaMe$

22.1   **22.2**   **22.3**

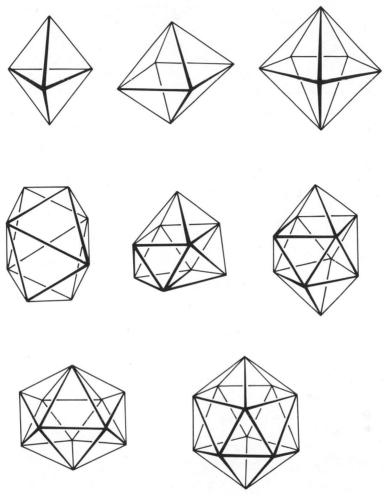

FIGURE 22.1. Deltahedra used by cage and cluster molecules. They are the trigonal bipyramid, the octahedron, the pentagonal bipyramid, the dodecahedron, the tricapped trigonal prism, the bicapped Archimedean antiprism, the octadecahedron, and the icosahedron.

compounds with other main group atoms. (b) Transition metal cluster compounds (22.4, 22.5).[1] These invariably have cyclopentadienyl or carbonyl groups coordinated to the transition metal. In many of these compounds the carbonyl groups are free to move about the surface of the metal cluster and the crystal structures of such species often show bridging CO groups. Similarly there are often low energy fluxional pathways associated with movement of hydrogen atoms in boranes and carboranes. In many of these transition metal containing examples, hydrogen atoms are often associated with the cluster[2] as in 22.6. (c) Metallacarboranes[7] (22.7–22.9). (d) Compounds that contain a small atom inside the polyhedron (22.10). Notice that the central carbon in this example is equally bonded to six Ru atoms.

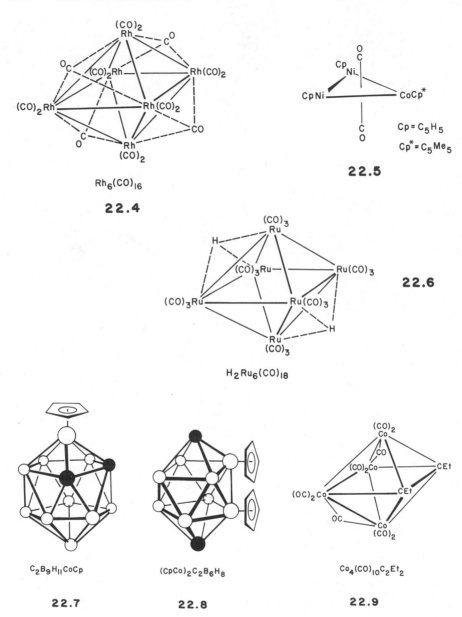

Rh₆(CO)₁₆

22.4

$Cp = C_5H_5$

$Cp^* = C_5Me_5$

22.5

H₂Ru₆(CO)₁₈

22.6

C₂B₉H₁₁CoCp

22.7

(CpCo)₂C₂B₆H₈

22.8

Co₄(CO)₁₀C₂Et₂

22.9

There are also examples of clusters where a hydrogen atom is bonded to from three (as in **22.6**) to six metal atoms. In addition the deltahedron may be either complete (a *closo* species) as in **22.1–22.10** or may have one vertex missing (a *nido* species) as in **22.11–22.13**, two vertices missing (an *arachno* species) as in **22.14, 22.15,** or perhaps more. In many of these cases the reader may be used to viewing these molecules, not as the remnants of a cage, but in a completely different way, perhaps.

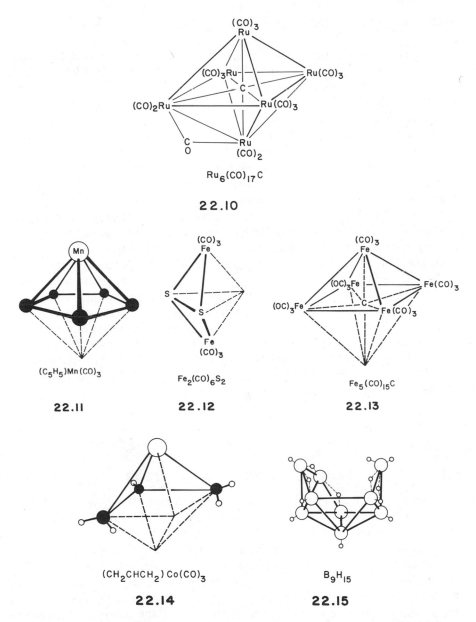

$Ru_6(CO)_{17}C$

22.10

$(C_5H_5)Mn(CO)_3$

22.11

$Fe_2(CO)_6S_2$

22.12

$Fe_5(CO)_{15}C$

22.13

$(CH_2CHCH_2)Co(CO)_3$

22.14

B_9H_{15}

22.15

22.14 might, for example, be regarded as being constructed from a π-allyl fragment bound to a transition metal carbonyl unit. 22.9 may also be viewed as an acetylene complex. In the previous chapter too we have encouraged the reader to be imaginative when considering molecular geometry. We will see that viewing these species as complete or fragmented deltahedra is a very profitable way to understand their structures.

22.2.   CLUSTER ORBITALS

One of the features of these molecules is that usually there are fewer electron pairs than close contacts (bonds) in the molecule. So the compound **22.1** has 6 B—H linkages and 12 B—B linkages requiring a total of 36 electrons if all were two-center–two-electron bonds. The molecule has only 26 electrons and because of this, such compounds are called electron deficient. This fact should not worry us since many of the molecules we have studied in this book are electron deficient when the same criterion is used. The list is long but H_3^+, Cp_2Fe, and many transition metal compounds are among them. For transition metal cluster compounds the 18-electron rule often fails too. Associated with each ruthenium atom in **22.6**, for example, there are a total of $8 + 6 = 14$ electrons from the $Ru(CO)_3$ fragment plus four more if each Ru–Ru linkage is assumed to contribute one electron to each metal atom. This makes a total of 18 electrons. However, in generating this figure we have ignored the two hydrogen atoms attached to the cluster which will give a total of $18\frac{1}{3}$ electrons per atom. In **22.4** similar counting gives $18\frac{1}{3}$ electrons per metal center too.

  We will derive a very general model for these systems by considering three types of orbital at each deltahedral vertex, assembled by considering the one s and three p orbitals at each center. Initially then we will focus on the main group examples **22.1–22.3**. There is an orbital **22.16** pointing away from each vertex of the deltahedron which will be exclusively involved in extradeltahedral bonds, for example, the B—H bonds in **22.1**. There are two types of orbitals involved in skeletal bonding,[5, 9–12] a single radial orbital **22.17** which points toward the inside of the cage and two tangential orbitals **22.18**. The orbital properties of the collection of inward

22.16   **22.17**   **22.18**

pointing radial orbitals are easy to visualize. We will assemble the *nido* trigonal bipyramid (this is just the tetrahedron) and the trigonal bipyramid itself by adding one or two capping orbitals to the triangular system. We readily recognize the orbital pattern of the triangular unit as being topologically analogous to the D_{3h}, H_3 system (Section 5.2). The resultant interaction diagram is shown in Figure 22.2. The horizontal dashed line indicates the energy of an isolated radial hybrid orbital. We have tilted the orbitals comprising the a_1 combination in the triangle during the addition of an extra atom so that they point toward the centroid of the tetrahedron. The energy changes will then not be exactly as we have shown but the general picture is clear enough. The level pattern for the radial orbitals of the tetrahedron is identical to that shown for tetrahedral H_4 in **4.27**. A strongly bonding combination of a_1 symmetry is formed. The antibonding combination meets the two nonbonding orbitals located on the triangle to form a t_2 set. When two extra atoms are added to the trigonal plane (to give the trigonal bipyramid) the orbital combination antisymmetric with respect to this plane finds no symmetry match with the plane

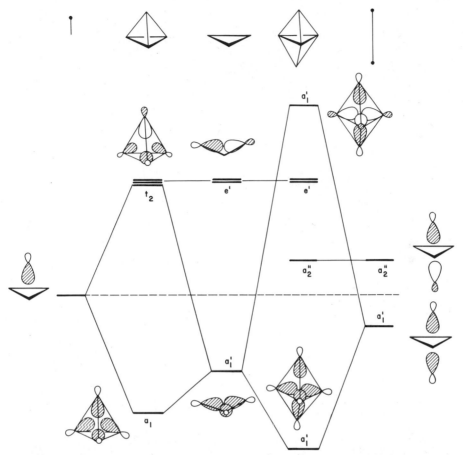

FIGURE 22.2. Derivation of orbital diagrams, built from the radial orbitals of the fragments, for the tetrahedron and the trigonal bipyramid, by capping a trigonal planar unit with one or two new fragments, respectively. Notice that the tetrahedron is a nido-trigonal bipyramid.

orbitals and so remains unchanged in energy. The symmetric combination (a_1') enters into bonding and antibonding combinations with the in-plane orbitals of the same symmetry.

Figure 22.3 shows a similar construction for the *closo* and *nido* octahedra starting with the square. The orbital picture is a very simple one to understand. The important result to note, which is easy to understand from these two diagrams, is that only one bonding orbital results. All the others are antibonding. For larger polyhedra the result is a little more involved. There is always one deep-lying bonding orbital, the in-phase combination of the radial orbitals **22.17**, but there may also be other bonding orbitals at higher energy.

The tangential orbitals are, in principle, no more complex to build up. Starting again with the *"arachno* trigonal bipyramid and octahedron" we need the tangential

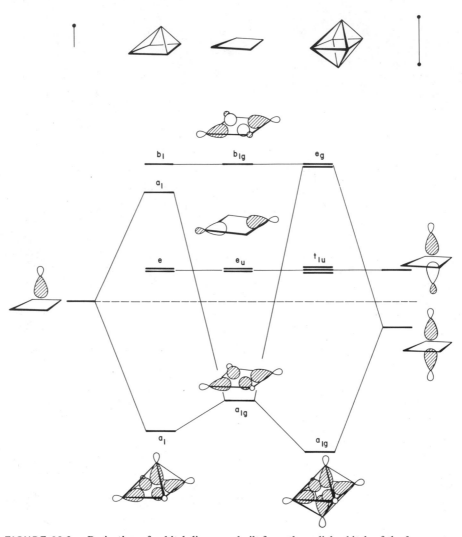

FIGURE 22.3. Derivation of orbital diagrams, built from the radial orbitals of the fragments, for the square pyramid and the octahedron, by capping a square planar unit with one or two new fragments, respectively.

orbitals of these species. They are of two types **(22.19),** one set parallel to the four- (or three) fold axis of the unit, and the other set perpendicular. The parallel orbitals are none other than the cyclic polyene $p\pi$ orbitals of Section 12.1, and the perpendicular orbitals of these fragments were derived in Chapter 5. Figure 22.4 shows the

22.19

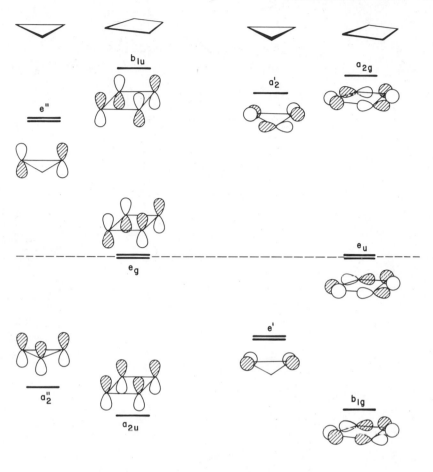

|| orbitals   ⊥ orbitals

FIGURE 22.4. The tangential orbitals of the square and triangle. These break down into two types. Those parallel (||) and those perpendicular (⊥) to the figure axis of the molecule.

parallel and perpendicular sets for the triangle and square. Note the total of six bonding or nonbonding orbitals for the square and three bonding orbitals for the triangle.

Figure 22.5 shows the generation of the orbitals of the octahedron and the *nido*-octahedron (square pyramid). The details of the origin of the new levels are a little difficult to follow since the "capping" orbitals may interact with both parallel and perpendicular sets of orbitals of the square. In brief, the two pairs of capping orbitals at the right-hand side of Figure 22.5 interact with the two pairs of doubly degenerate plane orbitals to produce a set of four orbitals, two of which are bonding and two of which are antibonding. Each pair has the same energy as a nondegenerate

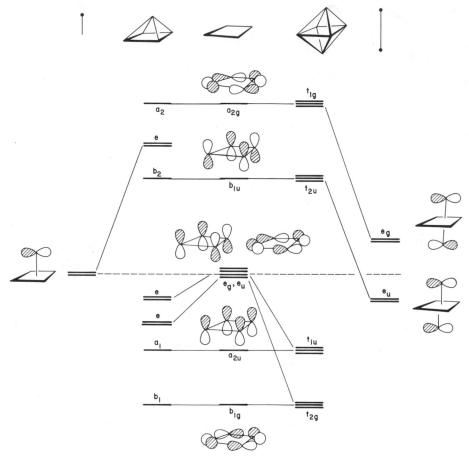

FIGURE 22.5. Assembly of the orbitals, derived from the tangential orbitals of the fragments, for the square pyramid and octahedron, by capping a square planar unit with one or two new fragments, respectively. Only one member of a degenerate orbital set is shown pictorially.

orbital of the plane. The net result is a total of four triply degenerate sets. For the square pyramid the three pairs of degenerate e orbitals (one from the capping atom, two from the square) lead to three new orbitals, two of which are bonding and one of which is antibonding.

Figure 22.5 shows that when the square is "capped" from above and from below to give the square pyramid and then the octahedron, a total of six low energy tangential orbitals are always present. (Recall that a total of six low energy orbitals of this type are found for the square too.) A similar result is obtained if the alternative *arachno*-octahedral structure **22.20** is included, although this is a more difficult result to derive as a result of the lower symmetry of this species.

The result for the trigonal bipyramid is easier to see (Figure 22.6) since e' and e'' orbital combinations on the capping atoms find symmetry matches with the orbitals

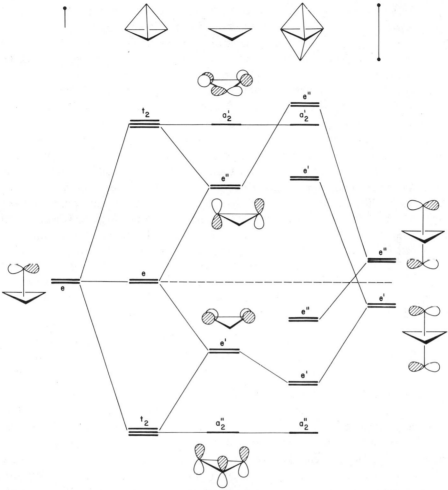

FIGURE 22.6. Assembly of the orbitals, derived from the tangential orbitals of the fragments, for the tetrahedron and trigonal bipyramid, by capping a trigonal planar unit with one or two new fragments, respectively.

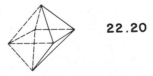

22.20

of the triangle and lead to two pairs of bonding and antibonding orbitals, one of each symmetry type. In the tetrahedron, the capping orbitals of e symmetry (in C_{3v}) mix with the two central atom e pairs to give a bonding, nonbonding, and antibonding trio. The top and bottom orbitals of the set are degenerate with the a_2' and

a_2'' orbitals of the trigonal plane and triply degenerate t_2 orbitals result. For the trigonal plane, capping by one or two extra units immediately produces a total of five low energy (bonding or nonbonding) tangential orbitals (Figure 22.6).

A general rule, which we may derive by extrapolation of these results, is that an n-vertex deltahedron has a total of n-bonding tangential orbitals and one deep lying bonding radial orbital, giving $n + 1$ low energy orbitals in all.[5, 8-12] This rule, very importantly, also applies to the *nido* and *arachno* cages where some of the n atoms are missing. The exception is the tetrahedron itself, which has a total of $n + 2$ skeletal pairs. In addition to being a deltahedron (it is the simplest one), it is a *nido* trigonal bipyramid and behaves electronically as such. Our analysis above has attempted to show in orbital terms how this comes about. It is not exact, since orbitals of the same symmetry, radial and tangential, can, and do mix with each other to change the picture a little. In particular, note for the octahedron that the combination of capping radial orbitals antisymmetric with respect to the square plane (Figure 22.3) will mix with the tangential a_{2u} orbital (Figure 22.5). Likewise there is some overlap, albeit small, between the tangential capping e_u pair and the radial e_u set. Similar situations exist for the other three clusters. In spite of such mixing we are still left with a total of $(n + 1)$ bonding orbitals for the n-vertex deltahedron. A more general group theoretical result, beyond the scope of this book, enables generation of this $n + 1$ rule in a much more elegant and general fashion.[11,12]

22.3.   WADE'S RULES[5]

The discussion above led to the conclusion that *closo*, *nido* and *arachno* species based on an n-vertex polyhedron required $(n + 1)$ skeletal electron pairs to fill all bonding orbitals and, by implication, lead to electronic stability. The application of this simple counting idea to cluster chemistry leads to a set of predictions known as Wade's rules.[5] The molecule **22.1** has a total of 26 electrons (13 pairs). Six pairs are used for the six external B—H bonds leaving seven pairs for skeletal bonding—just the right number to fill all the bonding orbitals of the octahedron. For the molecule **22.2** there are a total of 38 valence electrons (= 19 pairs). Nine pairs are involved in external B—H bonds and therefore 10 pairs remain for skeletal bonding, again just the right number to fill all 10 skeletal bonding orbitals of the tricapped trigonal prism. If there are bridging hydrogen atoms in the molecule then it is sufficient to regard them theoretically as H^+ ions coordinated to the heavy atom core. Thus each bridging hydrogen atom contributes one electron to the skeletal electron count. For a carborane with the general formula $[(BH)_a(CH)_b H_c]^{d-}$ containing c bridging hydrogen atoms the number of skeletal electron pairs is simply equal to $\frac{1}{2}(2a + 3b + c + d)$. Alternatively, we could regard each BH unit as contributing two electrons to the skeletal electron count and each CH unit as contributing three electrons, since an electron pair is absorbed by each BH or CH bond. For molecules containing skeletal atoms only, we must be careful to absorb two electrons per atom as lone pairs occupying the outward pointing orbital **22.16**. So for the *closo* trigonal bipyramidal structure of Bi_5^{3+}, each bismuth atom contributes two electrons to a lone-pair orbital and the rest ($5 \times 3 = 15$) to skeletal bonding. When the positive charge is taken into account there are then just the right number of electrons, six pairs, to stabilize the trigonal bipyramid. One interesting molecule in a historical context is the simple icosahedral species $B_{12}H_{12}^{2-}$ (22.21). With the -2 charge it has a total of

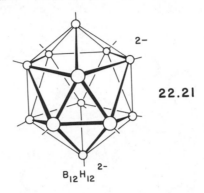

22.21

$B_{12}H_{12}^{2-}$

$\frac{1}{2}[(12 \times 2) + 2] = 13$ skeletal bonding electron pairs as befits the icosahedral geometry. Molecular orbital ideas using completely delocalized bonding as described here led[13] (in 1954) to the prediction of a -2 charge for this species, verified when the molecule was synthesized some years later. Table 22.1 lists the number of skeletal electrons contributed by each main group unit. It is easy to see how the number of skeletal electrons for the A, AH, AX, AL, and AH_2 cluster units are determined. If a vertex is occupied by a single atom A, then two electrons must be placed in the extracyclic bond **22.16**. Each unit then uses the one radial and *two* tangential orbitals (**22.17** and **22.18**, respectively) for skeletal bonding. If a vertex is occupied by an AH or AX unit, then the situation is similar except that only one electron needs to be placed in **22.16**. For AL (where L is a ligand carrying two electrons, such as NH_3) then no electrons need to be placed in this orbital. The AH_2 case is treated in exactly the same manner. The $2a_1$ orbital (see Figure 7.2) corresponds to **22.17**. The two tangential orbitals are b_1 and $1b_2$. The latter orbital is also in-

TABLE 22.1   The Number of Skeletal Electrons Contributed by Main Group Cluster Units

A	A	AH, AX[a]	AH_2, AL[b]
Li, Na	[-1]	0	1
Be, Mg, Zn, Cd, Hg	0	1	2
B, Al, Ga, In, Tl	1	2	3
C, Si, Ge, Sn, Pb	2	3	4
N, P, As, Sb, Bi	3	5	5
O, S, Se, Te	4	5	[6]
F, Cl, Br, I	5	[6]	[7]

[a] X = one-electron ligand, for example, halogen.
[b] L = two-electron ligand, for example, NH_3, THF.

volved in A—H bonding but we formally assign the electrons in it for skeletal bonding. Thus CH_2 contributes four electrons, rather than the two we may have initially
thought.

In the solid state there are many instances of infinite three-dimensional cage
structures. **22.22** shows the structure of CaB_6 where boron octahedra are linked

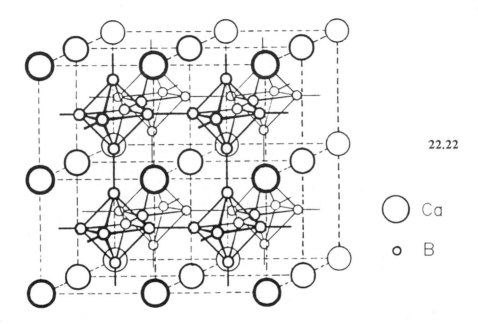

22.22

◯ Ca

∘ B

together via their vertices. Each B_6 octahedron needs a total of 6 electrons to form
two-center–two-electron bonds to the six adjacent octahedra and $2(n + 1) = 14$ electrons for skeletal bonding, a total of 20. The six boron atoms contain a total of
$6 \times 3 = 18$ valence electrons; the extra two come from the electropositive calcium
atom.[14]

For transition metal containing systems the generation of the number of skeletal
electrons contributed by various building blocks is a little more complex. In Chapter
16 we saw that in a single ML_n complex the metal valence orbitals split into a group
of $n\sigma^*$ orbitals and $(9 - n)$ nonbonding orbitals. We assume in our treatment here
that, out of this latter group, there are three[5,15,16] which are involved in skeletal
bonding just like the three orbitals **22.17** and **22.18**. The other $6 - n$ orbitals remain nonbonding orbitals and do not take part in skeletal interactions. It is a gross
simplification but one which is justified in the sense that it works well in practice.
It has strong links with the isolobal analogy of the previous chapter as we will
discuss later. We need to know how many electrons lie in this group of three orbitals.
For an $M(CO)_n$ fragment, the total number of d electrons is simply equal to v, the
number of valence electrons for the $M(0)$ metal. $2(6 - n)$ reside in the group of
nonbonding orbitals, leaving $v - 12 + 2n$ to participate in skeletal interactions. In
the case of a π-Cp unit, one electron is formally transferred to the η^5 Cp ring (effectively a triply coordinating ligand) so that each $M(Cp)_m$ unit has a total of $v - m$

TABLE 22.2   The Number of Skeletal Electrons Contributed by Transition Metal Cluster Units

Transition metal M	$M(CO)_2$	$M(\pi\text{-}C_5H_5)$	$M(CO)_3$	$M(CO)_4$
Cr, Mo, W	[−2]	−1	0	2
Mn, Tc, Re	−1	0	1	3
Fe, Ru, Os	0	1	2	4
Co, Rh, Ir	1	2	3	5
Ni, Pd, Pt	2	3	4	6

valence electrons and $6 - 3m$ nonbonding orbitals. This leaves $(v - m) - 2(6 - 3m) = v - 12 + 5m$ electrons for skeletal bonding. So for a $[M_a(CO)_b Cp_c]^{-d}$ cluster, the total number of skeletal electron pairs is simply $\frac{1}{2}[\Sigma v - 12a + 2b + 5c + d]$. Here Σv indicates the total number of valence $(s + d)$ electrons associated with the collection of a metal atoms, which need not all be the same. Just as with the case of a bridging H atom of a carborane it does not matter overall where the carbonyl groups are located. Whether they are terminal or bridging, their influence on the skeletal electron count is the same. A summary of the numbers of electrons contributed by some carbonyl- and cyclopentadienyl-bearing metal atom units is given in Table 22.2.

A comparison of the numbers of Tables 22.1 and 22.2 shows immediate links with the isolobal analogy of the previous chapter. So in Chart 21.1 CH_2 and FeL_4 are claimed to be isolobal species and accordingly they both contribute four electrons to skeletal bonding. Similarly CH, $Co(CO)_3$, and NiCp are isolobal fragments according to Chart 21.1 and indeed each contributes three electrons to skeletal bonding. In fact the one-to-one correspondence between the electron-counting ideas in this and the previous chapter is complete with a single exception. This concerns fragments derived from the square planar $d^8 ML_4$ **(21.10)** species. Recall that NiL_2 **(21.12)**, for example, is considered to be isolobal to FeL_4 **(21.8)** as a result of the derivation of the former from a 16-electron square planar d^8 ML_4 species. Note that NiL_2 is only isolobal with $Fe(CO)_4$ if the square planar coordination around Ni is retained. If this is not the case then NiL_2 is isolobal with C_{2v} $Cr(CO)_4$. In the molecules of this chapter, square planar coordination is never observed at any deltahedral vertex. As a result, in Table 22.2, $Ni(CO)_2$ contributes the same number of electrons to skeletal bonding as does $Cr(CO)_4$ (and not $Fe(CO)_4$). Recalling that a CH or CR group is isolobal and isoelectronic with a $Co(CO)_3$ or $Ir(CO)_3$ group the whole series of nido-trigonal bipyramidal structures (tetrahedra) are readily generated in **21.56–21.60**. These all have a total of $\frac{1}{2}(4 \times 3) = 6$ skeletal pairs. The $Co_2(CO)_6 C_2 R_2$ molecule is perhaps more often regarded as an acetylene complex (just as **22.9** is) but this series illustrates neatly the dominating electronic influence of the tetrahedral arrangement of heavy atoms. Along similar lines the molecule **22.7** is also known but with the CoCp unit replaced not only with $Fe(CO)_3$ but $BeNMe_3$, AlMe, or Sn, all of which contribute two electrons to skeletal bonding. The permutations of the building blocks of Tables 22.1 and 22.2 are almost endless.

Very similar ideas apply to the structures of *nido* and *arachno* cages.[5,6,8] Whereas *closo* species with n cage atoms require $(n + 1)$ skeletal pairs, *nido* and *arachno* species with n cage atoms have $n + 2$ and $n + 3$ skeletal bonding orbitals, respectively that is, an identical number of electrons as the parent *closo* species. Some examples are shown in **22.23** and **22.24** and include some rather unusual species as well as traditional molecules viewed in a rather different light. The series in **22.23** is iso-

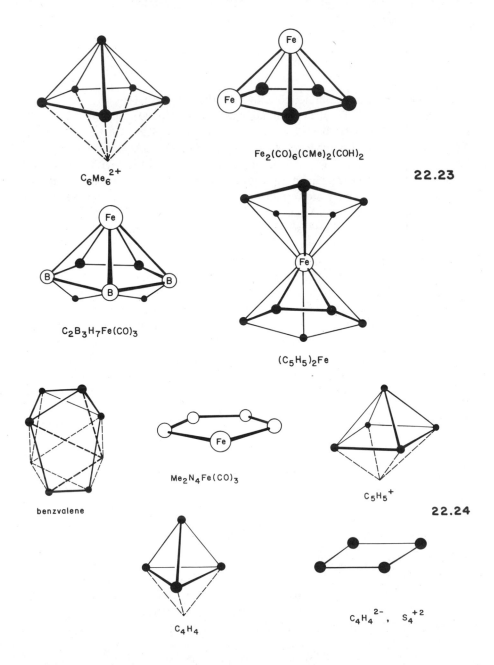

$C_6Me_6^{2+}$

$Fe_2(CO)_6(CMe)_2(COH)_2$

22.23

$C_2B_3H_7Fe(CO)_3$

$(C_5H_5)_2Fe$

benzvalene

$Me_2N_4Fe(CO)_3$

$C_5H_5^+$

22.24

C_4H_4

$C_4H_4^{2-}$, S_4^{+2}

electronic with ferrocene. $C_6Me_6^{2+}$ is simply a *nido* pentagonal bipyramid (number of skeletal pairs $= \frac{1}{2}[(6 \times 3) - 2] = 8$) with an apical CMe^{2+} unit; ferrocene contains the isolobal $FeCp$ unit in its place. Isoelectronic (Tables 22.1 and 22.2) $Mn(CO)_3$ may be used (22.11) to give $CpMn(CO)_3$ but BeX ($X = Cl$, Br, or Me) and PMe^+ are also known as apical groups. In 22.24 it is interesting to see that C_4H_4 (six pairs) is predicted to be a *nido* trigonal bipyramid (i.e., a tetrahedron) rather than the (Jahn–Teller unstable) square, which is the predicted structure for $C_4H_4^{2-}$ or S_4^{2+} (seven pairs). Although C_4H_4 is not known, the tert-butyl derivative is, and it does have a tetrahedral geometry. There are many examples isoelectronic and isostructural to those in 22.24. Other examples are $Co(CO)_3As_3$ and the tetrahedral structures found for elemental phosphorus (P_4) and arsenic (As_4).

However, there are exceptions. Although $C_5H_5^+$ is predicted and observed to have the geometry shown, isoelectronic $(PhC)_4(PhB)$ is a planar ring rather than a square pyramid. Similarly $(PhC)_4(BF)_2$, isoelectronic with $C_6Me_6^{2+}$, is also found as a planar six-membered ring rather than a pentagonal pyramid. The C_6H_6 molecule with $\frac{1}{2}(6 \times 3) = 8$ pairs should be an *arachno* D_{2d} dodecahedron. 22.24 shows the structure of benzvalene, a benzene isomer which is just this. The P_6^{4-} and $S_3N_3^-$ units of Section 12.4 are other exceptions. With eleven skeletal pairs a planar six-membered ring is unexpected.

The insertion of a small atom into the center of a cluster alters the counting rules since now there are extra orbitals to be included. A simplified picture, using the octahedral cluster is shown in Figure 22.7. Both the t_{1u} p orbitals and the a_{1g} s orbital find partners in the bonding octahedral cluster orbitals. The p orbitals overlap with a symmetry-adapted combination of tangential orbitals and the s orbital with the bonding radial cluster orbital. The result is a new set of four orbitals which are both skeletal bonding and central atom-cluster bonding. The cluster plus central atom is stable, therefore, for the same total number of skeletal electrons as the cluster alone. Thus the *nido*-octahedral species $Fe_5(CO)_{15}C$ (22.13) has 4 electrons (from the central carbon atom) + 10 electrons (two from each of five $Fe(CO)_3$ units) giving a total of 14 electrons ($= 7$ pairs) to be used in skeletal bonding—just the right number.

One interesting example of a structural change induced by a change in the number of skeletal electrons is shown in 22.25. With Fe^{2+}, Co^{3+}, or Ni^{4+} (formally a d^6 ion), then the symmetrical structure is stable. This can be envisaged as being isoelectronic with ferrocene where the C_5H_5 units are replaced by *nido*-$C_2B_9H_{11}^{2-}$ anions. Alternatively $X = M—C_2B_9H_{11}$ has just the right number of electrons to stabilize a *closo* icosahedral $XC_2B_9H_{11}$ unit. With two more electrons the metal slips off both rings. This may be simply regarded as a decrease in the hapto number of the metal as a response to the 18-electron rule demands or as the creation of a *nido*-13 vertex structure.

An interesting variant on the deltahedron of Figure 22.1 are the capped structures of 22.26. The capping group simply utilizes the existing outward pointing orbitals presented by the face and no new orbital requirements are made on the deltahedron. Such *n*-heavy atom species are then stable for a total of n skeletal electrons. So $Os_6(CO)_{18}$ (22.26) has a total of six skeletal pairs and six metal atoms and exhibits the capped structure but $Co_6(CO)_{15}^{2-}$ with seven skeletal pairs and six metal atoms exhibits the *closo* octahedral structure.

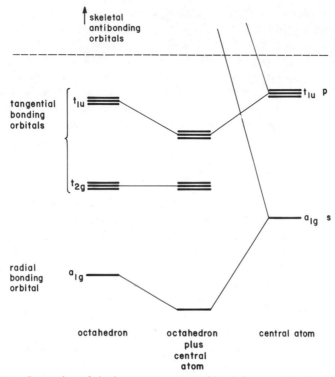

FIGURE 22.7. Generation of the lowest energy set of levels for an octahedron centered by a main group atom, from those of the octahedron and an isolated atom.

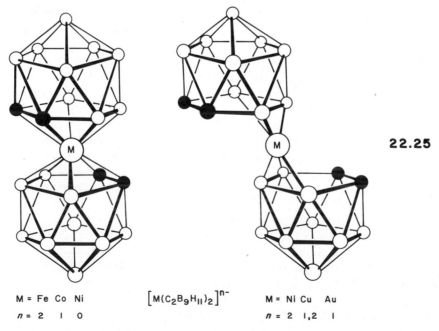

22.25

M = Fe Co Ni
n = 2 1 0

$\left[M(C_2B_9H_{11})_2\right]^{n-}$

M = Ni Cu Au
n = 2 1,2 1

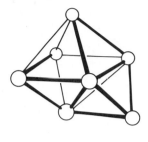

capped octhedron

$Rh_7(CO)_{16}^{3-}$

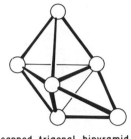

22.26

capped trigonal bipyramid

$Os_6(CO)_{18}$

Exceptions to this scheme are several. The molecule Ni_6Cp_6, for example, has $\frac{1}{2}(6 \times 3) = 9$ skeletal electron pairs but has the octahedral skeleton expected for seven skeletal pairs. The approach makes no comment on the relative stabilities of benzene, which does not fit into the scheme, and benzvalene **(22.24)** which does. Neither does it allow an *a priori* prediction conerning the best choice of *arachno* structures, that is, which pair of vertices to omit. So, for example, we have no way to distinguish energetically between the structure **22.20**, found for the seven-skeletal-pair molecule B_4H_{10}, and the square **(22.24)**, found for the isoelectronic molecule S_4^{2+}. In general more electronegative atoms are found in sites of lower coordination number as, for example, in **22.2** and **22.3**, although this choice is often mediated by the hydrogen atom locations. Overall however, the scheme is a remarkably successful one to view cluster chemistry. Counting rules have also been developed for more complex polyhedra.[17]

REFERENCES

1.   B. F. G. Johnson, editor, *Transition Metal Clusters*, Wiley, New York (1979).
2.   H. Vahrenkamp, *Struct. Bonding*, **32**, 1 (1977).
3.   P. Chini and B. T. Heaton, *Topics Current Chem.*, **71**, 3 (1977).
4.   E. L. Muetterties, editor, *Boron Hydride Chemistry*, Academic Press, New York (1975).
5.   K. Wade, *Adv. Inorg. Chem. Radiochem.*, **18**, 1 (1976).
6.   M. E. O'Neill and K. Wade, in *Metal Interactions with Boron Clusters*, R. N. Grimes, editor, Plenum, New York, (1982).
7.   R. N. Grimes, *Acct. Chem. Res.*, **11**, 420 (1978).
8.   R. W. Rudolph, *Acct. Chem. Res.*, **9**, 446 (1976).
9.   R. Hoffmann and W. N. Lipscomb, *J. Chem. Phys.*, **36**, 2179 (1962).
10.   R. Hoffmann and M. Goutermann, *J. Chem. Phys.*, **36**, 2189 (1962).
11.   A. J. Stone, *Inorg. Chem.*, **20**, 563 (1981).
12.   A. J. Stone and M. J. Alderton, *Inog. Chem.*, **21**, 2297 (1982).
13.   H. C. Longuet-Higgins and M. de V. Roberts, *Proc. Roy. Soc.*, **224**, 336 (1954).
14.   H. C. Longuet-Higgins and M. de V. Roberts, *Proc. Roy. Soc.*, **230**, 110 (1955).
15.   D. M. P. Mingos, *Nature (Phys. Sci.)*, **236**, 99 (1972).
16.   D. M. P. Mingos, *J. Chem. Soc. (Dalton)*, 133 (1974).
17.   D. M. P. Mingos, *Chem. Comm.*, 706 (1983).

Index

Slater type orbitals, 4
S_2N_2, 220
$S_3N_3^-$, 221, 437
S_4N_4, 224
$S_4N_4^{2+}$, 225
$(SN)_x$ chain, 245
Sn_2H_4, 166
Sn_2R_4, 166
Solids, 229
Spin state, inorganic compounds, 281
Spiro-antiaromatic, 227
Spiro-aromatic, 226
Spiro-heptatriene, 226
Spiro-nonatetraene, 227
Spiro-octatrienyl cation, 227
Spiro-pentane, 410
State energies, 110, 115
State wavefunctions, 110
Stereoelectronic control, 182
Sudden polarization, 159
Sulfonium ylides, 170
Sulfuranes, 269
Sulfur chains, 245
Symmetric direct product, 50
Symmetry, 39
Symmetry-adapted orbitals, 100
Symmetry-adapted wavefunctions, 42
Symmetry equivalent atoms, 42
Symmetry-forbidden reaction, 53
Symmetry operations, 40
Symmetry points of Brillouin zone, 252
Symmetry properties of integrals, 51
Symmetry species, of electronic states, 49
π systems, orbitals of, 71

Tangential orbitals, 72, 426
Tantalum tetrahalides, 249
Te_4^{2+}, 245
Tellurium chains, 245
Tetracyanoplatinate (TCP), 251
7,7,9,9-tetracyano-p-quinodimethane (TCNQ), 251
Tetrahedranes, 437
Tetrathiofulvalene (TTF), 251
Three-center bonding:
  four-electron, 86, 99
  two-electron, 86, 99
Three-center-two-electron bonding, 156
Three dimensional solids, 251
Through-bond coupling units, 198
Through-bond interactions, 107, 195
Tight binding method, 237

TiO, 150
Total interaction energy, 15, 16
Trans influence, 356
Transition metal carbenes, 398
Tricyclooctadiene, 193
Tricyclo-3,7-octadiene, 196
Trilithiomethane, 149
Trimethylenemethane-Fe(CO)$_3$, 390
Triple-decker-sandwich complexes, 390
TRIS(acetylene)W(CO), 299, 302
TRIS(ethylenediamine)Fe^{2+}, 278
Tungstenacylobutadiene, 348
Two-center-two-orbital problem, 12
Two-orbital-four-electron interaction, 15
Two-orbital-two-electron interaction, 15

Unit cell, 232

Valence band, 231, 237
Valence-shell-electron-pair-repulsion model
  (VSEPR), 101, 104, 263, 264, 266, 269, 289
van Hove singularities, 237
Variational theorem, 10
$V(CO)_6^-$, 123
$V_2(CO)_{10}$, 327
Vibrational frequencies, in N_2, N_2^+, CO, CO$^+$, 82

Wade's rules, 432
Walk arrangement, 415
Walsh diagram, 93
  H_3^-, 97
  H_3^+, 97
Walsh's rule, 94
Wavevector, 232
W(CO)$_4$CS, photochemistry, 318
Wolfsberg-Helmholz formula, 8

XCH$_2$CH$^-$, 177
XCH$_2$OH, 179
XeF$_2$, 262
XeF$_4$, 262, 270
XeF$_6$, 262, 264
Xenon fluorides, 262

Ziegler-Natta catalysis, 398
Ziegler-Natta polymerization, 356
Zeise's salt, 354
Zone center, 233
Zone edge, 233
Zwitterionic state, 160